Blade Design and Analysis for Steam Turbines

Blade Design and Analysis for Steam Turbines

Murari P. Singh, Ph.D.
George M. Lucas, PE

McGraw Hill

New York Chicago San Francisco
Lisbon London Madrid Mexico City
Milan New Delhi San Juan
Seoul Singapore Sydney Toronto

The McGraw-Hill Companies

Cataloging-in-Publication Data is on file with the Library of Congress

1 2 3 4 5 6 7 8 9 0 DOC/DOC 1 7 6 5 4 3 2 1

ISBN 978-0-07-163574-5
MHID 0-07-163574-2

Sponsoring Editor
Michael Penn

Editing Supervisor
Stephen M. Smith

Production Supervisor
Richard C. Ruzycka

Acquisitions Coordinator
Michael Mulcahy

Project Manager
Manisha Singh, Glyph International

Copy Editor
Patti Scott

Proofreader
Mohinder Bhatnagar, Glyph International

Indexer
Robert Swanson

Art Director, Cover
Jeff Weeks

Composition
Glyph International

Printed and bound by RR Donnelley.

This book is printed on acid-free paper.

About the Authors

Murari P. Singh, Ph.D., is the President of Safe Technical Solutions, Inc. (SAFETSOL). He has been involved in the design, development, and analysis of industrial turbomachinery for more than 30 years with Turbodyne Corporation, Dresser Industries, Dresser-Rand Company, GE CONMEC, and most recently GE Oil & Gas as Consulting Engineer. Dr. Singh has extensive knowledge and experience with fatigue and fracture mechanics, stress and vibration of structures, reliability, life analysis, and probabilistic analysis. His practical application experience features a variety of rotating equipment, including CAES, warm gas and FCC expanders, steam turbines, and centrifugal compressors. He developed the widely used SAFE diagram for reliability evaluation of turbine blades. Lately, Dr. Singh has been involved in developing and teaching applications of lifting strategy for many mechanical components. This includes subjects dealing with HCF, LCF, creep, and fracture mechanics. He has authored many technical papers and he is the coauthor of the book *Steam Turbines: Design, Applications, and Rerating,* Second Edition, published by McGraw-Hill.

George M. Lucas, PE, is a registered Professional Engineer with over 34 years of experience in the design, analysis, and operation of steam turbines and large rotating equipment. As Director of Engineering for Dresser-Rand's Wellsville Steam Turbine Operation, he was directly responsible for product design, steam path engineering, turbine performance, and manufacturing engineering for custom-engineered process drive steam turbines and steam turbine generators. Mr. Lucas was Design Project Leader for the joint Dresser-Rand/EPRI project that resulted in the successful commissioning of the 110-MW gas turbine/expander train at PowerSouth Electric Cooperative's McIntosh Compressed Energy Storage Plant. He received B.S. and M.Eng. degrees from Cornell University. Mr. Lucas holds two U.S. patents and has authored technical papers on topics including the use of advanced abradable seals in steam turbines, compressed air energy storage, and the development of advanced steam turbines for syn-gas compressor drive service. He is currently an independent consultant serving OEM and end-user clients in oil and gas, power generation, and manufacturing industries.

Contents

Preface

Turbine engineers and designers have made remarkable improvements in the efficiency and reliability of industrial steam turbines over the last 30 years. Remarkable improvements have been achieved for products that already had over 100 years of technical development behind them. For most of those first 100 years, the analysis of turbine blades had concentrated on the behavior of individual blades. A key change, and one of the most significant advances in turbine reliability, was the development and application of analytical techniques that make it possible to characterize and explain the behavior not simply of individual turbine blades, but of entire bladed disk assemblies.

Advancements in modal analysis and testing, fatigue analysis, creep analysis, fracture mechanics, aerodynamic theories, and the development of many new materials and manufacturing processes cleared the path for the design of more powerful, more efficient, and more reliable turbines. It became evident that design of blades is a multidiscipline activity. For a proper reliability assessment of a design, one needs to understand many fields of science and these must be applied as need be. These advancements helped designers to extend the capabilities of designs beyond past experience. This also helped to explain past successes and failures of components.

The simultaneous development of powerful and inexpensive computers has made it practical to quickly and efficiently carry out the calculations necessary to apply these advanced analytical techniques to the routine design of new and replacement blades and rotors for industrial steam turbines. Nowhere have these advances had a greater influence than on the design of critical service process compressor drives for the refining and petrochemical industries. Large drivers for ethylene and LNG processes exceeding 75 MW in power are in successful service. Older designs using double-flow exhausts with short, but very strong, blades have been supplanted in newer designs by single-flow exhausts with taller, but more reliable and aerodynamically sophisticated, stages. Inlet pressure and temperatures of 2000 psig/1000°F (140 barg/540°C) have become almost common in new process drive applications.

The purpose of this book is to introduce these advances in a concise volume and provide an easy-to-understand reference for practicing engineers who are involved in the design, specification, and evaluation of industrial steam turbines in general, and critical process compressor drivers in particular. This text has also attempted to present a unified view of concepts and techniques needed in the understanding of blade design. It includes some advanced concepts such as life estimation. One chapter is dedicated in introducing the reader and designers to the effect of uncertainty of input variables on the reliability of the design. Probabilistic-type analysis is introduced for reliability estimation, as it is said that every design decision has some risk associated with it and risk may be managed if it is known.

We would like to thank each person and the many industries whose works have been referenced in the book. We also take this opportunity to apologize to those whose work might not have been referenced by mistake. Thanks to the many associates during our employment and consulting work whose thoughts guided the selection of many materials. We hope these will help readers in their work or at least make them think. Last but not the least, many thanks to Seema Singh for reading the manuscript word by word and making numerous suggestions for changes that made the work better and more readable.

Murari P. Singh, Ph.D.
George M. Lucas, PE

Acknowledgments

Special thanks are extended to the following:

Advanced Turbomachine LLC	Steve Rashid
Arthaven	Barbara Lucas
Consultant	John Waggott
Dresser-Rand	Jay Scherbik, Randy Moll, Bob Voorhees, Jim Dello, Dan Flurschutz, David Nye, George Lentek, and Neeraj Bali
Elliott Co.	Robert Sloboda, Art Titus, and Brook Tolman
GE Oil & Gas	James Cencula and Leonardo Tognarelli
Mar-Test	Matt Webb
Metal Improvement Co.	Dave Massey

Blade Design and Analysis for Steam Turbines

CHAPTER 1

Introduction

1.1 Importance of Blades in Steam Turbines

Structural integrity of all rotating components is the key for successful operation of any turbomachinery. This integrity depends on the successful resistance of the machine parts to the steady and alternating stresses imposed on them. The challenge with rotating equipment, such as turbomachinery, is often more severe due to the significance of the alternating loads that must be carried to satisfy their purpose.

One of the major classes of rotating machinery is the mechanical drive steam turbines, i.e., steam turbines that drive pumps and/or compressors. These steam turbines are differentiated from those that drive generators in that they operate at variable speeds. Steam turbines may operate from 1 to 5 hp up to several hundred thousand horsepower; they may operate with steam that ranges from vacuum to thousands of pounds per square inch; and blade tip speeds can exceed the force of the most severe hurricane (a large, last blade row with an 8-ft tip to tip diameter operating at 3600 rpm will experience tip speeds in excess of 1000 mph).

One of the causes of blade deterioration is static stress which is primarily the result of steam bending and centrifugal loads. Alternating stresses are imposed due to the vibration of the parts in question, e.g., blades and disks. If the combined loads become too large, vibration-induced fatigue of the rotor blades or disks is a major concern. In addition to the imposed loads, these forces are subject to resonant amplification caused by coincidence with natural frequencies. To put the scope of this problem into context, one must realize that there may be thousands of blades in a steam turbine. For example, there may be 10 to 20 rows of different blade designs with the possibility of each blade row having different dynamic characteristics.

Steam turbines have been in operation for more than 100 years and have always faced this problem. As may be imagined, the technology in engineering and physics to support these designs has grown dramatically over that time; tools have been enhanced and technological developments incorporated.

1.2 Brief Historical Perspective of Technological Development

The current state of design, as represented in the API standards for this class of machinery, sets a life of 30 years for all components. In many cases, this translates into a design requirement for infinite life and may exceed the needs of a specific installation. This requirement may be driven by the actual desire for infinite life, limitations in analysis techniques, tools that have existed over the years, and/or an incomplete understanding of the tools that have appeared in the recent past and are currently fully or partially available.

A common cause of vibration-related failure in steam turbine blading is resonant excitation of the blading occurring at an integral order, i.e., multiples of the rotational speed, nozzle passing frequency, and multiples thereof. The associated mode of failure is high cycle fatigue. A primary feature of resonant excitation is that dynamic stress amplitudes rise as the exciting frequency approaches the resonant speed and the response decreases after passing through the resonant speed. Hence, it is necessary to identify resonant frequencies of the system.

It is impossible to include all the work done by the numerous researchers and designers of steam turbine blades. Effort is made to include some of those that describe the progress and current methodology for steam turbine blade design. Many textbooks were published on steam turbines during the last century together with many technical publications dealing with all aspects of turbine design, specifically blade design. Early publications by Stodola (1905) and Kearton (1922) are worth mentioning because these two books are credited with setting the stage for detailed vibration and reliability analysis for blades. In many different ways designers followed the processes and methods outlined in these books. As the turbine design matured and manufacturers gained experience, methods were adjusted to include new technical methods and lessons learned from field experience and each manufacturer has evolved its own methods and criteria to achieve successful design. Hence, methods and criteria should not be expected to be consistent across manufacturers.

Blade design has evolved from the analysis of spring-mass systems to a single cantilever beam to a band of blades to a bladed disk. In addition, steam turbines have included bands of blades on a disk as a system. Throughout the years many effects of turbine speed to increase blade frequency were found, and it gave rise to the term *centrifugal stiffening*. Campbell (1925), while examining the failure (bursting) of disks, concluded that blades were broken due to axial vibration. This publication reported the results of an investigation conducted at General Electric to understand the wheel failures, mostly in wheels of large diameter, that could not be explained on the basis of high stress alone. About this time certain types of vibrations of standing waves were investigated by means of sand pictures. This test was conducted by scattering sand over the wheel surface. Wheels were then excited by means of a magnet exciter, and the turbine wheel was placed in a horizontal position. An electromagnet was clamped with its poles close to the edge of the wheel, alternating current was passed through the coils of the magnet, and a series of pulls was exerted on the wheel. This resulted in deflecting the wheel in a transverse direction to the plane of the wheel. A variable-speed direct-current (dc) motor was used to drive the alternating-current (ac) generator and allowed the frequency of the pull of the magnet to be varied over a wide range. Frequency of excitation was varied until a sand pattern on the wheel appeared, and sand accumulated mostly in a radial line or pattern. When the frequency changed to some higher magnitude, a different sand pattern appeared on the wheel. These radial lines represented the location where the velocity of vibration was zero. The number of radial lines was always observed to be of an even number. These patterns, are known as nodal patterns, and two lines are taken as one diameter. It is now understood that the opposite radial lines might not be 180° apart. Frequencies at which these patterns are observed coincide with the natural frequency of the wheel in axial vibration associated with the mode shape represented by the sand pattern.

Figure 1.1 shows a picture of such a sand pattern. It is noticeable that sand has collected on certain portions of the wheel, and it forms a pattern showing four radial lines. This pattern is referred to as two nodal diameters mode. There are six radial lines in the pattern shown in Fig. 1.2. These modes are called three nodal diameters mode. Note that the radial lines pass through the balancing holes in the left picture while in the picture on right side these lines pass between the balancing holes. A detailed discussion of this phenomenon that forms the basis of bladed disk analysis is provided in Chap. 5.

Over time blades needed to be taller to accommodate the requirement of increasing power. This necessitated the blades to be joined together by a band of metal either at the tip or somewhere along the length of the blade. Kroon (1934) described a method to evaluate the effectiveness of such construction to reduce the dynamic response of the design under steam forces. Allen (1940) described design practices of blades in high-pressure and high-temperature stages. A detailed explanation for partial stage admission was included, as was one for full admission stage. Allen recommended limiting the number of blades per group to two for high-temperature service and argued that more blades in a group for high-temperature application tend to set up high stress. Two types of root attachment (axial entry vs. tangential entry) construction were explored, and the choice is dependent on the application, e.g., speed, power, and temperature. The effect of shrouding might be considerable for high-pressure blades. A reduction of 25 to 60 percent in bending stress may be achieved. The natural frequency

FIGURE 1.1 Four radial lines, two nodal diameters mode (Campbell, 1925).

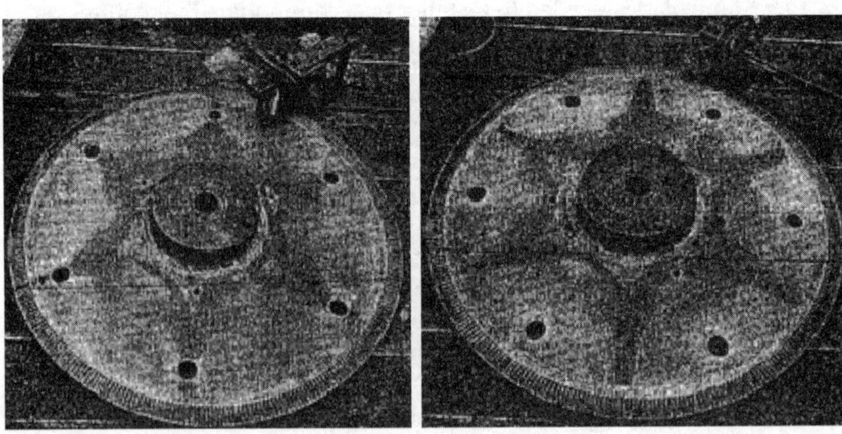

FIGURE 1.2 Six radial lines, three nodal diameters modes (Campbell, 1925).

for high and intermediate stage blades should be more than 5 times the running speed of the turbine. For low-pressure stage blades the natural frequency more than 4 times might be sufficient. Trumpler and Owens (1955) provided a detailed discussion of factors that affect the strength of the blade for full admission stage. Damping was measured during a test on partial admission stage. Values of logarithmic decrement varied from 0.02 to 0.06 for 12 to 13 percent chromium steel, and the approximate material damping of these materials is 0.02. The effects of fixity due to centrifugal force, surface finish between mating surfaces, geometric tolerance, and the length of operating time were not explored, however.

Problems of blade damage were encountered during World War II. At this time advances had been made in computers to handle large calculations. Weaver and Prohl (1956) demonstrated that blades are banded together, and blades in the band behave differently than single blades. Results showed that there were more frequencies and mode shapes of banded construction compared to what the analysis of a single blade provided. Large numerical calculations were performed for such a construction. The response of blades under excitation due to flow from nozzle, the natural frequencies of the banded blades, and the associated mode shapes were evaluated. In a companion paper Prohl (1956) described the numerical method and provided the equations that were used to estimate frequencies, mode shapes, and dynamic response of the banded blades. The basic beam equations for blades were developed where blades were coupled together through a band at the tip of the blades. To simplify the analysis and to make the results tractable, three key assumptions were made:

1. Series of identical blades were assumed to be parallel to one another, and they were uniformly spaced.
2. Principal axes of blade cross sections were assumed to be parallel to the tangential and axial direction of the rotor.
3. The shear center of the cross section was assumed to coincide with the center of gravity of the cross section.

It can be said that the above assumptions do not relate to the actual construction of bladed disks for steam turbines. For example, directions of blades' principal moments of inertia do not coincide with the tangential and axial directions of the rotor at least for reaction-type blades or exotic taper and twisted blades. However, they do coincide for impulse-type blades, which are not parallel to one another, but each is positioned in a radial direction on the disk. Spacing between blades changes from the base to the tip of the blade; however, the angle between them is equal. Shear center and center of gravity become coincident only for a circular cross section. Now with the development of finite element analysis (FEA), one does not have to make these assumptions. FEA made it possible to analyze the banded construction rather than a single blade for correct frequency and dynamic response determination. Notwithstanding these assumptions, results of this work provided the future direction for analysis. This was a milestone in the analytical development of blade vibration and the decision-making process for reliability.

Equations to estimate dynamic stress at the base of the blade due to loading from nozzles were provided. The static steam bending stress gets magnified during resonance. An assumption was made that energy supplied to the vibrating blade group by the exciting force is completely dissipated in damping.

$$\sigma_v = K(\pi/\delta)S\sigma_b$$

where σ_v = resonant vibration stress at blade root
K = resonant response factor
δ = logarithmic decrement of damping
$K(\pi/\delta)$ = amplification factor
S = stimulus (always less than unity)
σ_b = bending stress at root due to steam loading
$S\sigma_b$ = exciting stress

It was assumed that the stimulus is uniformly distributed along the length of the blade and that the phase between stimulus and blade motion is constant along the length of the blade. Figure 1.3 summarizes the results of the analysis for tangential vibrations for seven modes. The first six modes are considered to belong to the first bending of a single blade, and the difference among them is the phasing among blades. The seventh mode is the second tangential mode.

The analytical result for the dynamic response of blades was shown to be a function of (1) the number of harmonic excitation, (2) number of nozzle openings, and (3) number of blades per 360°. The shapes of these curves, which resemble that created by a bouncing ball, are functions of mode shapes of the blade group. These mode shapes can be seen to differ from one another by the difference in their phasing of blades and the results correlated with field data and test data from laboratory study. The result of this work created the need to analyze the blades as a group rather than as a single blade. The dynamic response of the group is due to the coupling between blades through the band. Frequency is a function of the relative stiffness of the band and the blade. The review of Fig. 1.3 indicates that with a proper selection of nozzle and blade pitching, it may be possible to considerably reduce the resonant stress even though operating precisely at the speed required to excite the natural frequency.

The next technological development was to include the disk in the analysis. It was recognized that a bladed disk, is a system and coupling

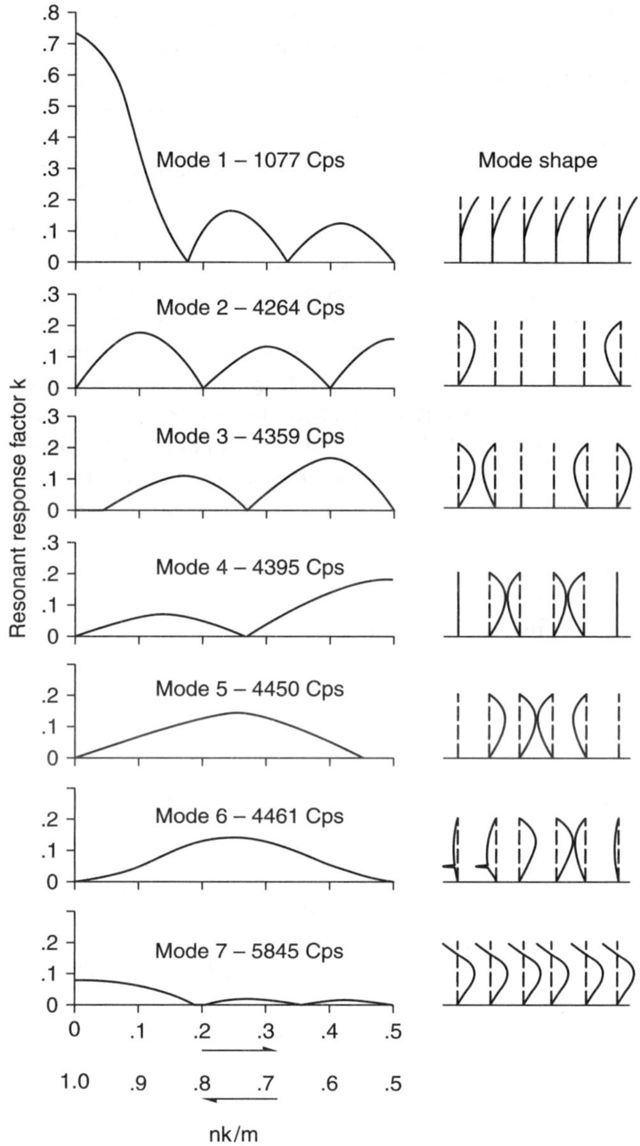

FIGURE 1.3 Resonant response factor (Weaver and Prohl, 1956).

between blades will be through the disk, and the consideration of the stiffness of the disk became important. In the tangential vibration of blades, stiffness of the disk may be considered very high, but it will be very small if there is any coupling through the disk. However, in the axial vibration, disk stiffness becomes a large contributor.

Including geometric variations among blades was the next logical advancement in the analysis of the dynamic response of blades. It has been shown that geometric variation influences mode shapes and frequencies and, in turn, the response of a bladed disk system. In a tuned system where each blade is identical, modes in general occur in duplicate. There are two modes that differ by a phase angle, but these modes have identical frequencies. However, when symmetry is disturbed through variation from blade to blade, these modes tend to split in two frequencies. Also, the shape of these modes gets distorted from the tuned case, which is a phenomenon called *mistuning*. Ewins (1969, 1970, 1973, 1976) has extensive discussions of this phenomenon, and the response of the mistuned case was found to be different from that of the tuned case. This was attributed mainly to a change in the mode shape and may pose a serious decision point for designers to account for variability among blades. Ewins (1976) specifically dealt with completely shrouded or unshrouded bladed disk constructions.

The next stage of advancement was made in the understanding and analysis of disks containing packets of blades. The blades of steam turbines more often are banded together in a packet. Singh (1982, 1988, and 1989) studied the dynamic behavior of packeted bladed disk construction. Figure 1.4 shows a comparison of a SAFE

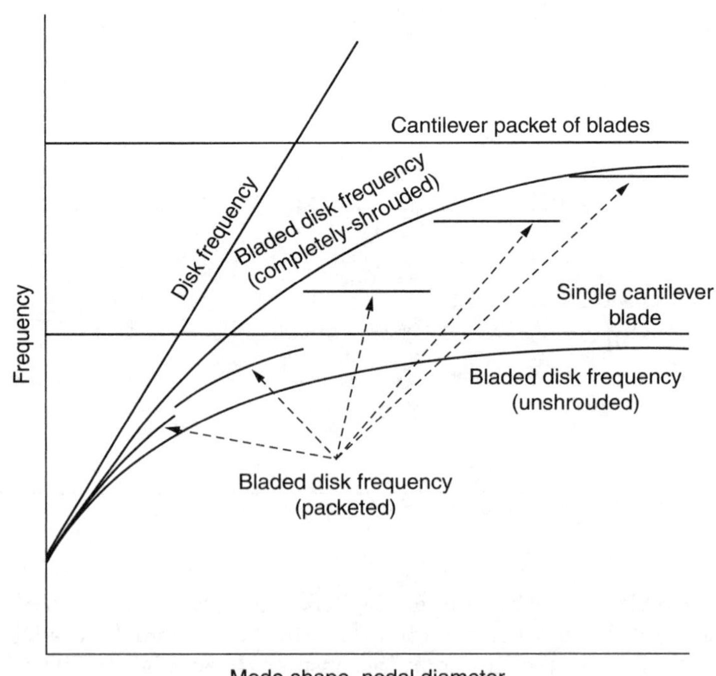

FIGURE 1.4 SAFE diagram for completely shrouded, unshrouded, and packeted bladed disk (Singh and Drosjack, 2008).

(Singh's advanced frequency evaluation) diagram for different types of construction for the same number of total blades mounted on the same disk. The vibration characteristic of packeted bladed disks is similar to that of the completely bladed disk, but it has some special features. This method of analysis is the focus of a large part of this book, and it is discussed in the appropriate sections.

Reliability evaluation of a mechanical component becomes a multidiscipline activity and has greater relevance in the design of steam turbine blades because of the blades' contribution to the reliability of the turbine. In addition, many technological developments have been made throughout the years for the design of gas turbine blades. Many of these analysis methods are applicable to steam turbine blades and should be used.

To calculate the resonance characteristics of blades, one needs to deal with the following key issues, which are not all-inclusive:

- Unsteady aerodynamics over a wide variety of flow conditions that may exist within the operating range
- Structural vibration characteristics, i.e., frequencies and associated mode shapes within the operating range
- Damping in the system (structural as well as aerodynamic)
- Estimation of material properties (fatigue properties, yield strength, ultimate strength, modulus of elasticity, etc.) at the temperature in the operating range
- Manufacturing and quality assurance processes

Each of these issues may require a separate technical discipline to support an evaluation. In the last few decades, significant advances have been made in several of these issues, even though additional analysis and test development are still needed. Utilization of the current state-of-the-art technologies and tools will provide a better evaluation of the issues, resulting in more accurate and more optimal designs.

Unsteady aerodynamic analysis can produce two necessary pieces of data: aerodynamic damping and the magnitude of the pressure acting on the blade (pressure profile due to flow variation). During each revolution, every blade experiences a variation in pressure or force and will experience the same force in the subsequent revolutions. Each blade experiences a periodic force with a time lag with respect to its neighbor. These aerodynamic forces depend on blade geometry (profile, stagger angle, angle of incidence), cascade solidity, and flow conditions (subsonic, transonic, and supersonic). In case of shock or flow separation, aerodynamic analysis becomes complex, and so the unsteady force is estimated.

The blade structural dynamic analysis must consider an assembly of blades and the disk as a system. The characterization includes natural frequencies in the operating range, associated mode shapes, and damping, which are the required input for estimating the forced response of the bladed disk assembly. Since the blades vibrate in a flow medium, an interdisciplinary approach that includes structural dynamics and unsteady aerodynamic analysis is necessary.

The amplitude of the dynamic stress is proportional to the damping. Damping is provided by a material's internal resistance as well as by the flow medium (aerodynamic damping). More is known about material damping, but aerodynamic damping is dependent on the characteristics of the flow. For example, a blade will experience greater resistance to vibration in a dense gas than in a less dense gas. In addition, a taller blade will experience greater damping due to large amplitude of vibration than a shorter blade having a small amplitude of vibration; i.e., damping can be wildly nonlinear.

Many new materials are now available that have been used in blade manufacturing. The advent of many superalloys has allowed blade design for higher loads. Many materials are suited for different environmental conditions, e.g., erosive and corrosive environments. Now materials are available that can be used for higher speed, higher steam pressure, and higher temperature. Some materials provide better fatigue properties while others provide better creep properties while still others provide better resistance to crack propagation. The choice of material depends on the condition and type of operational loads being imposed on the blade. This has allowed designers to balance the properties for a given application to achieve higher reliability. Less dense material, such as titanium alloys, has allowed blades to be designed for higher speeds that might not have been possible in the past. Many types of surface treatments, e.g., coatings, shot peen, and laser peen enabled the achievement of increased blade life, even in some adverse conditions.

Current and former successful results have been achieved by applying experience-based rules and limits, which can cause designs that may seem to run out of space (the limits set design boundaries that may not be analytically sound). The assessment of the useful blade life is discussed with respect to damage from high cycle fatigue and may impose overdesign requirements that may damage the economic viability of equipment designs cascading to projects. As there is always a risk associated with any design decision, a probabilistic concept will be presented that permits a quantification of the risk of a proposed design. The potential to apply probabilistic analysis to extend design capabilities is included.

Chapters 2 to 11 are arranged so as to permit the review of the pertinent technologies that have been applied in the design and evaluation of the reliability of steam turbine blades.

CHAPTER 2

Steam Turbine Design Process, Performance Estimation, and Determination of Blade Loads

Turbine blades only exist to convert the heat energy in steam to useful shaft work. In this context, when we use the term *steam turbine performance*, we are really discussing how (and how well) these blades convert energy in steam (known as enthalpy, or heat energy) to shaft power. In a specific turbine for a specific application, the final quantity, size, and shape of the blades are the result of a design process that follows a logical sequence which starts with the basic requirements of the application (power, speed, steam pressure, and steam temperature, to name a few of the important parameters) and ends with detailed designs for the individual components.

Since we are interested in how this design process affects the design and the reliability of the rotating blades, it is important to understand how the turbine design engineer designs or selects blades based on the application requirements, and how he determines the loads imposed on the blades in service. Because our focus is very narrow, specifically only the rotating blades, this is not intended to be a comprehensive discussion of steam turbine thermodynamics or flow path design. There are many excellent texts available that cover thermodynamics from introductory through advanced topics, and other texts cover the aero-thermodynamics of turbine flow path design in great detail, a few of which are listed as references. Instead, we concentrate on some of the most basic aspects of the turbine design process, to show how the application requirements drive decisions made by the turbine designer and how they influence the quantity, size, and shape of turbine blades. We conclude with a basic explanation of how the conversion of steam energy to shaft power imposes loads on the rotating blades in the form of steady and unsteady forces, moments, and pressures.

This is important because steam turbine blades operate in a very harsh environment. The stationary nozzles direct steam in a circumferential flow, in the process creating an extremely strong vortex in which the rotating blades must operate—in other words, an artificial tornado. As terrifying as natural tornadoes may be, the environment in a turbine stage is even more violent. Our artificial tornado has wind speeds up to 1000 mph, but with a fluid many times denser than air and, in some cases, at temperatures approaching 1000°F. Yet we expect the rotating blades to operate at high efficiency and with high reliability throughout literally years of continuous operation. The turbine design engineer must pay careful attention to blade design and to the determination of the loads imposed on the blades in service.

In the discussion that follows, we have adopted typical conventions and nomenclature commonly used in steam turbine practice. The velocity vectors and their components that represent the blade velocity (aka "wheel" speed) are referred to by *W*, while steam velocity vectors and their components are referred to by the letter *V*. Also, in keeping with common steam turbine conventions, flow path angles are referenced from the plane of blade rotation, rather than the turbine's longitudinal axis commonly used as the reference in compressor and gas turbine practice.

2.1 Turbine Design Process

The turbine design process, and more specifically the flow path design process, progresses through a series of steps from the very general to more specific, ultimately resulting in definition of the geometry for the stationary and rotating blade rows. The general sequence of this process includes

1. *Turbine specifications.* At their most basic, these include the rotational speed or speed range, steam pressures at the turbine's inlet and exhaust, steam temperature at the turbine inlet, and either the desired power output or the steam flow that is available.
2. *Determination of staging.* Based on the turbine specifications, the turbine designer makes some basic decisions on which

the flow path design is built—degree of reaction, desired blade peripheral speed, stage diameters, and number of stages in the turbine. At this stage, the number of rows of stationary and moving blades is established.

3. *Determination of optimum flow passage angles.* The turbine engineer creates velocity diagrams for each stage based on the mean diameter of the flow path to determine appropriate airfoil entrance and exit angles, for best performance at the design or "guarantee" point. This step determines the blade shapes.
4. *Detailed stage design.* With the turbine staging determined and appropriate airfoil shapes specified, the turbine designer determines the quantity and the size (i.e., the width or chord) of the blades, either at the mean stage diameter for "short" constant section airfoils, or at a series of radial stations for "tall" blades with twisted airfoil shapes.
5. *Reliability evaluation.* The blade designs are analyzed to ensure that steady steam bending and centrifugal forces are within acceptable limits. The vibratory characteristics of the blades are predicted and compared to the frequency and shape of unsteady forces from a variety of sources that act on the blades.

Much of this book is devoted to step 5 in this process—reliability evaluation. This chapter, however, is devoted to a brief overview of the other flow path design steps, to acquaint the reader with the decision-making process followed by engineers who design turbines. Relying on these basic concepts, the reader may gain some insight into the turbine designer's reasoning in selecting the shape, number, and size of the blades in a specific turbine based on its conditions of service—steam conditions, speed, power, and steam mass flow, among others.

2.2 Introduction to Steam Turbine Thermodynamics

The key thermodynamic property in all steam turbine performance calculations is enthalpy, which represents the heat energy in the steam, as distinct from the total energy which also includes pressure energy. While a pressure differential is essential to extract energy from the steam, the energy that can be extracted from each pound of steam depends on only the change in heat or enthalpy. Since steam does not behave as an ideal gas, the simple relationships that are used to calculate enthalpy and related thermodynamic properties for air and other ideal gases cannot be used for calculations involving steam. Instead, the thermodynamic properties of steam are characterized by charts (e.g., the well-known *H–S* diagram or Mollier chart for steam), tabulations of thermodynamic properties (the familiar ASME steam tables), or the equations developed by the International Association for the Properties of Water and Steam (IAPWS). For the casual user, steam property calculation programs are available from many sources, either as stand-alone programs or as add-ins to popular spreadsheet programs that can be used to simplify the thermodynamic calculations.

The basic calculation methodology for turbine performance is the same whether the calculation is for a complete turbine or for individual stages in a multistage turbine. The simplest way to illustrate the basic turbine performance calculation is by referring to the *H–S* diagram, or Mollier chart, for steam. In this discussion the thermodynamic parameters of interest include the enthalpy H, entropy S, the pressures P and temperatures T at the turbine inlet and exhaust, and specific volume of the steam at inlet and exhaust v. Specific volume is the reciprocal of density ($v = 1/\rho$) and is used by the turbine designer to determine the flow areas required for a given mass flow of steam. Subscripts are used to keep track of the location at which each parameter is measured—subscript 1 denotes inlet or initial steam parameters, while 2 denotes the exhaust or final parameters. Quantities with primes (′) denote intermediate steps in the calculation between initial and final parameter values.

Let's consider a turbine with inlet steam pressure P_1 and inlet temperature T_1, exhausting at pressure P_2, operating with flow path efficiency η. For the turbine inlet steam conditions P_1 and T_1, the corresponding enthalpy value H_1 is found from the intersection of the P_1 isobar and the T_1 isotherm lines on the Mollier chart, and initial entropy S_1 is read from the chart. Once the basic parameters are specified, the first step in the calculation is the determination of the total enthalpy available for the expansion process, shown as $H_1 - H_2'$ in Fig. 2.1. The total enthalpy **available** for conversion to work in the turbine equals the enthalpy change for an isentropic (i.e., no change in entropy) expansion from inlet P_1 and T_1 to exhaust pressure P_2. The quantity ($H_1 - H_2'$) represents the energy per unit mass steam flow that could be converted to shaft work by a perfect turbine with an efficiency of 100 percent.

However, for a turbine with known or predicted efficiency η, the final state for expansion lies on the P_2 constant-pressure line (or isobar), at its intersection with final enthalpy H_2. *Turbine efficiency* is defined as the decimal fraction of the total available enthalpy (i.e., $H_1 - H_2'$) that is actually converted to shaft work by the flow path (i.e., $H_1 - H_2$):

$$\eta = (H_1 - H_2)/(H_1 - H_2')$$

Final enthalpy H_2 is calculated by rearranging this definition of efficiency

$$H_2 = H_1 - \eta(H_1 - H_2')$$

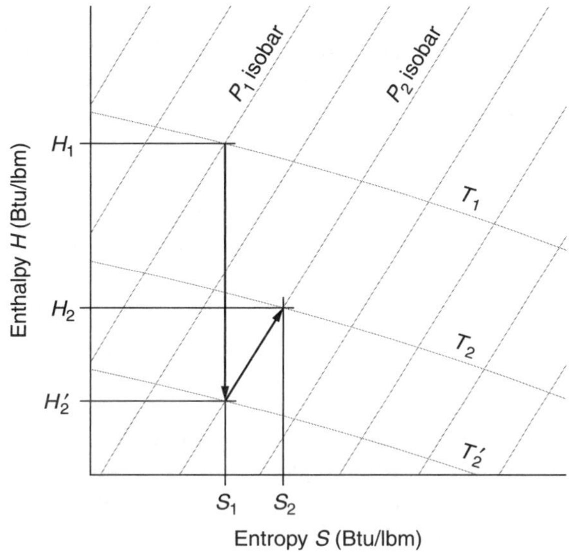

FIGURE 2.1 Basic enthalpy-entropy chart (*H–S* diagram) illustrating turbine calculations.

and the remaining thermodynamic properties (T_2, S_2, v_2) are easily found from the Mollier chart or the steam tables.

With the enthalpy change through the stage now known, the power Q produced in a turbine flow path passing M lbm/h of steam is given by

$$Q = M(\text{lbm/h}) \times (H_1 - H_2)(\text{Btu/lbm})/2545\ (\text{Btu/hp·h})$$
$$= M(\Delta H)/2545\ \text{hp}$$

or

$$Q = M(\text{lbm/h}) \times (H_1 - H_2)(\text{Btu/lbm})/3413\ (\text{Btu/kWh})$$
$$= M(\Delta H)/3412\ \text{kW}$$

Of course, in real turbines, one must account for bearing friction and other mechanical losses, which must be subtracted from Q to get the net power output W from the turbine. Nevertheless, this simple calculation is the basis for all steam turbine performance calculations. It is easy to see that these relationships may be rearranged so an engineer may determine the potential power output if M is known or specified, or to determine the required value of M to produce a given power output. Using only this basic relationship, some important information can be deduced:

- For example, if accurate values of the actual steam pressures and temperatures at the turbine inlet (P_1, T_1) and exhaust

(P_2, T_2) are known from field measurements, the remaining properties (S_2, v_2, H_2) are easily found from the chart or tables, and the actual stage efficiency can be calculated and compared to the design or guaranteed efficiency of the turbine.

- Conversely, if the power output can be measured or determined with sufficient accuracy, the value of M' can be calculated and compared to the design or guaranteed steam consumption of the turbine. (Elaborations from these two cases form the underlying basis for almost all steam turbine performance testing.)
- Even before the turbine has been designed in detail, the process engineer, armed with some knowledge of the range of flow path efficiencies possible for different levels of turbine complexity and cost, can estimate either the steam flow required to produce a given power or the potential power output from, say, a flow of waste steam from which energy is to be recovered with a turbine generator.

Example 2.1 Determine the steam flow rate required to produce 500 kW if a steam turbine with predicted efficiency of 75 percent, consumes steam at inlet pressure and temperature of 600 psia/750°F and exhausts at 200 psia. Assume 2 percent total mechanical losses.

Referring to the *H–S* diagram in Fig. 2.2, $H_1 - H_2' = 1379.8 - 1257.3 = 122.5$ Btu/lbm. For $\eta = 0.75$, $H_2 = 1379.8 - (0.075 \times 122.5) = 1287.9$. Note that the remaining properties at the end state—T_2, S_2—are now easily read from the Mollier diagram.

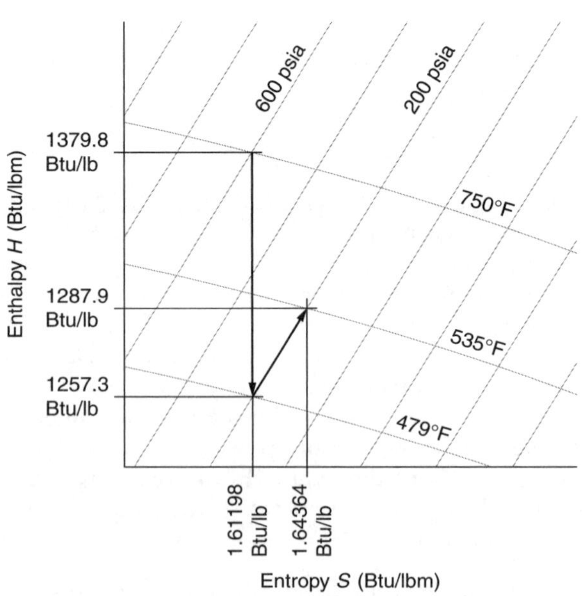

FIGURE 2.2 *H–S* diagram for example turbine calculation.

Power output Q is equal to the flow path power minus the mechanical losses L, which in this case equal 2 percent of the flow path power Q. Taking account of the mechanical losses yields

$$500 \text{ kW} = Q - L = Q - 0.02Q = 0.98Q = M(H_1 - H_2)/3413$$

which, rearranged to calculate M, yields

$$M = (3413)(500/0.98)/(1379.8 - 1287.9) = 18{,}948 \text{ lbm/h steam mass flow required}$$

2.3 Velocity Diagrams

2.3.1 Euler's Equation

The basic calculation above is the simplest approach to estimating turbine power output and steam flow, or to back-calculate turbine flow path efficiency. It relies on only some basic knowledge (or predicted values) concerning the steam conditions (pressures, temperatures, mass flow), and it provides helpful information that can be used at the preliminary stage of turbine design as well as in the overall process or plant steam system design. However, it is a "black box" calculation, providing virtually no direct information on the details of the turbine flow path. To arrive at a more complete understanding of a stage's actual performance and detailed flow path parameters, we turn to the velocity diagram.

Figure 2.3 shows the general concept of steam flow through a turbine stage. Steam expands through the stationary nozzles, leaving at

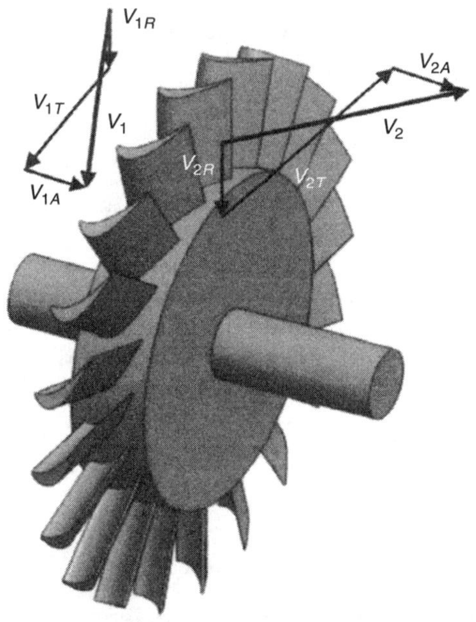

Figure 2.3 Turbine wheel illustrating general application of Euler's rule.

velocity V_1, passes through the rotating blades, and leaves the rotor at V_2. In this case V_1 and V_2 are the absolute velocities relative to the turbine's stationary components (nozzles, turbine case, etc.), each of which is composed of tangential, axial, and radial components, denoted by the subscripts t, a, and r, respectively. So V_{t1} and V_{t2} are the tangential components of V_1 and V_2, respectively. The rotor is spinning at angular velocity ω, which is equal to $2\pi N/60$ rad/s, where N equals the turbine rotational speed in rpm. Blade speed (again relative to a stationary reference frame) is $W_1 = \omega R_1$ at the point where steam enters the rotor and $W_2 = \omega R_2$ at the point where steam leaves the rotor.

If the mass flow rate entering the stage is M, the angular momentum about the shaft centerline of the steam entering the rotor blades is equal to $(M/g_c)(V_{t1}R_1)$, where g_c is the gravitational constant (32.2 ft/s^2 in USCS units) that relates force and mass in traditional units. The angular momentum of the steam leaving the rotor blades is similarly equal to $(M/g_c)(V_{t2}R_2)$. The rate of change of momentum through the stage is therefore

$$(M/g_c)(V_{t1}R_1 - V_{t2}R_2)$$

which has units of force times distance (that is, lb·ft or N·m) and is equal to the torque T about the turbine shaft centerline. The rate of work production w is equal to ωT (in units of lb·ft/s). Stage power Q of course equals $\omega T/550$ hp (in USCS units) or $\omega T/737.6$ kW (in SI units).

After combining, we arrive at the relation

$$550P = \omega T = Q = (M/g_c)(W_1V_{t1} - W_2V_{t2})$$

which can be further rearranged to give

$$Q/M = (1/g_c)(W_1V_{t1} - W_2V_{t2})$$

a relationship that is commonly known as Euler's equation.

Using the familiar equation for kinetic energy, the net rate of energy input to our generic turbine stage is equal to $(M/2g_c)(V_1^2 - V_2^2)$. Defining the stage efficiency as the ratio of work output to energy input, the efficiency of our turbine stage is given by

$$\eta = (M/g_c)(V_{t1}R_1 - V_{t2}R_2)/(M'/2g_c)(V_1^2 - V_2^2)$$

which simplifies to

$$\eta = 2(V_{t1}W_1 - V_{t2}W_2)/(V_1^2 - V_2^2)$$

It is important to note that V_1 and V_2 are three-dimensional vectors with radial, axial, and tangential components (V_{r1}, V_{a1}, V_{t1}, and V_{r2},

V_{a2}, V_{t2}, as noted), but since the radial and axial components do not contribute to the angular velocity and angular momentum change about the rotor centerline, only the tangential components V_{t1} and V_{t2} are of interest in calculating the shaft power output. However, it would be wrong to then assume that the axial and radial components are unimportant to the overall turbine design. For example, if the axial components all equaled zero, it would be impossible for steam to enter and exit the rotor blades at all. It is also clear from this discussion that the radial and axial components of V_1 and V_2 should be kept to the minimum necessary for turbine operation, in order to achieve maximum energy conversion efficiency. Finally, the axial and radial components of flow through the turbine are important factors in the calculation of thrust loads, rotor stability analysis, and even external loads imposed by the turbine on the plant structure and foundation.

In the field of steam turbine design, there are two fundamentally different approaches to flow path design, commonly referred to as *impulse* and *reaction staging*. The distinction between these concepts is important because the choice has profound effects on the blades' design, including the number of stages, number of blades per stage, airfoil size and shape, as well as the loads imposed on each row and the individual blades in the row. In simplest terms, in an impulse turbine, all the expansion (i.e., all the enthalpy change, with its resulting fluid acceleration) occurs in the fixed nozzles, and energy is transferred to the rotor solely by velocity change as the steam passes through the rotating blades. In a reaction turbine, the enthalpy change is divided between the fixed and moving rows, and energy is transferred to the rotor by both velocity change and additional fluid acceleration in the moving row of blades.

2.3.2 Impulse Turbine

Figure 2.4 is a vector representation of the pertinent steam and blade velocities overlaid on the nozzles and blades at the mean flow path radius of a typical impulse turbine stage. As we have already noted, in an impulse turbine all the enthalpy change takes place in the stationary nozzles. In practical terms, this also means that the entire pressure drop through the stage ($P_1 - P_2$) occurs across the stationary nozzles, and little or no pressure drop exists from the inlet to the exit side of the moving blades and the disk on which they are mounted. With no appreciable pressure drop across the disk, axial thrust loads on the rotor are very low, which is an important advantage for impulse stages.

In Fig. 2.4 steam exits the nozzles at velocity V at nozzle exit angle θ to the plane of blade rotation. Velocity V is calculated as if all the available enthalpy for an isentropic expansion from P_1 to P_2 were converted to kinetic energy of the fluid. Again referring to the *H–S*

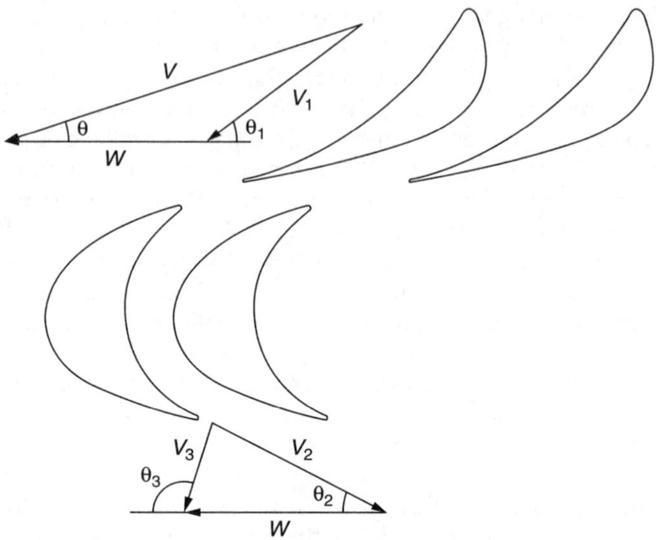

FIGURE 2.4 Impulse turbine cascade with associated velocity diagram.

diagram in Fig. 2.1, enthalpy available is $H_1 - H_2'$. Equating the enthalpy change to the steam's kinetic energy yields

$$M'(H_1 - H_2')(778.3 \text{ lb·ft/Btu}) = (M/g_c)(V^2/2)$$

Now M' drops out of both sides, leaving

$$V = [2(g_c)(778)(H_1 - H_2')]^{0.5} \quad \text{or} \quad V = 223.8\,(\Delta H)^{0.5}$$

(in traditional engineering units)

Of course, V is the absolute velocity of the steam relative to the turbine's stationary components. However, since the rotating blades are moving relative to these stationary components, the steam actually enters the rotating blades at relative velocity V_1, and at angle θ_1, which is the vector difference between V and W. Here, as before, W equals the tangential speed of the moving blades.

Steam exits the moving row at velocity V_2 and angle θ_2 from the plane of rotation. Neglecting friction and other losses for the moment, as the steam exits the moving row, scalar velocity $V_2 = V_1$. The vector difference of V_2 and W equals V_3, the absolute velocity of steam leaving the blades relative to the turbine's stationary components. The velocity diagrams at the blade entrance and the blade exit are often overlaid to create the composite velocity diagram in Fig. 2.5, which makes it easy to compare the values of V_{a1}, V_{t1}, V_{a3}, and V_{t3} directly from the diagram.

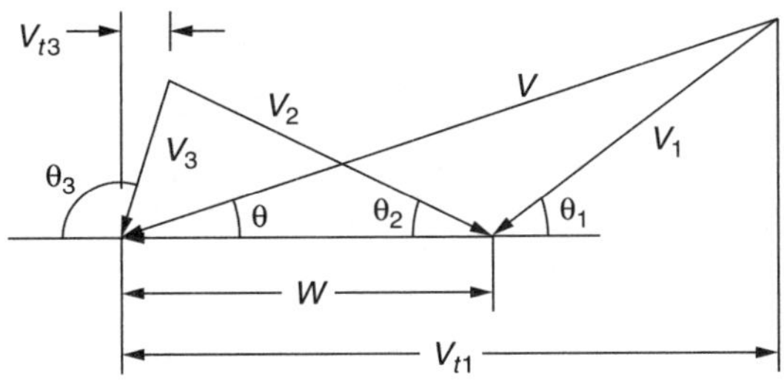

FIGURE 2.5 Impulse stage composite velocity diagram.

As long as the mean radius at the blade entrance and exit is the same, the efficiency equation reduces to

$$\eta = 2W(V_{t1} - V_{t3})/(V_1^2 - V_3^2)$$

since $V_{t1} = V_1 \cos\theta$, and if it is assumed that the energy represented by V_3 is not recovered, this further reduces to

$$\eta = 2W(V_{t1} - V_{t3})/V_1^2$$

Falling back on basic trigonometry, we can easily determine that

$$V_1 = (V\cos\theta - W)/\cos\theta_1 \qquad V_{t1} = V\cos\theta \quad \text{and} \quad V_{t3} = -(V_2\cos\theta_2 - W)$$

And as noted, if friction and other losses are neglected, $V_2 = V_1$.

Substituting and rearranging these relations yield the following expression for efficiency:

$$\eta = 2[W\cos\theta/V - (W/V)^2](1 + \cos\theta_2/\cos\theta_1)$$

From this simple relation, it is easy to show that for impulse stages, the optimum velocity ratio occurs at $W/V = \cos\theta/2$, and for the case where the blade entrance and exit flow angles are equal ($\theta_1 = \theta_2$), the maximum efficiency that can be attained is $\eta = \cos^2\theta$.

For the case where $\theta_1 = \theta_2$ (i.e., relative flow angles are equal at blade entrance and exit) and $\theta = 20°$ (a typical value for impulse stages), Fig. 2.6 shows a plot of predicted efficiency for an impulse stage as a function of W/V, alongside the predicted efficiency for the reaction stage discussed in Sec. 2.3.3. In real turbine stages, of course, actual efficiency will be lower due to the unavoidable losses from friction and other flow disturbances that have been neglected in this simplified analysis.

In practice, too many factors can influence actual stage efficiency to account for all of them analytically, so turbine manufacturers carry out extensive testing to develop empirical efficiency curves for their particular stage designs, to cover a range of velocity ratios. These

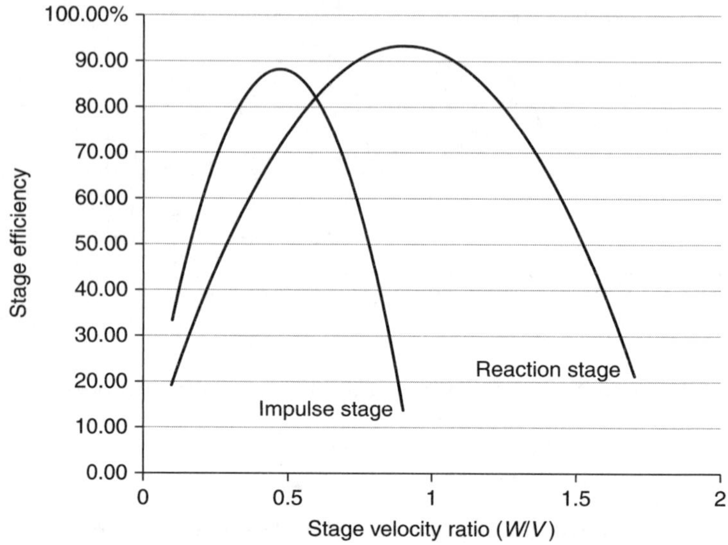

FIGURE 2.6 Efficiency vs. velocity ratio (W/V) for impulse and reaction stages.

proprietary efficiency relationships are then typically adjusted using additional empirically determined correction factors for such attributes as blade aspect ratio (i.e., airfoil height/chord width), the stage Reynolds number, and leakage losses, to list a few of many potential factors that can influence turbine stage efficiency.

2.3.3 Reaction Turbine

In contrast to the impulse turbine considered above, the enthalpy drop in reaction turbine stages is split between the stationary vanes and the moving blade row. The *degree of reaction* is defined as $r = \Delta H_{rotor}/\Delta H_{stage}$, which is usually stated as a percentage, 50 percent reaction, for example. Stages can be designed with a wide range of reaction values, depending on factors such as the available enthalpy and rotor length constraints as well as nontechnical factors such as relative cost or simply the turbine designer's preference. However, when most turbine designers discuss traditional reaction turbines, they are referring to a unit in which all the stages are designed for nominal 50 percent reaction. For the present discussion, a pure impulse stage is one designed with $r = 0$ or 0 percent reaction, while a traditional reaction stage is designed for $r = 0.5$, which is commonly described as a 50 percent reaction. An alternate definition known as pressure reaction is used by some turbine designers. It is approximately, although not precisely, the same value as reaction defined by the ratio of enthalpies.

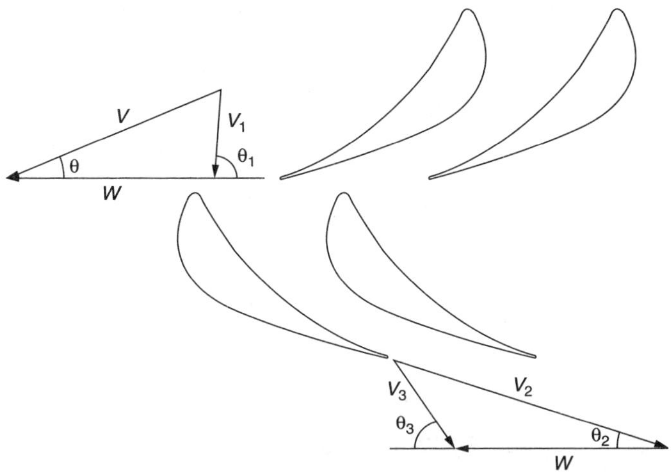

FIGURE 2.7 Reaction turbine cascade with associated velocity diagram.

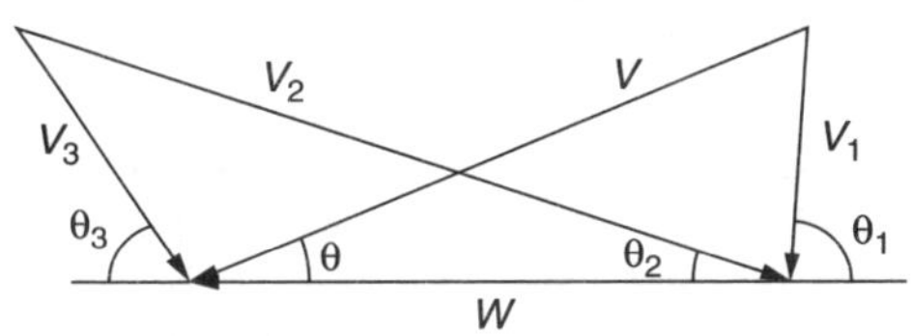

FIGURE 2.8 Reaction stage composite velocity diagram.

Figure 2.7 shows a reaction turbine cascade, with its associated velocity diagram, and Fig. 2.8 shows the composite velocity diagram for this same stage. At first glance, the velocity diagram appears very similar to the impulse stage diagram. The fundamental difference lies in the way the steam velocities are calculated. In this velocity diagram at the exit of the stationary nozzles, the vectors V, W, and V_1 are all calculated in exactly the same way as in an impulse stage, except that the value of ΔH_1 is now one-half of the total enthalpy available for the stage (based on 50 percent enthalpy reaction). For 50 percent reaction stages, symmetric blading may be used, which in Fig. 2.7 means that blade and flow angles $\theta = \theta_2$ and $\theta_3 = \theta_1$.

The enthalpy change in the moving row (that is, $\Delta H_2 = 0.5\Delta H$) affects the exit velocity from the moving row. Ignoring friction and other flow losses, the kinetic energy $V_1^2/2g_c$ of the steam entering the moving row is augmented by the conversion of enthalpy ($\Delta H_2 = H_1$). Ignoring friction and other losses, the energy balance across the moving row of blades is given by

$$(V_1^2 - V_3^2)/2g_c = (778)(\Delta H/2) \quad \text{or} \quad V_3^2 = V_1^2 + 778g_c(\Delta H) \text{ USCS units}$$

The rate of work output in the moving row (again, only the tangential velocities count in this calculation) is given by

$$Q = (1/g_c)(WV_{t1} - WV_{t3}) = (1/g_c)[W(2V\cos\theta - W)]$$
$$= (1/g_c)V^2(W/V)(2\cos\theta - W/V)$$

Using the definition of efficiency as the ratio of work output to energy available in the stage (that is, $\eta = Q/\Delta H$) and combining the above relationships yield

$$\eta = V^2[(2W/V)\cos\theta - (W/V)^2]/(V^2 - V_3^2)$$

If we again assume that the energy represented by V_3 (equal to $MV_3^2/2g_c$) is not recovered, the stage efficiency reduces to

$$\eta = (2W/V)\cos\theta - (W/V)^2$$

For this case with symmetric blading, since once again flow losses and leakage were neglected, the optimum velocity ratio is equal to $W/V = \cos\theta$, or just less than $W/V = 1$ for small values of nozzle and blade exit angle θ, and also approximately twice the optimum velocity ratio for a pure impulse stage. The maximum efficiency attainable is $\eta = \cos^2\theta$. As noted previously, Fig. 2.6 shows a curve of efficiency vs. velocity ratio (W/V) for a 50 percent reaction stage with symmetric blading and nozzle angle $\theta = 25°$, along with the efficiency curve for the impulse stage analyzed in the previous section. The reader must note that these efficiency values are based on some rather unrealistic assumptions about blade path losses (i.e., we have ignored flow losses altogether) so practical efficiencies are significantly lower in real stages. The quantity $V_3^2/2g_c$ is referred to as *kinetic energy carryover*, and the recovery of this energy to the greatest extent possible is of great concern to turbine design engineers. So our simplifying assumption that the energy represented by V_3 is wasted adds a little pessimism in our efficiency relations for both reaction and impulse stages. However, this pessimistic assumption only slightly offsets the far more optimistic assumptions that friction and other flow losses equal zero.

There are a few important conclusions the reader should draw in comparing the efficiency curves for impulse and reaction stages. First, the peak efficiency for impulse stages occurs at a value of W/V equal to $0.5\cos\theta$, while the peak efficiency for a reaction stage occurs at a value of W/V equal to $\cos\theta$. This implies that for identical tangential blade speeds, the energy converted per stage by a reaction turbine is one-half that converted per stage by an impulse turbine. In practical terms, this means that a reaction turbine requires twice as many stages to convert the same available enthalpy as an impulse turbine. However, the peak efficiency for a reaction turbine is slightly higher than that for an impulse turbine doing the same work.

2.4 Application Requirements and Conditions of Service

In turbine design as in most engineering design problems, the first step in the process is the determination of the application requirements. The most basic application requirements in this case include the power required by the driven machine, the speed range over which it must operate, and sufficient information regarding the steam inlet and exhaust conditions to determine the energy available from the steam. The purpose of identifying the range of service conditions for the turbine is to ensure that the turbine designer knows, and can design for, all the required combinations of speed and shaft power that a successful application requires. Since the turbine converts heat and kinetic energy in the working fluid (steam in this case) to shaft torque and power, it is also critical to know the complete range of inlet pressure, inlet temperature, and exhaust or discharge pressure over which the turbine must operate. In general, the designer is looking for those sets of concurrent conditions that determine the required flow path heights and areas, as well as the sets of concurrent conditions that result in the highest imposed loads on individual turbine components. The imposed loads due to rotational speed (i.e., centrifugal forces), demand for power (i.e., steam or gas bending forces), and pressure differentials acting on the casing, stationary flow path, and rotating blades are among the most critical.

With a basic understanding of turbine stage design and an understanding of the fundamental thermodynamics involved, most of the key conditions of service are easily identified. For example, the turbine designer must ensure that the turbine can produce rated power output using steam supplied at the minimum specified inlet pressure and temperature while operating with the highest specified exhaust backpressure. Since this combination of conditions results in both the least energy available per pound of steam and the lowest fluid density at the turbine inlet, this set of conditions is used to determine important design parameters such as minimum inlet area, control valve sizes, and ultimately the required heights and passage sizes of the stationary and rotating turbine blades.

Similarly, the designer must determine the operating point or points that result in the worst-case combination of imposed loads on the rotating blades. For multistage turbines, this may, in fact, encompass several distinct operating points, depending on the specific stage being considered. The designer must determine, stage by stage, which of all the potential operating conditions produce the worst-case combination of steam forces, centrifugal forces, and pressure loads on the rotating blades. For a typical process drive steam turbine, some of the critical operating points that the turbine designer must consider might include the following.

1. Rated output at minimum speed. When specified, this operating point may result in the highest torque produced by the

turbine rotor and consequently the highest steam bending forces acting on the individual rotating blades. When the driven machine is a process compressor, it may be impossible to ever attain this torque level at less than full speed, but most turbine designers (and many turbine design specifications) take a conservative approach and design the turbine for this condition anyway. The engineer must exercise judgment before applying such a conservative assumption, as it can lead to unnecessarily high costs and may actually force the turbine designer to accept undesirable compromises in efficiency or rotor stability.

2. Maximum flow at maximum inlet steam pressure and temperature and minimum exhaust pressure, while operating at maximum continuous speed. This set of concurrent conditions almost always results in the highest combined loading on the rotating blades. If the turbine is designed to produce rated power at minimum steam conditions that are significantly lower than normal steam conditions, at maximum flow the turbine may well be able to produce higher powers (and potential blade loadings) than the driven machine is capable of utilizing. Again, the turbine designer has to decide whether to design the blades for this impossible-to-achieve condition or to define a more realistic maximum load case. Many designers, as well as many industry specifications, take a conservative approach and design for this admittedly fictitious condition.
3. Control stages are a common exception to item 2 above. For most control stages, maximum blade loads occur when the turbine operates with the minimum arc of admission (e.g., first valve wide open for multivalve turbines) at maximum steam inlet pressure and temperature, coupled with minimum exhaust backpressure, and operating at maximum continuous speed. This usually results in the highest combined loading on the control stage blades.
4. Control stages are also often evaluated at the first valve wide-open point while operating at minimum continuous speed as well. This potential operating point usually results in the highest potential steam bending forces on the control stage blades. (This set of conditions may be more likely to occur in real-life service conditions than some other common conditions at which blade loading is evaluated.)

Many, if not most, process drive turbines are designed in accordance with API machinery specifications, in particular API 612, Special Purpose Steam Turbines, or API 611, General-Purpose Steam Turbines. These specifications are very useful as they define a common set of terms and definitions that are used by end-users, purchasers, plant and process design engineers, and turbine design engineers to specify the conditions of service for process drive steam turbines. They specify in some detail how to determine the range of conditions for which process drive steam turbines must be designed including, for example, maximum and minimum inlet pressures and temperatures, the design speed range for the turbine, and important concurrent combinations of speed and power required by the driven machine. These definitions, in conjunction with the API data sheets, are extremely helpful to the turbine designer in determining the combinations of speed, power, and steam conditions that define critical loading for each stage in the turbine.

2.5 Preliminary Turbine Design

As noted before, turbine design proceeds through a series of steps from high-level "global" design decisions to successively more detailed definition in later steps. A key step in this process is the decision regarding reaction levels and number of stages.

2.5.1 Number of Stages

Although single-stage turbines are occasionally used to drive process compressors, the vast majority of process compressor drive turbines are multistage turbines. A multistage turbine breaks up the available pressure and enthalpy drop across a number of stages. By selecting an appropriate combination of number of stages, stage diameters, and pressure drops (i.e., apportioning enthalpy changes to each stage), the turbine designer can produce a turbine that operates at near-optimum efficiency. The first step in this process is to determine the number of stages required.

At this point in the design process, the designer needs to know the total energy available based on inlet and exhaust steam conditions, the rated speed or speed range for the application, and the power required by the driven machine. Based on these data, the designer knows (or can calculate) the important thermodynamic quantities (H_1, H_2', S_1, S_2' as well as specific volume at the inlet conditions).

To determine the number of stages required, the first important decision is the level of reaction for which the unit will be designed. Steam turbine manufacturers self-classify themselves as either impulse turbine builders or reaction turbine builders. In real-world practice, few current-day turbine manufacturers actually build either pure impulse or pure reaction turbines, so this line is blurred at best; but this distinction provides a useful place to start this discussion.

Impulse Turbine—Number of Stages

For a known available enthalpy (that is, $\Delta H = H_1 - H_2'$), the engineer's first problem is to determine an appropriate number of stages in a turbine to achieve as near-optimum efficiency as possible. An appropriate blade or wheel speed W is usually chosen by the designer based on some knowledge of the application. For example, a relatively low-speed compressor drive turbine (e.g., cracked gas compressor drive, or a propylene refrigeration compressor drive) is usually designed for moderate blade speeds in the range from 400 to 700 ft/s, measured at the mean flow path diameter D_m for most of the stages in the turbine. On the other hand, high-speed compressor drives [e.g., those that process light gases such as methane, hydrogen, or synthesis gas ("syngas")] often must be designed in a manner that limits the number of stages in order to meet rotor length constraints, and therefore often use blade speeds of 1100 ft/s for high-pressure stages, and up to 1700 ft/s or more for condensing exhaust stages.

For impulse turbines, we have shown that peak efficiency occurs at $W/V = \text{Cos}\,\theta/2$, which for practical nozzle angles will occur slightly below $W/V = 0.5$. This can be rearranged as $V_{\text{optimum}} = 2W/\text{Cos}\,\theta$, where W is blade speed, θ is the nozzle discharge angle, and V_{optimum} is the desired steam velocity at the nozzle exit.

Example 2.2 If the engineer chooses $W = 500$ ft/s and $\theta = 15°$, then $V_{\text{optimum}} = (2)(500)/\cos 15 = 1035$ ft/s.

Since it has already been shown that $V = 223.8\sqrt{\Delta H}$, it is easy to calculate the optimum ΔH for this stage:

$$\Delta H_{\text{stage}} = (1035/223.8)^2 = 21.4 \text{ Btu/lbm}$$

This approach is easily generalized to the entire turbine to determine the number of stages required:

$$N_s = \Delta H_{\text{total}}/\Delta H_{\text{stage}} \qquad \text{(where } N_s \text{ equals the number of stages)}$$

Example 2.3 An impulse turbine driving a compressor at 4000 rpm is supplied with steam at 600 psig/750°F, exhausting at 150 psig. The turbine designer wants to determine the optimum number of stages to compare different turbines using 700- and 800-mm mean blade diameters, respectively.

From the H–S diagram, $\Delta H_{\text{total}} = H_1 - H_2' = 1376.4$ Btu/lbm – 1226.8 Btu/lbm = 149.6 Btu/lbm.
For the turbine with 700-mm wheels, $\Delta H_{\text{stage}} = [(2 \times 481 \text{ ft/s})/(223.8 \times \cos 15)]^2 =$ 19.80 Btu/lbm.

$$N_s = 149.6/19.8 = 7.55 \text{ or 8 stages if rounded to nearest integer}$$

For the turbine with 800-mm wheels, $\Delta H_{\text{stage}} = [(2 \times 550 \text{ ft/s})/(223.8 \times \cos 15)]^2 =$ 25.88 Btu/lbm.

$$N_s = 149.6/25.88 = 5.78 \text{ or 6 stages if rounded to nearest integer}$$

It is worth noting that while we have rounded to the nearest integer in this example, many designers might opt for a different rounding rule, depending on their knowledge of the application and other design objectives, such as minimizing first cost, or optimizing efficiency at more than one known set of operating conditions. For example, since the efficiency vs. W/V curve has a relatively broad top, rounding down to the next-lower number of stages results in lower overall cost and shorter bearing span at a small sacrifice in peak efficiency.

Reaction Turbine

For the same set of conditions, the reaction turbine's peak efficiency occurs at a W/V value equal to $\cos\theta$, which means for the same nominal blade speed, designing for peak efficiency drives the solution to more stages, each converting less energy per stage. Using the same example as for the impulse turbine above yields $V_{optimum} = W/\cos\theta$. Because of the lower overall velocities, the airfoil shapes typically found in reaction stages have more open exit angles to produce adequate exit flow areas while still providing sufficient blade quantities to adequately guide the flow. For this reason, nozzle exit angles of 25° are typical, and angles less than 20° are rarely seen in practice. For this example, nozzle exit angle $\theta = 25°$ is assumed, so if $W = 500$ ft/s, then $V = 500/\cos\theta = 500/\cos 25° = 551.7$ ft/s.

Therefore $\Delta H_{nozzle} = (551.7/223.8)2 = 6.08$ Btu/lbm, and $\Delta H_{stage} = 2 \times \Delta H_{nozzle} = 12.16$ Btu/lbm.

For the same total enthalpy available $\Delta H_{total} = H_1 - H_2' = 149.6$ Btu/lbm. For a similar 700-mm mean blade diameter, $W = 481$ ft/s and

$$\Delta H_{stage} = 2[481/(223.8\cos 25)]^2 = 11.25 \text{ Btu/lbm per stage}$$

For an 800-mm mean blade row diameter, $W = 551$ ft/s and

$$\Delta H_{stage} = 2[551/(223.8\cos 25)]^2 = 14.76 \text{ Btu/lbm per stage}$$

So, $N_s = \Delta H_{total}/\Delta H_{stage} = 14$ (13.29 rounded up) stages at 700-mm mean diameter, and $N_s = 11$(10.13 rounded up) stages at 800-mm mean diameter.

As these examples show, for similar available enthalpy and nominal blade speed, a reaction turbine will have roughly twice as many stages to process the same available energy as an impulse turbine designed for the same conditions of service.

2.5.2 Airfoil Section Shape

The velocity diagrams for impulse and reaction turbine stages reveal some key differences in the desired airfoil shapes for each category of turbine. Figure 2.9 shows a side-by-side comparison of typical

Figure 2.9 Impulse and reaction airfoil shape comparison.

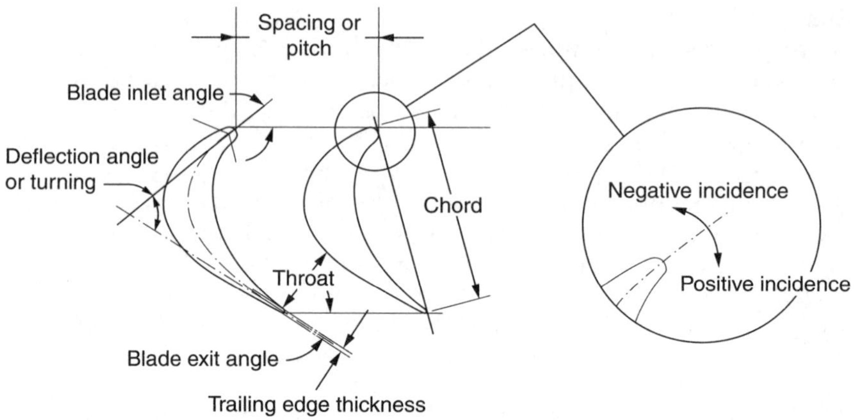

FIGURE 2.10 Basic airfoil geometry definitions.

turbine cascades for impulse (Fig. 2.9*a*) and reaction (Fig. 2.9*b*). Turbine designers use the velocity diagram for a stage to determine the best shape for the airfoils, rather than, as a layman might assume, use the airfoil angles to produce the velocity diagram. Designers either choose or design an airfoil with entrance and exit angles that match the calculated flow angles, within a reasonably small tolerance. At the entrance to the moving row, the designer wants to match the angle of the steam velocity vector to the pressure surface of the airfoil. Referring to Fig. 2.10, the best case is with an incidence angle (as defined in Fig. 2.10) equal to zero. But since this is rarely possible, an airfoil with a slight positive incidence angle is usually chosen, since small positive incidence entails negligibly small losses, where a negative incidence angle produces much larger flow losses, negatively impacting the efficiency of the stage.

The angular difference between the steam entering the moving blades (V_1) and the steam leaving the moving row (V_2) is referred to as *turning*, and it is obvious that the turning in an impulse row is significantly greater than that in the moving row of a reaction stage. As one would logically expect, this must be reflected in the shape of the airfoil. In addition, since the relative velocity between the steam and the moving blades is necessarily greater in an impulse stage, the steam velocity through the blade passages is greater, with correspondingly smaller entrance and exit area requirements. In turn this tends to produce smaller entrance and exit angles as well, which may also benefit the stage efficiency at the design point.

For 50 percent reaction stages, the airfoil angles can be made symmetrical. That is, the same airfoil section shape can be used for both the stationary nozzle partitions and the rotating blades, a situation shown in Fig. 2.9*b*. In some applications, both the stationary and the moving blades can be produced from the same precision rolled airfoil stock, which can significantly reduce the cost per row.

Besides the relationship between airfoil section shapes and steam flow velocities, the modulus of the airfoil in bending, airfoil mass (affecting the centrifugal force on the disk and root), and such mundane considerations as manufacturing and assembly processes for blades and turbine rotor disks all must be considered in selecting or designing the final airfoil shape for a specific stage.

Twisted Blades

In discussions so far, we have considered the velocity diagram only as it appears at the mean blade path diameter. This is the normal design approach for "short" stages, or stages with relatively large hub/tip ratios (that is $D_h/D_t = 0.85$ or greater), as typically found in the high-pressure stages of both impulse and reaction turbines. However, in low-pressure stages, taller blade heights are required to create stage areas sufficient to pass the large volume of steam flow. In condensing stages of both industrial turbine generators and process compressor drives, hub/tip ratios as low as 0.5 (and sometimes even slightly lower) are often found in the last or exhaust stage of the turbine. In the case of tall blades, the blades' tangential velocity, and thus the velocity ratio, varies along the blade height. Since blade velocities are smallest at the root and highest at the blade tips, the lowest velocity ratios (W/V) are found at the hub, and increase as one considers stations at larger radii.

To accommodate this, modern turbines are designed with twisted blade shapes so that the total turning and exit angles match the velocity triangles, while ensuring radial equilibrium along the length of the blade. Figure 2.11 shows an example of a tall blade and its associated velocity diagrams at the base, mean diameter, and tip of the blade. At the base, the blade section and its associated velocity diagram may look very much like a typical impulse stage, while at the mean flow path radius and at the tip the airfoil appears more typical of reaction stages. To avoid significant flow disturbances that can affect reliability as well as stage efficiency, the designer must be careful to ensure that recompression (sometimes referred to as negative reaction) does not occur at the base section.

As one may imagine, the design of tall or twisted blades is a complex topic, and massive collections of literature exist on various aspects of their design. For more comprehensive discussions on the design of twisted airfoils, the interested reader is encouraged to look in this book's reference list, which contains several texts with excellent discussions covering both basic and advanced topics in twisted airfoil design.

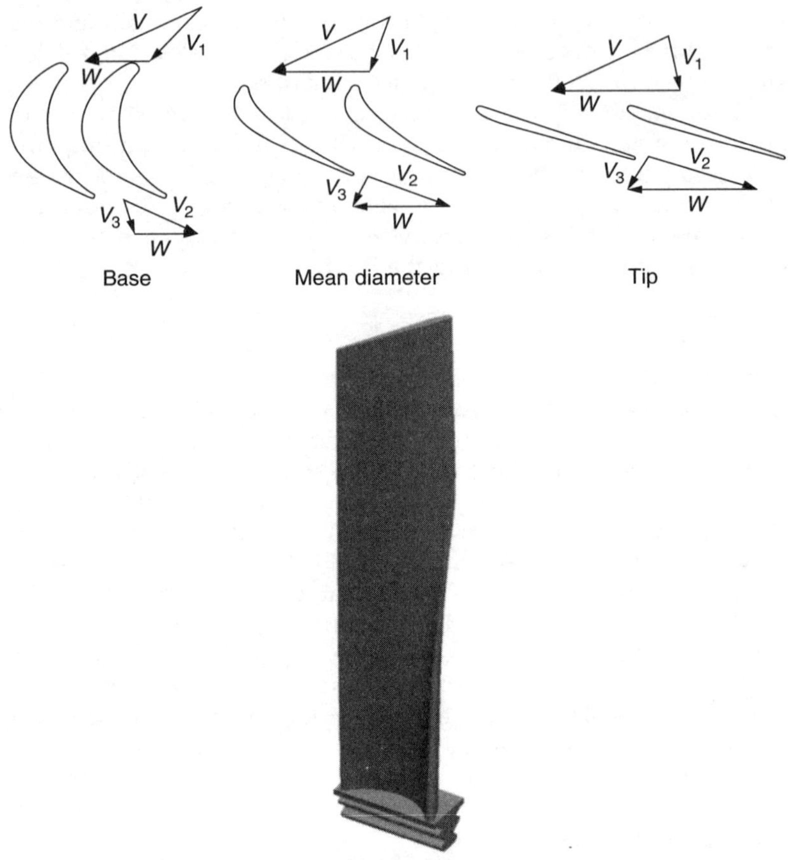

FIGURE 2.11 Twisted blade airfoil shape and its associated velocity diagrams at base, mean diameter, and tip.

2.5.3 Number of Blades

The number of blades in each row, and in the moving rows in particular, is closely linked to the airfoil shape. Most turbine manufacturers rely on a library of standard airfoil section shapes that are selected to match the desired flow angles determined by the velocity diagram for each stage. Once the basic airfoil shape has been selected, the turbine designer can then "scale" the airfoil to an appropriate size for a particular stage. In this way, the turbine designer can fine-tune the number of blades in the row and the strength of individual blades and can optimize the proportions of the flow passages bounded by the blades. Note that both the axial and tangential directions must be scaled by the same factor, or the basic airfoil section shape (importantly, the inlet angle and exit angle) will be changed.

While the turbine designer is selecting or designing the airfoil shape, his real design objective is to optimize the passages bounded by the blades. The blades' width and section thickness may be scaled in size, but the blade-to-blade spacing must also be scaled to maintain appropriate passage proportions. Optimum blade spacing (sometimes called the *blade pitch*) is defined by a parameter called *solidity*. Solidity is defined as the ratio of blade chord to spacing or pitch, as defined in Fig. 2.10. The optimum solidity for a stage using a specific airfoil shape depends primarily on the inlet and exit angles of the airfoil, but also on the passage area and relative flow velocities. The effect of solidity can be seen intuitively by considering a stage with very low solidity, that is, narrow blade chords coupled with wide blade spacing as in Fig. 2.12*b*. Too low solidity results in insufficient guidance for the flow, particularly at the blade exit, allowing the exit flow angle to deviate significantly from the desired exit angle. Conversely, a stage with very high solidity (see Fig. 2.12*a*) does an excellent job guiding the flow, but the extremely long narrow passages result in excessive friction losses along the passage walls. Once the optimum solidity is determined for a selected airfoil, the number of blades in the row can be calculated by dividing the stage circumference at mean blade diameter by the optimum pitch.

Depending on the application requirements, it is often desirable to deliberately choose the narrowest blade chord that will meet the minimum strength criteria for the stage, since it generally will result in the shortest axial length for the stage and shortest overall rotor-bearing span. It is almost always desirable to minimize the rotor length to help achieve a stiff and unresponsive rotor-bearing system, allow use of the smallest possible rotor diameter, and minimize the overall size and cost of the turbine.

The turbine designer must juggle several factors when determining the minimum chord and scale factor for a given airfoil. The airfoil must have sufficient section modulus to keep peak bending stresses within appropriate design limits. Fortunately, the bending strength of the airfoil is proportional to the scale factor cubed, while the load per blade is inversely proportional to the number of blades. Since the number of blades is also inversely proportional to the scale factor,

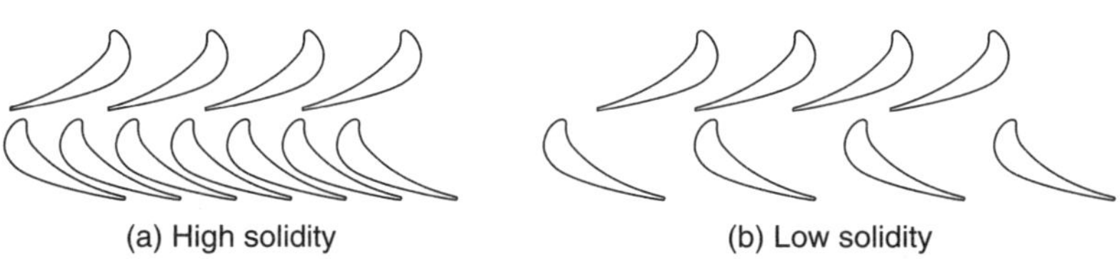

FIGURE 2.12 Comparison of blade rows with (*a*) excessively high solidity and (*b*) low solidity.

relative bending strength of a given airfoil section increases with the square of the scale factor, so a small increase in airfoil chord results in a large increase in bending strength.

Another important consideration is the shroud attachment. Especially for blade rows with riveted shroud attachments, the airfoil thickness must be great enough to accommodate a tenon of sufficient size to attach the shroud. For stages operating at moderate to high speed, or requiring heavy shrouds, the required tenon diameter may become a critical factor in determining the minimum blade chord.

With both the optimum solidity and the minimum airfoil section chord/width determined for the selected airfoil shape, the number of blades in the row can be calculated by dividing the stage circumference at the mean blade diameter by the optimum pitch.

$$N_b = 2\pi D_m / s$$

where N_b = number of blades
D_m = mean blade path diameter
s = blade spacing (pitch)

For reasons that will become clear in later chapters, it is sometimes important that the final blade count be evenly divisible by a specified number of blades under each shroud segment. The engineer may find it necessary to adjust blade size and/or blade pitch to achieve an acceptable shroud packeting scheme.

2.6 Blade Loading

Blade loading, as the term is used in this context, encompasses all the external loads that are applied to the rotating turbine blades as a result of the demands imposed on the turbine by the driven machine, plus loads that arise internally to the turbine during operation. Regardless of their origin, blade loads all fall into two broad categories—steady loads and unsteady loads.

2.6.1 Steady Loads

Steady blade loads are a result of the conversion of thermal energy in the working fluid to mechanical shaft work. Through the mechanism of momentum change as it passes through the flow path passages bounded by the rotating blades, the working fluid imposes forces on the blades. To ensure that the blades are designed to resist these forces, their magnitude must be known or predicted based on the power demanded by the driven machine and knowledge of the turbine stage design.

To show how stage energy relates to power developed by a turbine stage and ultimately to the loads imposed on the turbine blades, consider the simplest example of a turbine with one stage (i.e., a single-stage turbine), for which some basic parameters are known:

P_1 = inlet pressure

T_1 = inlet temperature

P_2 = exhaust pressure

M_1 = steam mass flow

D_m = blade path mean diameter

X_b = number of blades on disk

N = turbine rotational speed, rpm

Since the loads on individual blades must be predicted to analyze how the blades respond to the load imposed, the turbine stage parameters to be predicted include

$\Delta H = H_1 - H_2$

w = shaft power (rate of work or energy conversion), hp or kW

T = shaft torque

F = total tangential force acting at D_m

f_b = force on individual blade acting at D_m

To apply this simple example to determine the load on the blades in a hypothetical stage passing mass flow M_1 (lbm/h), the stage power can be determined by multiplying the enthalpy change $(H_1 - H_2)$ Btu/lbm by the steam mass flow rate for the stage (M_1), which results in

$$M_1 \times (H_1 - H_2)/C_e = w$$

where C_e is a proportionality constant relating stage energy to power output. In USCS units C_e = 2545 Btu/hp·h and w = shaft work.

Turbine torque T is calculated from w and speed N:

$$T = wC_t/(2\pi N)$$

where C_t is a constant relating torque to power. In USCS units (lb, ft, s) C_t = 33,000 lb·ft/min.

To produce torque T, the tangential force required at the mean blade path diameter is $F = T/(D_m/2)$, and the force on each individual blade is therefore $f_b = F/X_b$ or $f_b = 2T/(D_m X_b)$, which is the equivalent force acting on each individual blade at the mean blade path height, required to produce the output torque T and power P_1.

This value can then be used by the turbine designer to carry out blade loading calculations such as shear and bending stress in an airfoil, using the airfoil's section properties, or combined with other known or calculated loads (e.g., centrifugal force) to predict combined or total stress in the blade.

This simple calculation carried out at the mean blade path diameter is often sufficient for short blades in light to moderately loaded stages. For more critical stages, especially taller blades in stage with hub/tip ratios less than 0.90, a more sophisticated steady blade loading prediction method is called for. Traditionally, this is a streamline calculation that predicts pressures on the blades' pressure and suction surfaces at a series of stations from the base or hub diameter to the blade tip. Figure 2.13 shows the plotted pressures predicted from a streamline calculation of an airfoil. Streamline calculations have an advantage in that they have been in use for many years and consequently have been validated by both field experience and dedicated test data. Streamline codes are now being supplanted by computational fluid dynamics (CFD) routines for the same purpose, and CFD will probably completely replace the older streamline calculations as the experience base of CFD builds and results are validated by measurement and field experience.

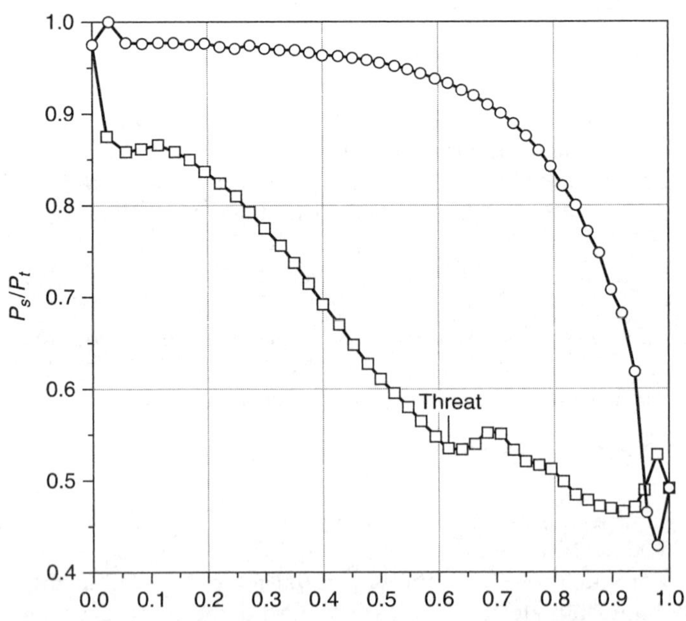

FIGURE 2.13 Streamline calculation output for pressure and suction surface pressure distributions. (*Courtesy of Advanced Turbomachine, LLC.*)

2.6.2 Unsteady Blade Loads

Unsteady loads are, as the name implies, variable loads imposed on the blades from a variety of sources. Variable loads arise from variations in imposed load during normal and abnormal operation of the driven machine, variations and disturbances in flow through the blade passages, and forces imposed on the turbine from external sources. In general, the life of mechanical components is directly related to the magnitude and quantity of repeated loads imposed on the blades. Sufficient repetitions of even relatively small loads can lead to failure of most metals. Turbine blade reliability is similarly highly dependent on the magnitude and frequency of the unsteady loads to which it is subjected. The ability to identify the entire range of unsteady loads on a particular stage is critical to the design of reliable turbines.

Turbine design engineers are, of course, interested in both the frequency and the magnitude of unsteady forces. Unfortunately, while the frequency can usually (though not always easily) be determined from limited knowledge of the turbine construction and conditions of service, the magnitude of the force is much harder to determine with any degree of accuracy. In most cases, turbine designers have to satisfy themselves with knowledge of frequency and then rely on conservative (usually worst-case) assumptions about the magnitude of the imposed forces.

Nozzle Passing

Unsteady forces due to effects from the rotating blades passing the stationary nozzle partitions are possibly the most important, and consequently the most studied, source of excitation for blade vibration in axial flow turbines. These forces arise mainly because the nozzle vanes' or partitions' trailing edges have finite thickness, and because the steam is discharged toward the moving blades via discrete nozzle passages that are subject to a variety of small variations in geometry. In turn, these produce a variety of flow nonuniformities and disturbances at the nozzle exit, immediately upstream from the moving blade row. Designers have several design options available that they use to minimize the effect of nozzle disturbances on the adjacent rotating blades. These include minimizing nozzle trailing edge thickness to reduce the size and effect of trailing edge blockage and wakes, and increasing the setback dimension (i.e., the distance from the nozzle exit to the leading edges of the moving row) to attenuate secondary flow disturbances as much as possible before the steam enters the moving blade passages. Figure 2.14 illustrates those features of a typical turbine stage.

The flow downstream from the stationary nozzles contains wakes from the trailing edges, nonuniform pressure distributions due to variations in the discrete nozzle passages, and secondary flow disturbances in the form of vortices or eddies at the inner and outer endwalls. Turbine designers also do their best to avoid transonic

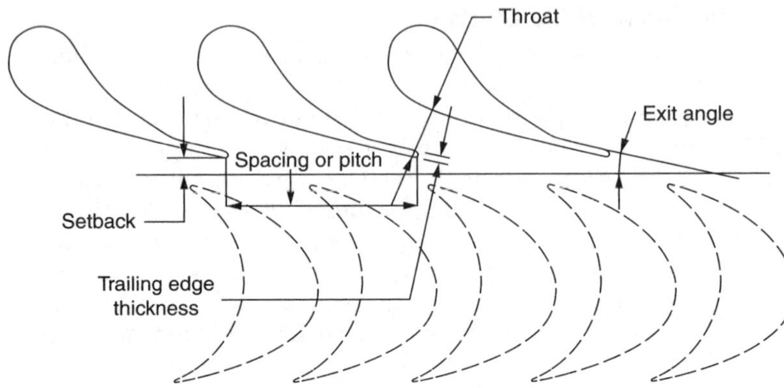

Figure 2.14 Nozzle geometry definitions.

velocities in normal operation, but turbines may also experience off-design operating conditions where nozzle exit velocities are high enough to create shocks in the flow immediately downstream from the nozzles. Regardless of the source of the flow field nonuniformities, the rotating blades are exposed to nonsteady forces at nozzle passing frequency, equal to the number of discrete nozzle passages times the rotational speed in revolutions per unit time, or to low multiples of nozzle passing frequency.

These same disturbances that influence the moving blades may also have a negative impact on stage losses and efficiency, so the turbine designer has an additional incentive to minimize these effects. For example, turbine design engineers strive to produce trailing edges that are as thin as possible to minimize their effects on steam flow, i.e., wakes and flow blockage, and their subsequent effect on the rotating blades. However, manufacturing limitations and simple strength considerations often impose practical constraints on how thin the nozzle trailing edge can be produced.

Trailing edge thickness depends on several factors, including the size or scale of the vanes. Intuitively, trailing edge thickness depends on the height and width of the row, as well as on the forces due to differential pressure imposed on the vanes. Basic strength considerations lead to thicker trailing edges as the nozzle height increases, and as the stage pressures and temperatures increase. In the end, this means that the blades with the highest imposed loading due to stage power levels also experience high excitation forces due to nozzle passing. In addition to nozzle strength considerations, the manufacturing process specified for the nozzles also influences the trailing edge thickness, depending on the precision and accuracy of the method used to produce the vanes. For example, cast vanes' trailing edges are typically thicker, and vary more in thickness, than fully machined or drawn (precision rolled or extruded) vane sections.

The effects of these disturbances usually die out quickly with distance, as the wakes and zones of disturbed flow mix out, and designers often take advantage of this phenomenon by deliberately leaving additional distance, or setback, between the trailing edge of the stationary nozzle vanes and the entrance to the moving blade rows. However, while increased setback reduces the unsteady blade forces due to nozzle passing, excessive setback may also reduce stage efficiency. For this reason, turbine designers usually limit the setback to a relatively small value—the drive to maximize turbine efficiency means that the distance between the nozzle's trailing edge and the blade immediately downstream is too small to afford much opportunity for these disturbances to mix out.

If the nozzle assemblies were perfect, with each partition perfectly placed so there were no variations in nozzle exit area, nozzle spacing, and setback distance, the designer would have to consider only the fundamental nozzle passing frequency (that is, 1X nozzle passing) and its low multiples (2X, 3X, and so on)—pure tones in other words—when analyzing the frequency response of the rotating blades. However, since these nozzle characteristics always exhibit variations as a result of manufacturing tolerances (refer to Fig. 2.15), the designer must consider their influence on the rotating blades. The effects of variations in nozzle exit area, circumferential pitch or spacing, and setback modify our otherwise pure signal at 1X nozzle passing frequency, and superimpose other harmonic frequencies on the predominant 1X nozzle passing frequency "signal" or forcing function as illustrated by Fig. 2.16. Although this discussion is primarily concerned with variations from design values due to manufacturing

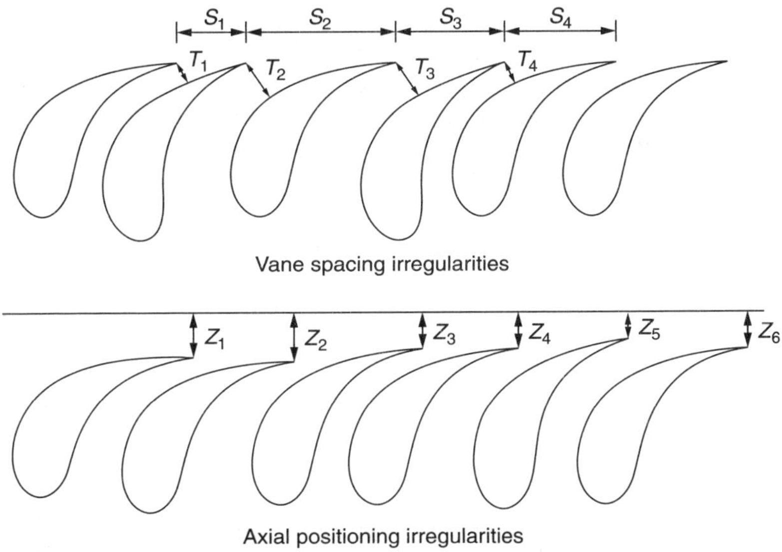

FIGURE 2.15 Nozzle manufacturing variations.

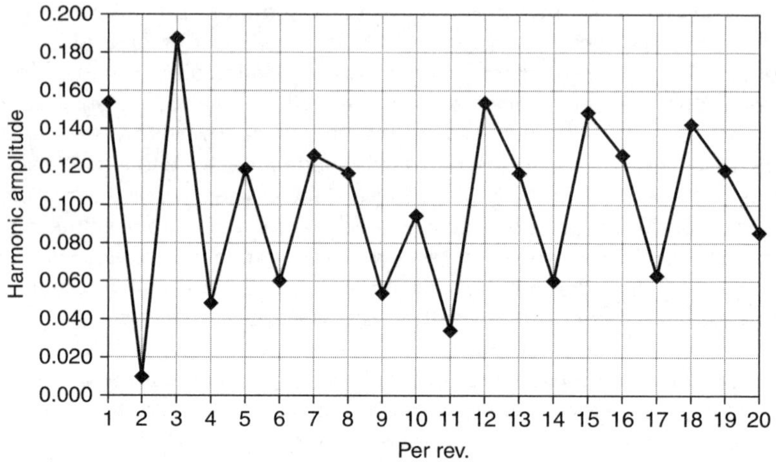

FIGURE 2.16 Harmonic stimulus from vane spacing irregularities.

tolerances, the inevitable variations that originate from erosion, wear, and foreign object damage during turbine operation can also produce potentially destructive unsteady forces on the rotating blades.

Virtually all turbine manufacturers and major turbine service organizations have developed procedures to evaluate and deal with the effect of manufacturing variations on the vibratory behavior of the rotating blades. In general, they all fall into three broad categories:

1. *Dimensional and area tolerance criteria.* This is the simplest approach, where maximum allowable tolerances are specified for passage area, spacing, and setback for the nozzle or diaphragm assembly in question. The specified tolerance limits are based on field experience with a great number of turbine stages, and they may be further bolstered by more detailed analyses to help the engineer create a conservative standard tolerancing process.
2. *Statistical criteria.* The application of automated inspection and data collection to document the variations in nozzle construction was followed closely by the application of statistical methods to evaluate whether these variations are acceptable. Statistical criteria are often used in conjunction with simple tolerance criteria for area, pitch, and setback measurements. One common measure of consistency in the measured parameters that is used in many statistical evaluations looks at the standard deviation of area, pitch, and setback for the nozzles. Since a nozzle or diaphragm assembly has a finite number of nozzles, another important measure looks at the extreme values (large or small) of area, pitch, and setback, either in absolute terms or in comparison to the mean value or standard

deviation of each parameter. Based on knowledge of the design or the manufacturing processes involved, the engineer may also wish to statistically evaluate specific features of the nozzle assembly. Examples of such features could include the nozzle passages immediately adjacent to the split line on horizontally split diaphragms, or the gaps between the individual arcs of the segmented nozzle rings that are used on many high-pressure and temperature control stages.

3. *Forcing function criteria.* A number of more sophisticated methods have been developed to evaluate the effect of nozzle variations, not simply on a statistical basis, but to determine if the specific distribution of these variations around the circumference of a specific nozzle assembly can unintentionally excite one or more resonant frequencies of the moving blades or of the bladed disk assembly. Turbine manufacturers and most major turbine service organizations have developed proprietary analytical routines to carry out this task. While there are differences in the details of the inspection routines and analytical processes for each manufacturer, they all rely on the same basic methodology. Almost all these methods employ a Fourier series analysis to determine the frequencies that are embedded in the nozzle passing "signal" or forcing function, along with their relative magnitude. The same calculation usually provides guidance to specify appropriate remedial operations for each specific nozzle passage within the assembly, and to evaluate the results of those remedial operations.

Partial Admission

Several special cases arise in the design of turbine stages that affect blade loading. One common cause of unsteady loading arises from the use of partial admission stages. All the cases considered so far have assumed that a full 360° arc of nozzles direct steam into a full 360° arc of moving blades. There are cases, however, when a partial arc of steam admission is required. In other cases partial admission stages may be used to achieve lower costs, increased operating flexibility, or improved part-load efficiency.

Where and why are partial admission stages used? By far, control stages represent the most common application for partial admission stages. As the name implies, control stages are used to control steam flow to the turbine, to achieve the driven machine's demand for speed, power output, or torque. Since the turbine is part of a larger system, steam flow is often also controlled to affect other system parameters, such as minimum boiler header pressure, extraction header pressure, extraction or exhaust mass flow rates, and a host of other possible control parameters.

Figure 2.17 schematically illustrates a common configuration for control stages with several short nozzle arcs, each controlled by a

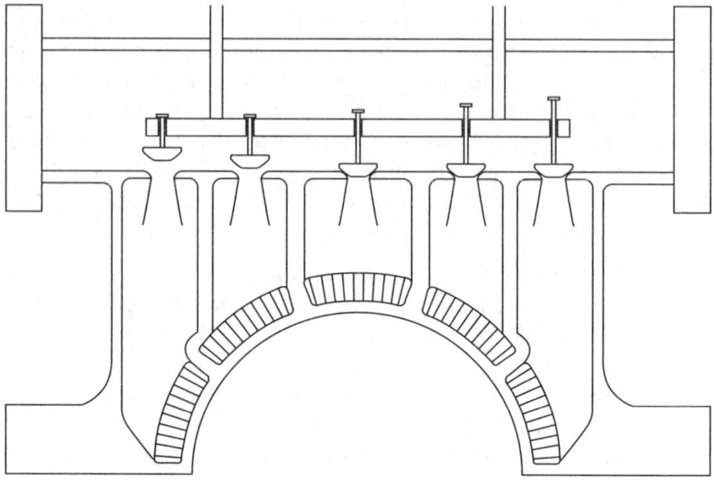

FIGURE 2.17 Schematic view of control stage for a multivalve turbine inlet showing partial arc nozzles, each controlled by individual valves opened in sequence.

separate valve. Valves are opened in a specific order by a mechanical or hydraulic mechanism. Depending on the application, a variety of mechanical, pneumatic, hydraulic, or electromechanical mechanisms may be used to open the valves in sequence to admit more or less steam to turbine. Figure 2.18 shows one common control configurations for process drives.

FIGURE 2.18 Control stage valve actuator mechanism—cam-operated valves. (*Photograph courtesy of Dresser-Rand Company.*)

Partial arc stages sometimes pose difficult design challenges for two main reasons. First, as already pointed out, all the power produced in the stage (which may be considerable in control stages) is produced in a fraction of the blades in the row. In addition, in the course of every revolution, each blade in the row goes from a zone where it is subjected to zero steam force to a zone where it is acted on by the maximum steam force, then back to zero. The blade thus sees the maximum stress range possible, resulting in a much higher alternating stress range than a full admission stage. In fact, since the blades are subject to drag during the inactive portion of each revolution, the minimum torque may actually be negative (i.e., a drag torque).

The primary difference for partial admission is that the stage power is produced in only the working blades—those on which steam is acting instantaneously, basically the blades that are directly downstream from the admitting arc of nozzles. In the previous section, the loading on individual blades was shown to be $f_b = T/(D_m/2X_b)$. For partial admission stages, the number of blades X_b must be changed to reflect the blades actually producing power. Individual blade loading may be arrived at in one of two ways:

1. *By nozzle count.* In this approach, the designer counts (or specifies) the number of active nozzles and divides the total nozzle count (as if the stage were full admission) by the number of active nozzles, and then uses this ratio to adjust the individual blade loading, with the following results: $f_b = 2TX_b/(D_m X_a)$.
2. *By arc of admission.* In this approach, the designer specifies the angular arc of admission, then adjusts the individual blade loading by the ratio of $360°/\theta$, where θ is equal to the active arc of admission in degrees. While not as precise as the actual nozzle count, this method is usually accurate enough to ensure a reliable design, and it is often used for preliminary design calculations, when the final nozzle count may not yet be specified. In this case, $f_b = 720T/(\theta D_m)$.

For multivalve control stages, there is an additional consideration. Critical loading usually does not occur at maximum flow or maximum stage horsepower, but at the point where the first valve to open reaches its fully open position. At this operating point, there is full pressure immediately upstream of the nozzles while the total flow through the turbine is still relatively small, so the stage pressure after the first moving row is still relatively low. This set of conditions makes the maximum energy available to the control stage, while all the power produced is concentrated in the smallest possible arc of admission. This operating point—"one valve wide open"—defines a critical design loading condition for control stages. To ensure that it truly represents the worst case loading, it is always calculated at the maximum continuous inlet pressure and temperature (or maximum extraction pressure for extraction control stages), in conjunction with minimum backpressure on the turbine.

Turbine Structure

Other potential sources of unsteady blade forces arise from various turbine structures that are necessary to support and align various internal components, including seals, bearings, and stationary flow path components as well as the turbine pressure casing itself. In fact, just about any structure upstream or immediately downstream from the bladed disk assembly has the potential to impose a force of some magnitude on the blade or bladed disk. Examples of such structural components include structural supports and struts in extraction and exhaust passages. Figure 2.19 shows how struts are used in a typical condensing exhaust to stiffen and support large, relatively flat, thin walls against differential pressure loads, and to support the inner casing that supports the stationary flow path components. Struts such as these present obstructions to flow and cause flow disturbances that may propagate upstream to affect the last-stage rotating blades. Depending on the number and location of structural supports, the turbine designer can estimate potential frequencies of interest based on the number and spacing of the struts, splitters, or other structural components found in the flow passages.

Variations in stage backpressure can also be caused by nonuniformities in the shape and size of casing passages in extraction and exhaust casings. Extraction openings in casings are an example of a structure that can potentially cause variations in stage backpressure around the periphery of the moving blade row immediately upstream of the extraction opening. In most cases, the variation in pressure is

FIGURE 2.19 Condensing exhaust showing internal struts. (*Photograph courtesy of Dresser-Rand Company.*)

small enough to present no danger, but the engineer must be aware of the potential for trouble, and must ensure that the passages are designed to prevent destructive variations in stage backpressure.

An example of this may be found in many older process drives (pre-1980s), where it was common to use double-flow and even triple-flow exhaust staging. By dividing the flow into two streams, shorter (one-half as tall!) last-stage blade heights can be used, simplifying their design, enabling the turbine design engineer to use stiffer low-pressure blades that are less susceptible to destructive vibration than longer blades. With modern blade analysis, multiflow exhausts are not widely specified for new turbines, but there are still many double-flow exhausts in service. A potential problem with multiflow exhausts occurs in the stage immediately ahead of the multiflow section. Steam is transferred from this stage to the double-flow section via one large or several smaller crossover passages that bring steam out from the main turbine casing and back into the multiflow section (see Fig. 2.20).

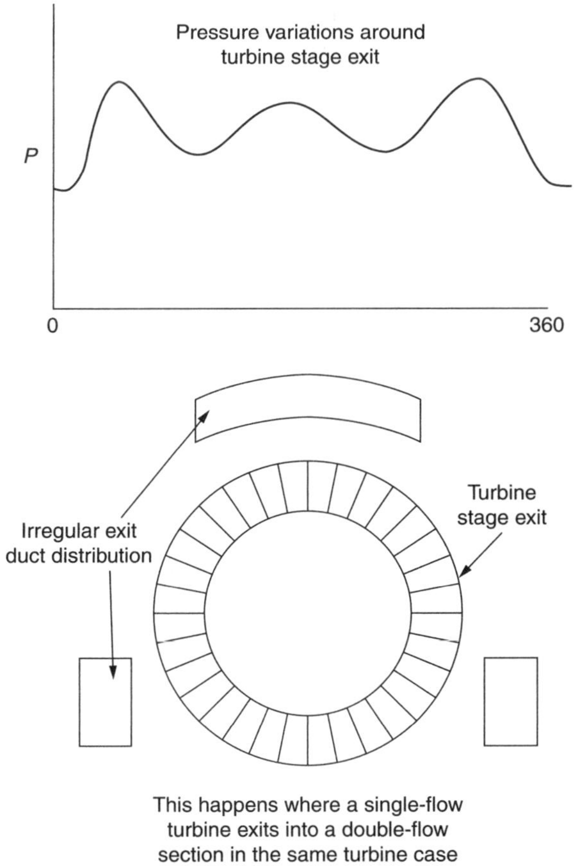

FIGURE 2.20 Pressure variations due to irregular duct spacing.

Since these passages are necessarily discontinuous around the periphery, there are variations in backpressure and potentially significant flow patterns around these crossover passages. The turbine design engineer must recognize the potential impact of double-flow passages and must either design the blades in the stage ahead of the multiflow section to withstand high unsteady forces or take steps to reduce the magnitude of these forces. A number of different construction features have been developed by turbine manufacturers to attenuate these forces, including diffusers, specially contoured exit turns, and internal baffles that shield the blade from backpressure variations as the steam exits the moving blades ahead of the crossover passages.

Although far from exhaustive, these few examples of how internal turbine structures give rise to unsteady forces are illustrative of other potential sources of unsteady forces. As readers consider how these structural features give rise to potentially destructive unsteady forces on the moving blade rows, they should be able to easily identify other potentially significant sources of blade excitation and to take reasonable precautions to analyze these forces, reduce their magnitude, or avoid them altogether.

CHAPTER 3

Turbine Blade Construction, Materials, and Manufacture

The first steam turbines were applied to generators, ships' propulsion systems, and other mechanical drive applications in the early 1880s. In the intervening 130 years, steam turbines have been applied to every conceivable type of service requiring shaft power. Today, steam turbine designs for industrial, marine, and utility services range in size from a few kilowatts in output to multicasing giants producing over a gigawatt. Industrial steam turbines are still used in a wide variety of processes, but today their major applications are as turbogenerators for in-house power and cogeneration and as mechanical drives for pumps, compressors, and fans in the power generation, refining, and petrochemical process industries.

Process drive steam turbines represent a specialized subset of industrial steam turbines that are mainly used by the refinery and petrochemical industries to drive compressors in critical services. These machines range in size from a few kilowatts at the small end to perhaps 100 MW at the large end of the range. They are usually characterized by variable-speed operation (as distinct from the constant synchronous speed of generator drive turbines), sustained operation at high speeds and outputs, and a requirement for high reliability. The varied demands of such a vast application range have given rise to an amazing number of mechanical variations in the design of all turbine components, and especially in the construction details of the rotating blades, a few of which are shown in Fig. 3.1.

Fortunately, when the discussion is limited to industrial steam turbines, the variations in blade size and design are manageable. For industrial steam turbines, the range of blade heights under consideration starts

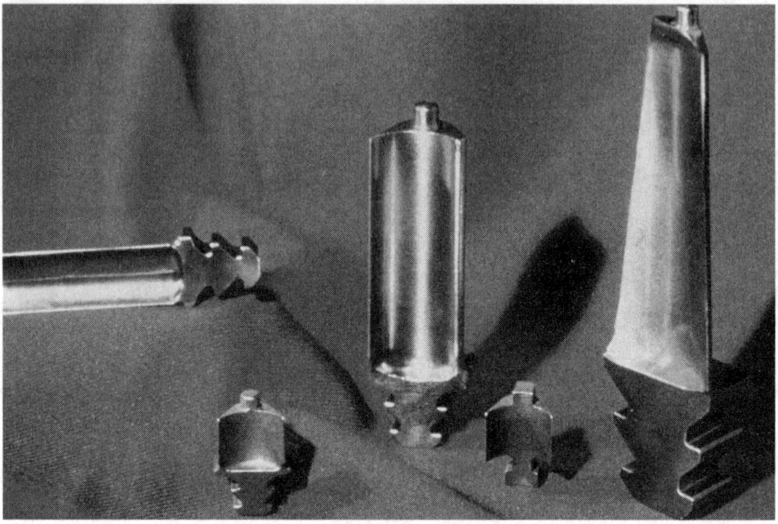

FIGURE 3.1 Blades for process drive steam turbines come in a wide variety of sizes, shapes, and styles. (*Photograph courtesy of Dresser-Rand Company.*)

around 12 mm (0.5 in) and goes up to a maximum of 750 mm (30 in). In addition, while the number of *potential* materials is extensive, the list of practical, widely used materials is limited to a handful of martensitic stainless steels, titanium, and a few specialty alloys. Similarly, the range of potential processes and tools that can be applied to manufacture turbine blades is very broad, but in practice most blades are produced using only four basic processes—milling, grinding, roll forming, and forging.

A brief overview such as this cannot adequately cover every one of the materials, design variations, and manufacturing processes that are currently employed for industrial steam turbine blading. Surveying the wide variety of designs that are found in successful operation around the globe makes it obvious that many equally good solutions have been developed to meet the demands of any steam turbine application imaginable. Therefore, the inclusion (or omission) of specific features in our discussion should not be interpreted as an endorsement of those features over alternative designs, or of one turbine supplier's solution over another supplier's solution.

While geometric detail dimensions may vary widely, just about every turbine blade can be broken down into a common set of basic features—the airfoil that interacts with the steam flow, the root or disk attachment, the platform that forms the inside-diameter surface of the flow path; and the shroud that forms the outside-diameter flow path surface (see Fig. 3.2). In addition to these basic elements, some blade designs incorporate specialized structures such as retainers,

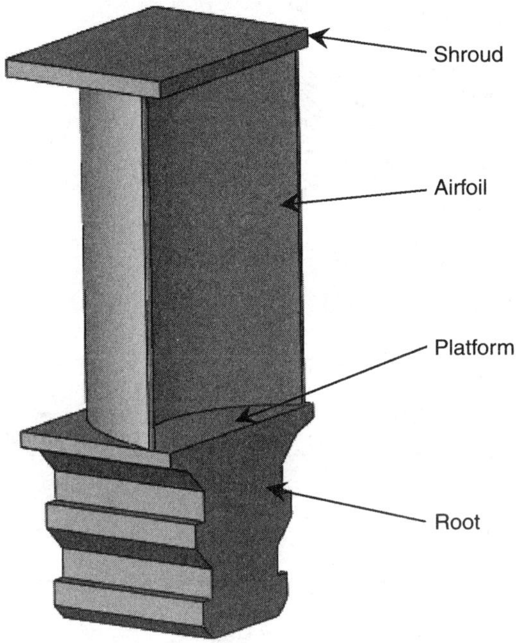

FIGURE 3.2 Common blade design elements.

lacing wires, or auxiliary dampers. These elements may appear as part of monolithic individual blades or as separate component parts that are assembled to form a complete blade system. As we will see, the designer has a wide choice of construction features from which to tailor the blade for a specific application. The final choice of blade design details is often influenced by available machine tools and manufacturing processes, available materials, and relative part cost considerations.

3.1 Airfoils

If we take these common blade construction features in order, it is logical to start with the airfoil, because the airfoil does all the work. The airfoils of the rotating blades represent the only components in the entire turbine that convert energy in the steam to shaft torque and power. Obviously, the performance and reliability of the blades are critical to successful process drives. Critical attributes of the airfoil include the profile accuracy of the airfoil surface contours, blade-to-blade spacing, and blade path exit area after installed in the wheel. Figure 3.3 illustrates the common terminology applied to airfoils, defining some key airfoil parameters such as chord, entrance and exit angles, exit area, and others.

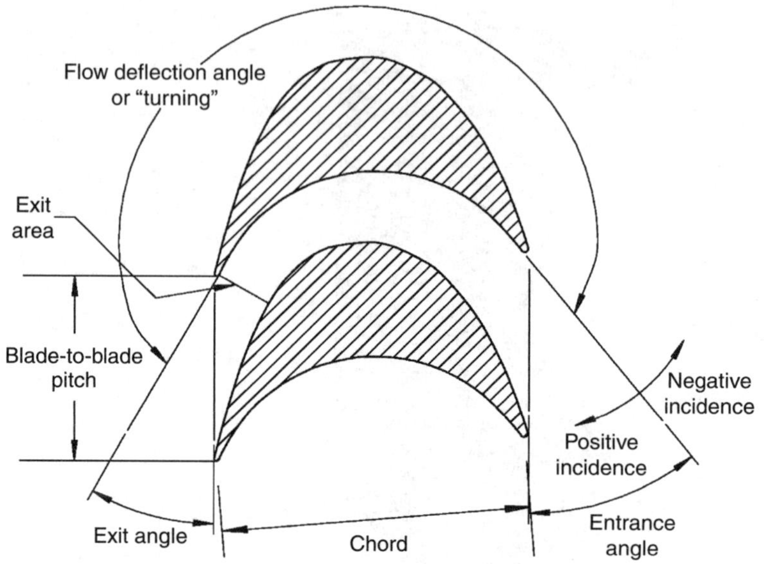

FIGURE 3.3 Airfoil geometry definitions.

3.1.1 Impulse and Reaction Blades

As noted in Chap. 2, there exists a major division in steam turbine types between impulse turbines and reaction turbines. To review, in pure impulse turbines the entire stage pressure drop is taken across the stationary nozzles, converting heat energy in the steam to kinetic energy at the nozzle exit. That kinetic energy is then transferred to the shaft purely by momentum change (i.e., by turning of the velocity vector only) in the moving blade row, with no significant additional pressure drop.

In contrast, in reaction stages, only a portion of the stage pressure drop is used to convert energy in the steam to kinetic energy in the stationary nozzles, while the remainder of the total stage pressure drop is taken across the moving blades. The momentum change in the moving blades is therefore contributed by conversion of the kinetic energy at the nozzle exit, plus the reaction forces from acceleration of the steam. This acceleration occurs as the remainder of the available stage energy is converted to kinetic energy in the moving blade row. As you might expect, these differences turn out to have important influences on the blade shapes used in each type of machine.

3.1.2 Impulse

Since peak efficiency occurs at low W/V [i.e., at $W/V = (\cos\theta_1)/2$] for impulse stages, the blade speed relative to the steam exiting the nozzles is lower, and the steam enters the moving blade row at a greater

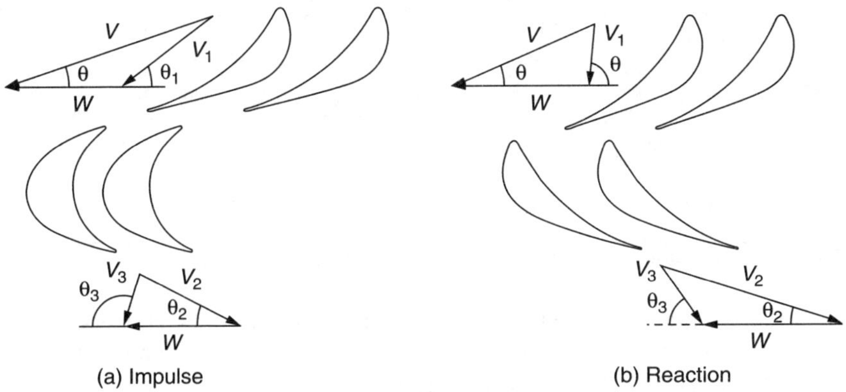

FIGURE 3.4 Comparison of (*a*) impulse and (*b*) reaction cascades and their velocity diagrams.

angle to the turbine axis (refer to Fig. 3.4*a*), it must be turned through a greater deflection angle before exiting the moving row. Airfoil sections for impulse stages are characterized by greater turning, and thus a greater total angle, between entrance and exit sides of the moving row. In addition, as we have already seen, to maintain the relatively low W/V values necessary for peak efficiency, the impulse turbine design engineer usually opts for a small number of stages with higher energy per stage, which tends to result in greater steam forces acting on the blade row. Once an airfoil section is selected with the desired total turning and appropriate inlet and exit angles, higher steam loading drives the designer to scale up the airfoil proportions, resulting in fewer, but more robust, blades in the moving row.

3.1.3 Reaction

In reaction stages, where peak efficiency occurs at high W/V, blade speed is relatively low compared to the steam and consequently enters the moving row at a relatively small angle to the turbine axis (refer to Fig. 3.4*b*). In contrast to the case of an impulse turbine, the reaction turbine design engineer usually opts for more stages with higher W/V, and lower energy per stage, resulting in relatively lower steam forces on the blades of each moving row. In counterpoint to impulse stages, reaction stages tend to have a greater number of narrower, less robust blades in each moving row.

Figure 3.5 is comparison of typical turbine cascades for impulse and reaction turbines for the same services. A few additional key features of each turbine type stand out: (1) In the reaction turbine cascade, the stationary nozzle vanes and the rotating blades both utilize mirror images of the same airfoil profile (i.e., symmetrical blading). (2) The impulse stages, due to the higher loading in each stage and on individual blades,

FIGURE 3.5 Comparison of impulse and reaction turbine cascades.

are wider than their less highly loaded reaction counterparts. The reaction turbine may be slightly longer than the impulse cascade, but not in direct proportion to the number of stages.

3.1.4 Twisted-Tapered Airfoils

In Chap. 2, the need for airfoils of varying section from base to tip for taller reaction stages and for the low-pressure stages of impulse turbines was explained. Such airfoils are often referred to as twisted or twisted-tapered airfoils. Figure 3.6 shows several versions of blades with twisted

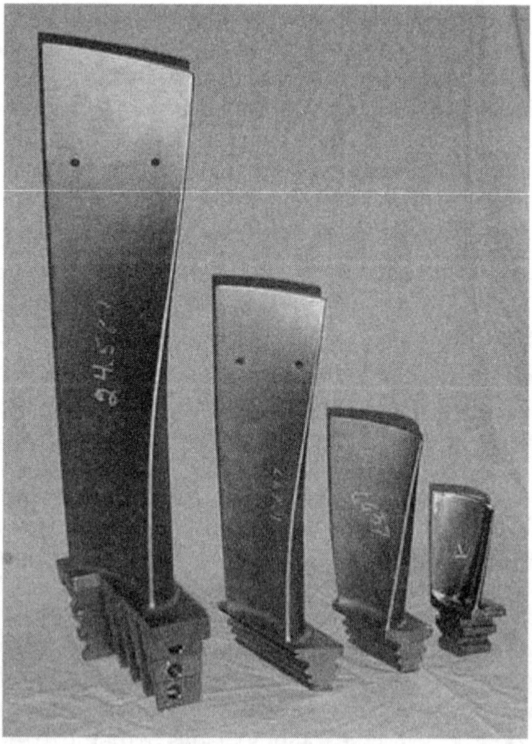

FIGURE 3.6 Tall blades with twisted airfoils. (*Photograph courtesy of Dresser-Rand Company.*)

airfoils, illustrating a few important aspects of blades with twisted airfoils. From Figure 3.6 it is evident that all these blades have airfoil shapes with a great deal of turning at the base section, and very little turning at the tip, as we noted in Sec. 2.5.2. In fact, the tip section of the tallest blade is very nearly flat in cross section, denoting virtually no turning, only acceleration of the flow inside the moving blade passages.

3.2 Roots

Turbine blade roots come in a vast array of shapes, sizes, and attachment geometries. The root is the primary mechanical attachment between the rotating blade and the turbine disk. The root must restrain the blade against centrifugal forces to keep it attached to the periphery of the disk and at the same time transmit steam forces from the blade to the disk to provide shaft power.

Roots in common use can be divided into three broad classifications—dovetail or circumferential roots, axial or fir tree roots, and pinned root attachments (Fig. 3.7). The turbine design engineer must understand the advantages, limitations, and common applications for each root design.

3.2.1 Circumferential or Tangential Dovetail Roots

As their name suggests, circumferential root attachments are circumferential in form. They are the most widely used root form in industrial steam turbines and process drives because they are easily machined on the rotor by turning in a lathe or machining center. The ability to turn the form in the rotor means that the rotor machining is

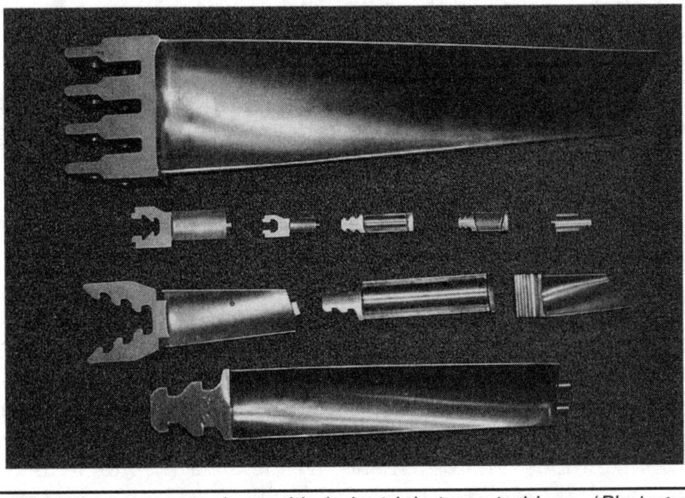

FIGURE 3.7 Roots commonly used in industrial steam turbines. (*Photograph courtesy of Dresser-Rand Company.*)

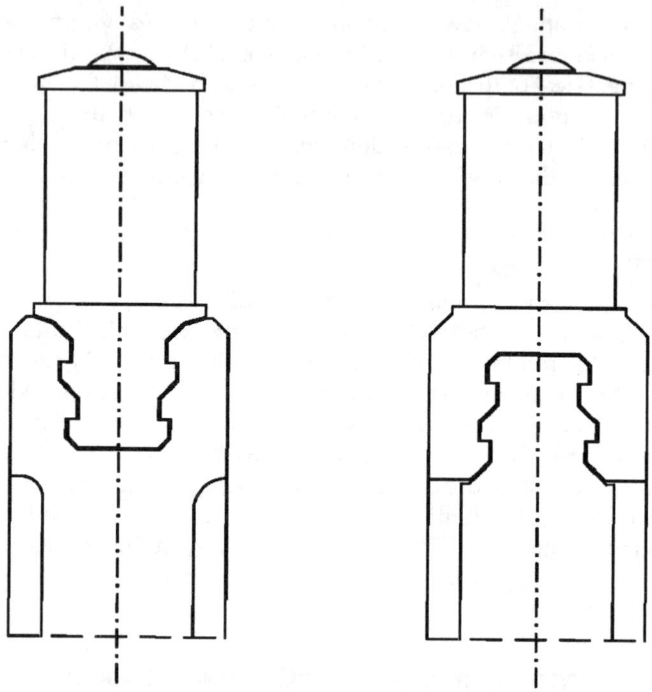

FIGURE 3.8 Dovetail- and straddle-style circumferential roots.

much simpler and less expensive than alternate root forms such as axial fir trees. Figure 3.8 shows the basic configuration of the circumferential root form. The internal dovetail root style and the straddle root style are typical of tangential roots in wide use today. The simplest root form is that shown in Fig. 3.9, which is a basic one-tooth form. A single tooth on each side of the root engages a matching single tooth in the wheel form. Without multiple contact points on each side of the root, the close tolerances between contact surfaces required of multiple-tooth root forms are unnecessary. A single-tooth root form is always less expensive than multiple-tooth root forms.

Multiple-tooth forms (refer again to Fig. 3.8) are used when centrifugal and steam bending loads are moderate to high, resulting in loads that exceed the capability of a single tooth. If the load is shared between additional contact points, both contact and root bending stresses are reduced, enabling higher powers and speeds than are possible with single-tooth root forms. At the same time, multiple contact points impose more demanding tolerance requirements on the wheel and the blade roots to ensure that the loads are shared more or less equally among multiple contact points.

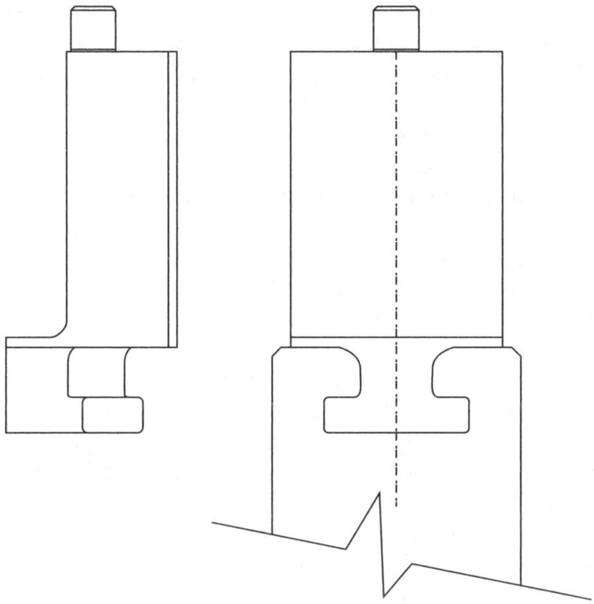

FIGURE 3.9 Single-tooth root geometry.

To ensure load sharing among multiple contact surfaces, blade manufacturers employ matched sets of cutters, produced to very close tolerances, for the blade root and the wheel groove form to ensure correct fit. Figure 3.10 shows typical milling cutters for producing the root form on blades with dovetail roots. The milling cutters are intended to be used in a milling machine or flexible machining

FIGURE 3.10 Matched dovetail root cutter set. (*Photograph courtesy of Dresser-Rand Company.*)

center of the type discussed in the section on machining. The wheel root form is machined using a matching form cutter that is procured as a set with the blade root milling cutter to ensure that the tolerances between parts are very closely held. Even with matched cutters, it is sometimes difficult to ensure that the blades and the matching wheel grooves achieve sufficient contact on all contact points. It is common practice to produce one of the components, then fine-tune the setups and cutters for the other component to ensure that the desired contact between the components is achieved. Contact between the mating parts can be inspected and verified using machinists' bluing or a similar dye transfer process to verify the extent of contact between the parts.

While circumferential roots are perhaps the most widely used type of blade attachment, they have some limitations, especially for units with high speeds and high centrifugal loads, such as, tall last-stage blades or extremely robust control stage blades. In addition to the contact stresses and tensile stresses in the root, the centrifugal force on the root results in bending deflection of either the wheel rim or the straddle root of the blade. Figure 3.11 illustrates this problem. Wheel rims and blade platforms can be made wider to resist this bending and reduce deflection, but there are practical limits to this solution. The disk stress

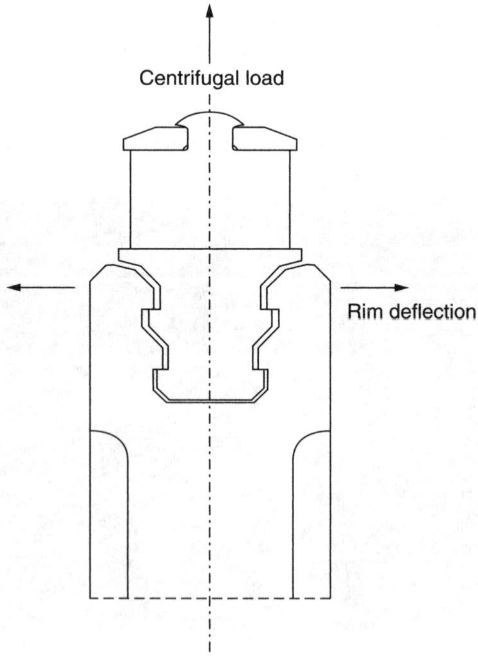

FIGURE 3.11 Rim bending under centrifugal loads.

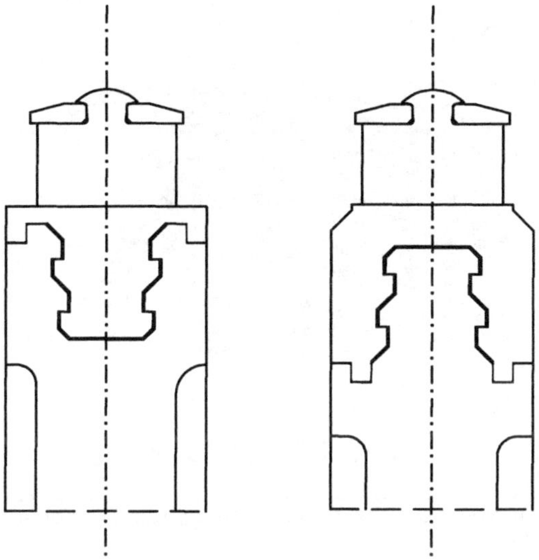

Figure 3.12 Interlocking platform on dovetail root form to resist rim bending.

due to the large mass required at the rim may exceed the disk material's strength limits, or the bladed disk assembly's width may become too great for the available rotor span. Root designs have been created to resist this bending deflection using interlocking forms between the rim and the blade (e.g., Figs. 3.12*a* and *b*), that extend the application range of tangential roots to larger blades and higher speeds. Even with interlocking geometry, however, there is a limit where the rim or blade width becomes too wide to be practical.

3.2.2 Axial Roots

As the name implies, axial roots consist of profiled slots arranged parallel (or at small angles) to the longitudinal axis of the turbine shaft or rotor, extending completely through the periphery of the disk, with profiles that interlock with matching profiles on the blade root. Figure 3.13 shows a common type of axial root used in many industrial steam turbines. Figure 3.14 defines the important features of the most commonly used form, known as a fir tree root for its resemblance to a Christmas tree.

For modern industrial steam turbines, practically all axial fir tree roots are produced by milling each slot individually in the turbine disk (Fig. 3.15). This is a consequence of the nearly universal application of integral rotors—i.e., shaft and disks machined from a single large integral forging—to industrial turbine generator sets and

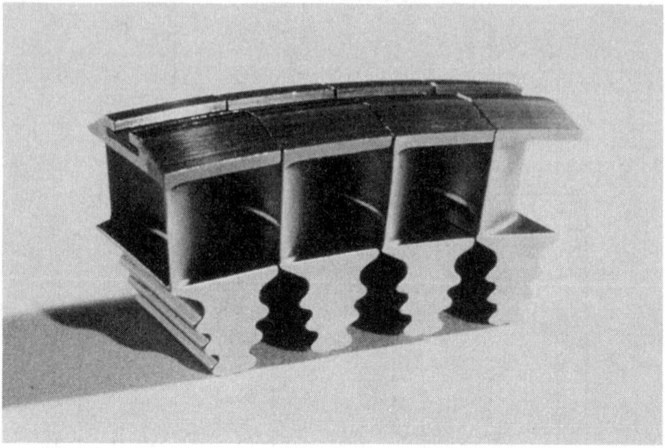

Figure 3.13 Turbine blades with axial fir tree roots. (*Photograph courtesy of Dresser-Rand Company.*)

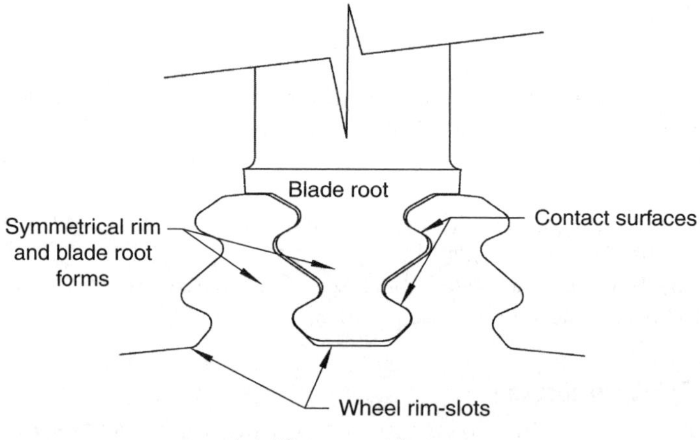

Figure 3.14 Fir tree root key features.

process compressor drive steam turbines. (In contrast, most gas turbines with separable disks have axial entry roots that are produced by broaching.) The accuracy of the root profile as well as the angular indexing of each slot to the next must be held to a very close tolerance level. Axial entry roots are more expensive when compared to circumferential roots, but for high-speed, critical service stages they are often the logical choice.

Figure 3.15 Turbine disks with milled axial roots. Note the narrow space between adjacent disks which limits the diameter of cutters and therefore the width of the root form. (*Photograph courtesy of Dresser-Rand Company.*)

Fir tree roots are a good choice for high-speed stages with high centrifugal loads because the blade root and the disk root forms are essentially mirror images of each other. In effect, the imposed loads and stress levels in each of the two parts can be made virtually the same. If the blade and disk materials have significant differences in strength, the width of the disk roots and blade roots can be adjusted slightly to balance the stress levels, recognizing the relative material strength available in each part. Inherent in their design, axial roots also do not suffer from the characteristic bending stress that tries to open up the sides of circumferential dovetail roots. By taking rim bending out of the total and equivalent stress calculations, axial entry roots are able to withstand higher centrifugal and radial tensile loads than is possible with circumferential dovetail roots. Typical applications in high-speed process compressor drives include condensing last-stage blades and highly loaded first-stage blades. Titanium blades coupled with axial fir tree roots are the ultimate choice in last-stage blades for process drives, allowing the use of significantly taller, more highly loaded last-stage blades than are possible with dovetail roots and stainless steel material.

3.2.3 Pinned Roots

If locking blades and locking spacers can be retained by pins through the blade and wheel rim, it is logical that pins can also be used as the primary retaining mechanism for turbine blades. In fact, pinned roots have been used since early in the development of steam turbines. Pinned roots have been used extensively to retain tall low-pressure blades in medium and large utility condensing turbine generator sets, in addition to low-pressure blades in industrial steam turbines. Proper function of axial and circumferential roots depends upon the precision with which the mating surfaces are produced, requiring very close tolerances to be held on the mating surfaces. Pinned roots, on the other hand, depend on the assembly process. While the wheel and blade root must still be produced to tolerances measured in thousandths of an inch (0.001 in), this is still far less demanding than the precision required for axial or tangential roots where tolerances of ten-thousandths of an inch (0.0001 in) are required.

Pinned roots derive their strength from multiple load-bearing points. Figure 3.16 shows how the root takes the form of multiple

FIGURE 3.16 Pinned root concept—pins attach blades with one or more projecting forms that match interlacing forms on the wheel rim.

projections interlaced between similar grooves on the wheel rim. Each pin shares the load on several shear sections. The area in contact with the pins is roughly equal for both blade and wheel rim, resulting in roughly equal contact stresses in the mating parts. The designer may adjust the ratio of contact and tensile areas of the blade and wheel rim to make best use of the relative strength of the materials.

Multiple pins are often used, resulting in even higher load-carrying capability for the overall assembly. As with the retaining pins for locking buckets, a close side-to-side fit between the blade roots and the wheel is critical to ensure that the pins are loaded primarily in shear. Elastic deformation of the blade and disk materials inevitably results in bending stress in the pin. A close fit between the pin and its board minimizes this bending stress and helps to ensure that the centrifugal and steam bending loads are shared between multiple pins. The assembly process is therefore critical to successful pinned root designs.

Because these root forms do not automatically determine the blade location, the assembly process must correctly locate each blade in relation to the turbine rotor at the correct radius to the turbine centerline and with the airfoil stacking axis in the correct angular relation, not only to the rotor but also to adjacent blades in the stage.

Depending on the tools available and the preference of the process engineer, blades and wheel rim may have pilot holes predrilled, or the drilling may all be done in its entirety after the blade is fixtured in its correct location in the wheel. In either case, the final holes are align-reamed through both blade and wheel rim simultaneously at very close tolerance, ensuring proper fit of the pins. To ensure close fit, in some cases the pins may be ground to match the size of individual bores. In many cases, pins that are sized for a slight interference fit must be cooled to facilitate assembly after align-reaming. After align-reaming, the disk, blades, and pins become, in effect, a matched assembly. In this case, if they are removed for any reason, the blades and pins must be reassembled in their original locations on the disk.

Pins must also be positively retained in their bores. Thermal transients during startup or large load changes may result in looseness even in interference-fit assemblies. Coupled with vibration and pressure loads on the wheels, pins will "walk" out of their engagement in the bladed wheel rim if they are not positively retained. A common means to retain the pins involves staking or upsetting the material adjacent to the end of the pins in several locations at each end, to prevent axial movement in the wheel, but they may also be retained by mechanical fastenings of one sort or another. Figure 3.17 illustrates a typical staked pin and root assembly.

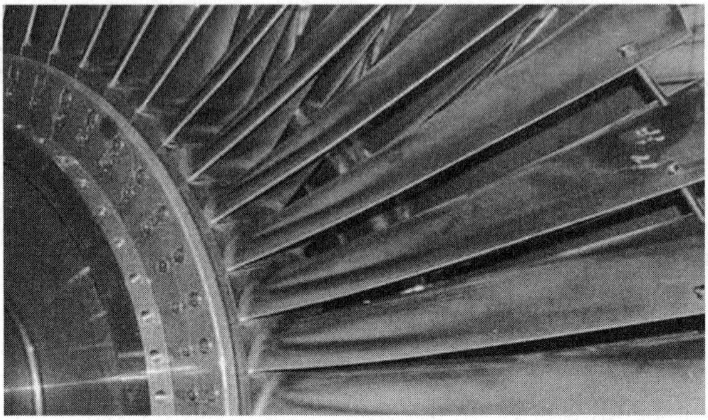

Figure 3.17 Staked pins in bladed disk assembly. (*Photograph courtesy of Dresser-Rand Company.*)

3.3 Shrouds and Auxiliary Dampers

Although some process drive steam turbines are built using shroudless, freestanding blades, mainly low-pressure, condensing stages, a majority of blade rows encountered in service will be equipped with shrouds or auxiliary dampers of some sort.

Shrouds take many forms, but regardless of the form, they are all designed to contribute structural stiffening, damping, or a combination of both to reduce service stress levels and the vibratory response of the blade. In addition, shrouds are often important elements in the aerothermodynamic design of the stage, forming the outside-diameter surface of the flow path and helping to minimize over blade tip leakage and secondary flows. To the extent that the stage depends on the shroud to reduce the amplitude of vibration or to change the frequency of vibration, they are critical elements for reliable operation of the turbine.

3.3.1 Riveted Shrouds

Although other designs have important applications, the most commonly used shroud design in service today is still the riveted shroud, where a separate band of material is attached to the tips of each blade's airfoils by one or more peened tenons. Figure 3.18 illustrates the most common type of shroud made from a series of shroud segments, each covering several blades. Shrouds are usually installed to create discrete packets of blades that are tied together by the shroud. The length and number of packets are chosen by the stage designer to modify the vibratory behavior of the bladed disk and to avoid potentially dangerous specific mode shapes and frequencies of vibration for the bladed disk assembly. The importance of packeting and its effect on the bladed disk's vibratory behavior are covered in detail in later chapters.

FIGURE 3.18 Photograph of packeted shroud bands typical of process drive turbines. (*Photograph courtesy of Dresser-Rand Company.*)

Longer shroud segments (so-called long arc shrouds) are difficult to produce, due to the difficulty of installing them without making the holes too large or bending the blades during installation. If long arc shrouds or even a completely shrouded stage is specified by the designer, a double-layer shroud design is often used, made up of two partial thickness bands with their ends staggered or overlapped to create a continuously shrouded stage.

3.3.2 Integral Shrouds

Unlike riveted shrouds, which are separate parts attached by peened tenons after the blades are assembled in the turbine disk, integral shrouds are produced as integral parts of the individual blades. The complete shroud is therefore formed from the segments of the shroud that are attached to each blade tip. Figure 3.19 shows a turbine rotor equipped with several stages of integrally shrouded blades.

Figure 3.19 Turbine rotor with integrally shrouded blade rows. (*Photograph courtesy of Dresser-Rand Company.*)

3.3.3 Z-Lock Shrouds

A special version of integral shroud is the Z-lock shroud, consisting of interlocking shroud segments that are loaded against one another in service to ensure contact at the abutment surfaces. Figure 3.20 shows a typical Z-lock shroud as installed in an assembled turbine wheel. Z-locks are almost always used on twisted airfoil sections because they are intrinsically activated by the twisting of the airfoil that occurs with increasing speed, as Fig. 3.21 illustrates. The Z-lock

Figure 3.20 Bladed disk assembly with Z-lock shrouds. (*Photograph courtesy of Dresser-Rand Company.*)

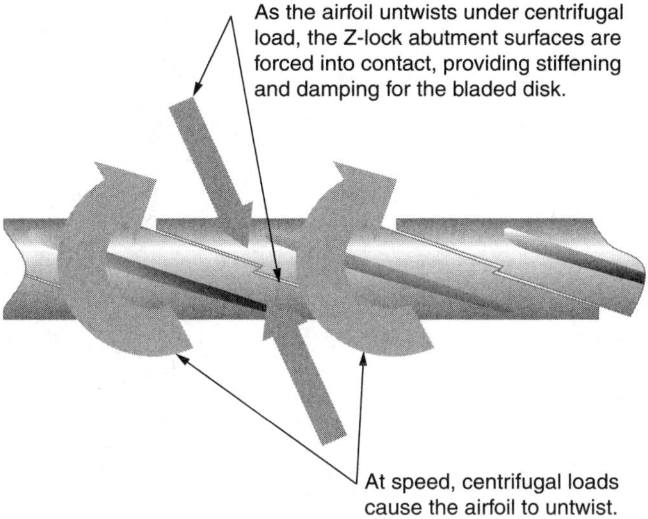

FIGURE 3.21 Self-actuating design of Z-lock.

surfaces are oriented to ensure positive contact at the abutment surfaces at speed. To be effective, the positioning and preload of Z-lock shrouds must be carefully controlled. The Z-lock shrouds are almost always used with axial fir tree root designs because the axial root design locates the blades and prevents relative twist between adjacent blades that could lead to variations in contact force, defeating the purpose of the interlocking shrouds.

The abutment surfaces are often treated with some type of surface modification to prevent wear and fretting on the abutment surfaces. The abutment surfaces may move relative to one another in service, and this rubbing between the surfaces can lead to unacceptable wear of Z-lock contact points. A variety of coatings have been used to combat this wear. A typical surface treatment consists of a Stellite 6 weld overlay, which is easily deposited with either manual or automated equipment and is compatible with virtually all 400 series stainless steel blade materials.

3.3.4 Auxiliary Shroud Dampers

For additional structural support or to ensure good coupling between adjacent blades with integral shrouds, several types of auxiliary shroud damping devices may be used. One common form is a wire inserted in a slot in integral shrouds and retained either by the form of the wire and the slot themselves or, in the case shown in Fig. 3.22*b*, by upsetting or rolling the material on either side of the wire.

For tall blades, high centrifugal loads may preclude the use of continuous shroud. In these cases the blade designer must choose

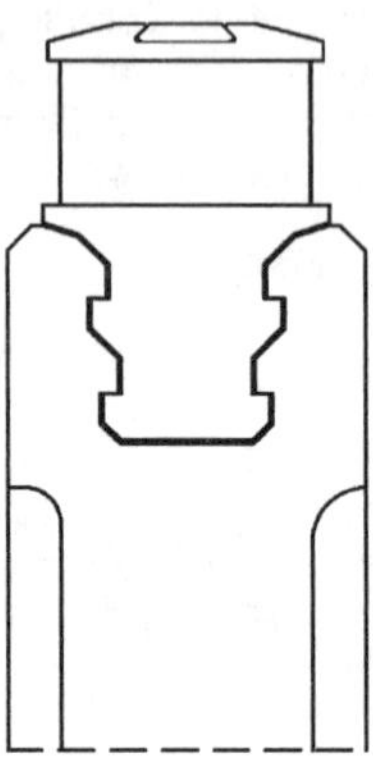

(a) Friction band shroud with "loose" wire retained by slot form

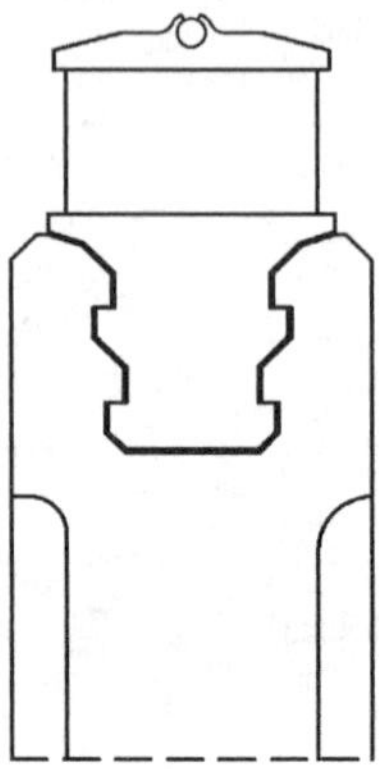

(b) Inserted wire retained by rolling shroud material next to wire

Figure 3.22 Inserted damping wires in integrally shrouded blades.

between a freestanding blade or specifying a lightweight damping device that spans the gap between individual blades. The most common device is either a wire that forms continuous 360° ring or individual dampers that span the individual gaps from blade to blade, similar to that shown in Fig. 3.23.

Figure 3.23 Blade-to-blade damping wires. (*Photograph courtesy of Dresser-Rand Company.*)

3.4 Blade Materials

With the wide range of services and the conditions of service that are encountered in the design of industrial turbines, one might expect an equally broad range of materials that could be specified for turbine blades. In fact, blades for modern industrial turbines and process drives, regardless of the original manufacturer, are almost all manufactured from a rather small set of widely used materials. An ideal turbine blade material has both high tensile strength and high fatigue strength. In addition an ideal blade material is corrosion-resistant and erosion-resistant, exhibits high ductility, is easily formed and machined, is widely available from multiple sources, and can be obtained at reasonable cost.

The high rotational speed of modern turbines means the blades are subject to large centrifugal forces, in addition to the potentially high repetitive bending forces imposed by the steam. One would therefore think that high tensile and fatigue strengths are the paramount requirements for turbine blades. While strength is a critical attribute of turbine blades, a high degree of ductility is also a critical property in any good blade material. In spite of closely held dimensional tolerances and careful assembly, the mating surfaces of blade and disk root structures never share perfect contact. On initial start-up therefore small zones of plastic deformation occur that result in improved contact between mating surfaces. This process is so critical to successful long-term reliability that turbine design engineers are willing to sacrifice some degree of tensile strength in return for improved ductility.

Nearly all steam turbine blade materials in common use are wrought materials. Webster's Dictionary defines *Wrought* as "worked; to fashion or create a useful or desired product by expending labor or exertion on" (see Merriam-Webster Unabridged). Force is applied by various forming processes such as forging, pressing, and rolling to reduce large sections to smaller shapes. In the process, material properties are changed and enhanced. Most forming processes also create directional variations in material properties as a result of the process. Forging and rolling operations tend to align and elongate the metal's grains in the direction of maximum elongation of the material and to narrow the grains in the direction of maximum reduction in section. Wrought materials usually exhibit some degree of directionality in properties such as tensile (and fatigue) strength, modulus of elasticity, and Poisson's ratio. Designers take advantage of this directionality by aligning the long axis of the grains with the direction of the maximum alternating bending stress, for example.

Stocks of blade materials are available in a variety of forms, including hot-rolled and cold-rolled bar stock, forged bars, as well as "bar" stock that is actually produced by reducing (cutting or slitting) larger forms into smaller sizes. This last approach can significantly reduce the cost of blade material, but the designer must be aware of how the directionality of rolling and forming affects the material properties. For most forgings and bar stock, the directionality is obvious from the shape of the material. For some forms, such as machining blanks that are cut from larger bars, it is important to keep track of the grain's direction to ensure that the axis of the material's grain flow coincides with the radial axis of the finished blades.

3.4.1 Stainless Steel

Type 403 and 410 stainless steels are among the most widely used materials for steam turbine blades. These alloys are very similar in chemical composition. Both are martensitic stainless steel alloys that can, through appropriate heat treatments, exhibit strength levels covering a wide range, depending on process variables such as the quenching and tempering temperatures, among others.

As used in blade applications, these materials are usually heat-treated to moderate strength levels, allowing the material to retain relatively high ductility, in the range of 20 percent or greater. This combination is ideal for turbine blades that are subject to high centrifugal and bending loads. The ductility is a very important characteristic as it allows, e.g., root contact surfaces to udergo slight plastic deformation to better share service loads, without the risk of cracking or failure. This same property gives turbine blades the ability to absorb minor damage in service from erosion, rubs between stationary and rotating parts, and foreign object damage without failure.

Microscopic material defects and inclusions well known to reduce fatigue strength act as starting points for cracks in steel materials, and therefore have a severe impact on the reliability of turbine blades. To avoid these problems, special care must be exercised to ensure that blade materials are as defect-free as possible. Vacuum melting processes produce "cleaner" steels than air melt processes. While single vacuum melt materials are widely used, for maximum reliability, the use of double vacuum melt or VIMVAR (vacuum induction melt, vacuum arc remelt) material is recommended for all critical process drive blading.

Type 403 stainless steel exhibits a significant decrease in usable material strength at elevated temperatures. Because of this, turbine design engineers often specify alternate materials for high-temperature stages. Type 422 SS is often specified for high-temperature, high-pressure stages because of its higher tensile strength and ultimate strength at high temperatures, as well as higher creep resistance at high operating temperatures. Type 422 is not commonly used for low- and moderate-temperature applications, since it is more expensive than 403 or 410, and the ductility of 422 at low temperatures is significantly less than either 403 or 410 SS materials. However, at service temperatures above 850°F, its ductility is comparable to those of 403 and 410 SS.

Typical applications include first-stage blades in high-temperature applications, and as an alternate to 403 and 410 SS in stages with moderate service temperature and high stress levels.

3.4.2 Titanium

Titanium is an excellent material for low-pressure steam turbine blades. The usable strength of titanium falls in a similar range to that of 403 and 410 SS, but its density is only 60 percent of that of commonly used martensitic steels. When Ti is substituted for 403 SS, for example, the centrifugal stress is reduced due to the lower density. With reduced stress level from centrifugal load, the blade is able to sustain higher steam bending stresses without exceeding the material's maximum allowable principal stress or shear stress criteria. Taller blades, with increased flow limits, are possible for high-speed applications.

The ratio of titanium's elastic modulus to density (E/ρ) is very close to the ratio exhibited by 403 and 410 martensitic stainless steels, which exhibit an elastic modulus in the range of 29 to 30×10^6 psi. As noted above, titanium's density is only 60 percent that of stainless steel. Coupled with an elastic modulus of approximately 16.8×10^6 psi, this results in a ratio of elastic modulus to density that is virtually the same as that for 400 series martensitic stainless steel. This produces fundamental blade frequencies and mode shapes that are nearly identical when the same blade geometry is produced in either titanium or 400 series stainless steel. In many cases therefore titanium can be treated as a direct substitution for stainless steel in highly stressed stages. This characteristic provides an opportunity, when considering upgrades or unit reapplication studies, to significantly increase the turbine's speed and output while retaining the original rotor. Titanium is, however, limited to low- and moderate-temperature applications, so it is not a substitution candidate for stages with high operating temperatures.

For LP stages, titanium's greater inherent resistance to water droplet erosion also means that it can be applied at higher erosive potentials than stainless steels without additional erosion protection. In many applications, eliminating erosion protection treatments offsets a significant portion of titanium's higher raw material cost.

Ti-6Al-4V is the most commonly used alloy for blades, although other alloys in the Ti-6Al family have also been used. It is available in a range of forms and sizes. Smaller titanium bars are often stocked by metals suppliers, but larger sizes may require longer lead times.

Table 3.1 compares properties of the most common blade materials. In general, the cost of finished blades increases as you progress from 403 and 410 SS to 422 SS, and finally to Ti, because of increasing raw material cost, coupled with higher forming, machining, and finishing time and costs. As one would intuitively expect, stronger

	403 SS	410 SS	422 SS	Ti-6Al-4V
Tensile strength, ultimate, ksi	110	120	150	120
Tensile strength, yield, ksi	85	105	125	110
Elongation, percent	23	21	19	15
Elastic modulus (10^6 psi)	29	29	28.9	16.8
Hardness (Brinell)	170	237	275	319
Relative machinability, percent	100	75	62	55

Table 3.1 Comparison of Common Steam Turbine Blade Materials

materials are harder to machine and therefore more costly to process. (The reader is cautioned that the values in this table are average handbook values, and the properties of specific materials and particular heats of a specific material may vary significantly depending on the actual process parameters.)

3.4.3 Other Blade Materials

While 403 and 410 SS are the most widely used stainless alloys, there are a few materials that see limited use in process drives for special situations. Austenitic 300 series SS materials have been used in very limited numbers of specialty turbines, particularly in applications where the steam is likely to contain corrosives. The strength of most austenitic stainless steels cannot be increased by normal heat treatment processes, so the material can only be applied at very low stress levels. A286 alloy is special case of an austenitic superalloy that can be strengthened by precipitation hardening. This property makes A286 an appropriate choice for blades and disks in process hot gas turbines operating at temperatures of 1000°F and above, and it has been applied in some steam turbines operating at high temperatures.

Other superalloys have also seen very limited application in steam turbines. IN718 can be used as an alternate to 422 SS in very highly loaded, high-temperature (≈1000°F) first-stage blade applications. IN718 is an excellent material for high-temperature applications, but its significantly higher raw material cost and difficulty in machining limit its application to special cases.

There are also some alloys that see limited use in low-pressure turbine stages because they exhibit good erosion resistance without additional surface treatment or overlays. Vasco M252, for example, sometimes called *Jet Hete,* was one of the earliest superalloys developed for gas turbines. It is now sometimes used for last-stage blades in erosive applications. It has the advantage of higher inherent strength than either titanium or 400 series stainless steel, and it exhibits a high degree of inherent erosion resistance. It has been used successfully in last-stage blade applications without additional surface treatments, but its use is limited by its higher material cost and high relative machining difficulty.

3.4.4 Material Forms

Blade materials are available in a variety of raw forms. In practice, the manufacturing process chosen for a specific blade row determines to a very large extent the required material form. The vast majority of blades for process compressor drives are fully machined from bar stock. Some variations on this process include the following:

Bar Stock

The most common material form used in blade manufacture is bar stock. The 403, 410, and 422 martensitic stainless steels are all widely available as both hot- and cold-rolled bar stock. Almost any square or rectangular shape up to an approximate 3 in width can be procured as bar stock. Bar stock shapes are created by rerolling larger shapes or billets to smaller size, ensuring that all the material is subject to significant deformation, resulting in uniform properties and ensuring that the resulting material has the desirable properties of a wrought material.

Blanks for blade machining are often produced by cutting or slitting larger material forms to produce smaller blanks of appropriate size. Since many of the desirable properties inherent in wrought material depend on the directionality of the final forming process, it is critical that the resizing process take this into account to ensure that desirable directional effects are preserved in the stock used to produce blades.

Forged Bars

The center of large (greater than 3 in) bar stock may not experience the degree of deformation necessary to impart the material enhancements expected from wrought materials. A material form known as *forged bar* is often used to provide improved control of properties in completely machined blades requiring stock with thickness greater than about 3 in.

Envelope Forgings

Envelope forgings are rough forgings of the airfoil with 1 mm to 3 mm stock allowance over the entire surface, depending on the size of the blade and the specific machining process to be used. Strategically placed blocks of material is provided in the forged shapes from which roots, tenons, and shrouds are formed by machining. Additional material is often added to provide temporary fixturing points or lugs to facilitate stock holding on the machine tools. These auxiliary fixturing points are removed once they have served their purpose. Due to the relatively relaxed tolerances required in the forging, envelope forgings require less expensive tooling and per piece cost than net shape forgings.

Envelope forgings that are later fully machined have largely replaced the use of net shape airfoil forgings in small and medium steam turbines. Forgings are often used for medium to large blade sizes to minimize total machining times. Envelope forgings require less expensive tooling than net shape forgings and can be produced on a wider range of forging presses. Lead times for envelope forgings are typically about one-half those for final net shape forgings. Although envelope forgings must be completely machined, the cost of finished blades is similar to that of net shape forged blades, because the total amount of material removed is minimal. Blades produced from envelope forgings, because they are fully machined, can be produced with finer dimensional tolerances on the airfoil contours, which in turn produce closer control of stage areas, more accurate surface profiles, and consequently improved aerodynamic performance. Figure 3.24 illustrates an envelope forging and the finished blade produced from that forging.

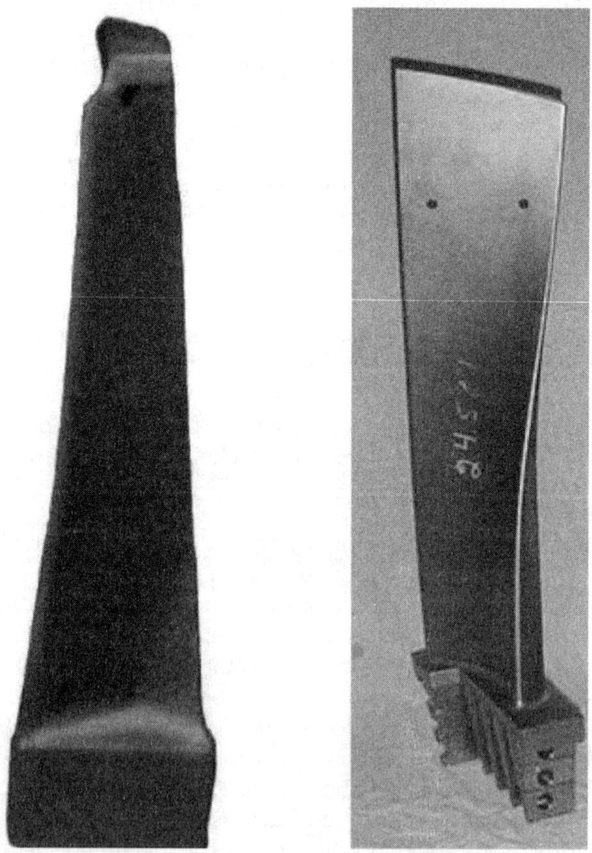

FIGURE 3.24 Example of envelope forging and completed blade. (*Photograph courtesy of Dresser-Rand Company.*)

3.5 Manufacturing Processes

Blades with Constant Section Airfoils

Blades with constant section airfoils are usually fully machined from bar stock. The airfoil surface is produced by a milling process in which the airfoil contour is formed by a milling cutter that traces the airfoil contour on a CNC machining center. Figures 3.25 through 3.28 illustrate some of the typical steps involved in production of a turbine blade from bar stock. Some variations on this process include the following:

1. For blades with riveted tenons, the airfoil is formed by tracing around the entire airfoil shape (pressure and suction surfaces) with a side milling cutter. For maximum accuracy and speed, the largest cutter diameter is chosen that can produce the minimum concave radius of the pressure surface. The blade is then cut to length and the tenon (round or kidney) is produced on the blade tip by milling.
2. For blades with flat-top integral shrouds, the cutter that forms the airfoil must also incorporate the fillet radii at the junction of the airfoil with the top of the platform and the underside of the shroud.

FIGURE 3.25 CNC machine tool for blade machining. (*Photograph courtesy of Dresser-Rand Company.*)

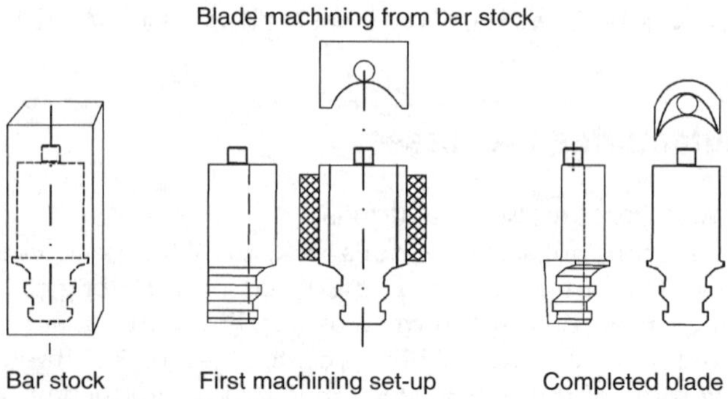

Figure 3.26 Blade machining steps.

Figure 3.27 Bar stock "blank" for turbine blade. (*Photograph courtesy of Dresser-Rand Company.*)

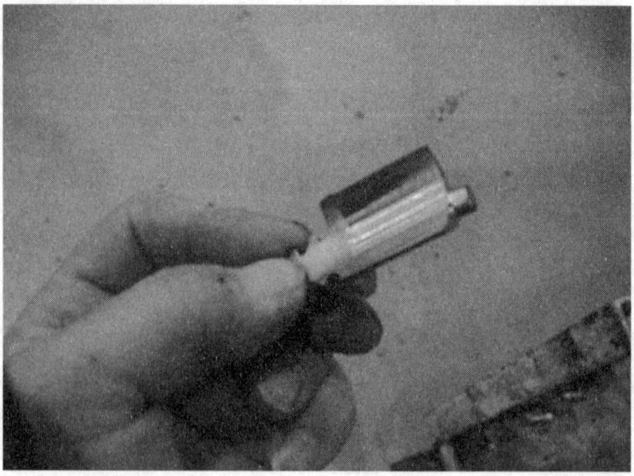

Figure 3.28 Completed turbine blade. (*Photograph courtesy of Dresser-Rand Company.*)

3. When a slant-top integral shroud cannot be avoided, the area under the shroud must be finished using a ball end mill. This is a time-consuming and expensive process, since the mill cannot exceed the diameter of the fillet radii between the airfoil and underside of the shroud.

Blades with Taper-Twisted Airfoils

Twisted and tapered airfoil sections are used in reaction turbine stages, including the low-pressure stages of impulse turbines. The airfoil section varies continuously from the platform to the blade shroud or tip, to account for the changes in relative velocities of blade and steam flow.

Twisted airfoil sections are produced by three main methods. For small (a relative term) blades the airfoil may be produced completely by machining from bar stock. The airfoil surface is produced by a milling process in which the airfoil contour is by a milling cutter that traces the airfoil contour under CNC. A typical machining process uses a four-axis or five-axis milling machine or machining center equipped with a ball end mill to trace the three-dimensional surface of the airfoil. This machining process produces a high-accuracy surface, but leaves small ridges of material where successive passes of the milling cutter overlap.

These ridges are removed by a subsequent grinding and polishing process to reach the final airfoil shape. This may be a manual process using belt sanders and polishing wheels, although industrial robots are taking over this process to produce a more consistent result. The choice of manual vs. robot is driven by the number of blades to be produced, i.e., higher quantities justify robot setup and programming. This process is capable of producing blades with minimal dimensional tolerances.

Airfoil surfaces may also be produced by forging processes. Envelope forgings are formed to the airfoil contours with several millimeters of excess stock. This stock is removed by the same machining and polishing process used for bar stock.

The forging process can also produce an airfoil forged to net shape, requiring only grinding and polishing to finish the airfoil. Machining stock is provided for root and tip machining operations.

The airfoil is numerically modeled by defining a series of sections at regular intervals along the stacking axis from platform to shroud. Although as few as four sections can be used, a smoother and more accurate profile can be obtained with more sections. Ten sections, defined by the profiles at 11 stations, are typical. The points that define these sections are used to create a three-dimensional spline representation of the airfoil surfaces. The same solid model created by these boundaries is used for engineering analysis (e.g., aero and stress), and then to program the CNC machining center and

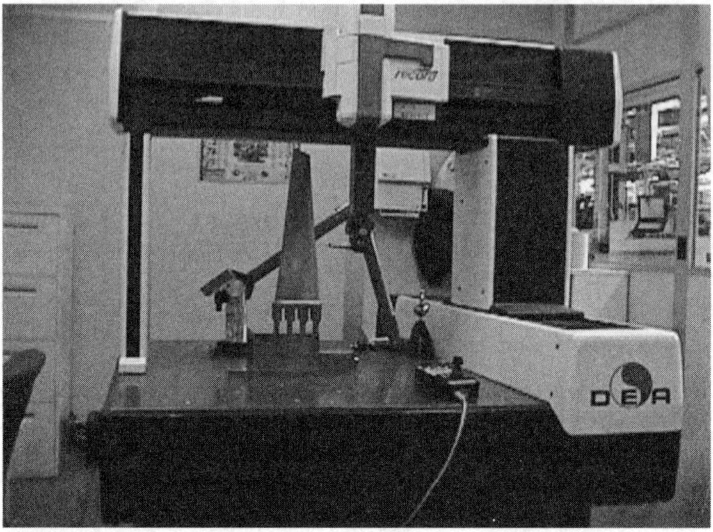

Figure 3.29 CMM inspection of blade with twisted airfoil. (*Photograph courtesy of Dresser-Rand Company.*)

the CMM (as illustrated in Fig. 3.29) for automated inspection of the finished blade.

3.6 Erosion Protection—Condensing Stages

Low-pressure stages in condensing steam turbines present another special case that the designer must consider. For condensing turbines, power produced in the last stage is a strong function of the exhaust backpressure. Low-pressure stages in condensing turbines, and in the last stage in particular, are other special cases that the designer must recognize and contend with. The term *condensing* is used for turbines with subatmospheric exhaust pressure, where the exhaust flow discharges directly to the service condenser (a heat exchanger that condenses the steam). The liquid condensate is then reused as feedwater, recirculating through the system many times.

The minimal losses of water for this closed system are replaced by makeup water that is demineralized before entering the system. In a closed system, control of water chemistry is critical, to prevent the buildup of harmful constituents. Dissolved oxygen, e.g., is corrosive, affecting not only the turbine materials, but also the boiler, plant piping, and any other equipment in contact with the steam. Other potential contaminants include minerals such as silica that may accumulate on turbine blades and of nozzles, reducing flow areas and ultimately reducing machine capacity. So why is this important to the discussion?

In condensing turbines, as the enthalpy of the system of steam is reduced, some of the steam may condense, so that flow through the low-pressure stages is actually composed of two phases, saturated steam and liquid condensate. Water-soluble minerals, chemicals, and gases tend to concentrate in the liquid fraction, which may result in very high concentrations of contaminants in the liquid phase. These contaminants may be corrosive or erosive, leading to damage in low-pressure turbine components. As steam starts to condense in the low-pressure stages, centrifugal force causes the raw condensate to collect at the outside periphery of the flow path. As the water condenses on the surfaces of the stationary flow path, droplets coalesce into water that streams from the trailing edge of the nozzles directly into the path of the moving blades. The liquid droplets move at much lower velocity than the vapor phase, so the leading edge of the blades smash into the slow-moving liquid droplets at high speed. This can be highly erosive to untreated blade materials. As a minimum, the loss of blade material to erosion leads to performance degradation and, in severe cases, compromises the mechanical integrity of the blade. Since velocities are greatest near the tips, the most severe erosion occurs in the outer third of the airfoil and the shrouds, if they are present. To combat erosion, a wide variety of treatments have been devised to combat water droplet erosion in low-pressure turbine stages.

For a given unit, the potential of erosion increases with increasing blade speed and with the amount of liquid phase present. In general therefore large tip diameters and higher rotational speed increase susceptibility to erosion. This is an important subject in turbine design, and over the years, many researchers have worked to characterize the erosion problem and have developed a number of approaches to predict erosion potential based on stage operating parameters. Virtually every turbine manufacturer has developed their own empirical erosion protection guidelines based on field experience with their particular stage designs. In general, erosion is dealt with in two main ways—remove the water from the flow path and use erosion-resistant materials or material overlays.

Untreated 403 or 410 stainless steel is quite susceptible to liquid droplet erosion and should only be used in low-speed stages with relatively small amounts of liquid condensate present. However, since these materials are strong, durable, and widely available, a variety of surface treatments have been developed to improve their erosion resistance.

One of the simplest approaches is flame hardening of the leading edge. In this process, an oxy-fuel torch heats the leading edge of the blade, immediately by a spray quench. The process is usually automated to ensure repeatability in heating and quenching for all the blades in a stage set. The objective in the flame hardening process is rapid heating of the surface area to be protected, followed by rapid quenching. This results in a thin layer of stronger, harder material on the surface with a core of ductile material underneath. Flame hardening is quite effective for stages with moderate erosion potential. Where it is appropriate, it has significant advantages over more aggressive treatments. Since the ductile core is maintained in all but the thinnest blades, flame hardening is a low-risk process. Flame hardening is inexpensive and requires short production cycle times, and the retention of the ductile core makes it a very low-risk process. If erosion in service proves to be a problem, there are repair techniques available to salvage blades with moderate erosion damage and to upgrade erosion protection in future operations.

However, flame hardening is not applicable to all blades. Very thin blades, are susceptible to through-hardening, which completely eliminates the ductile core. And conventional flame hardening may not provide adequate erosion protection in stages with high erosion potential. One alternative in such cases is laser hardening. Laser hardening relies on the same concept as flame hardening, except that it uses a high-powered laser to heat a very thin surface layer very rapidly, and it quenches the hot metal just as quickly, resulting in a harder but thinner layer at the surface for conventional flame hardening, and the harder surface layer protection provides in stages with high erosive potential.

One of the earliest approaches to prevent leading-edge erosion was to undercut a portion of the leading-edge contour and insert a thin layer of hard, erosion-resistant material such as Stellite or similar alloys. In this process a thin layer of Stellite strip or bar stock is attached to the undercut by brazing or welding. The Stellite bar is then blended to using abrasive belts or grinders to restore the desired airfoil surface contours.

Stellite is extremely erosion-resistant, but it is common to observe blades with long service where the untreated blade material adjacent to the Stellite inserts has eroded away around and even underneath the insert. Stellite bars are still widely used today, but more common today is a weld overlay process to deposit a layer of Stellite that more closely conforms to the surface of the blade. This reduces the total amount of material required and reduces the time and cost of blending the Stellite overlay to the final airfoil profile. This process permits its use on many blades that are either too thin or have contour changes too sharp to use a flat insert. Incidentally, the same process can also be used to build up and restore eroded blade surfaces on blades exhibiting erosion damage after service.

Some available blade materials in service today are also naturally resistant, to a greater or lesser extent, to water droplet erosion. Titanium blades, as we have already noted, are used in many high-speed process drives, since the material has strength similar to that of stainless steels but weighs only 40 percent as much. Titanium is also significantly more erosion-resistant than 400 series stainless steel and is therefore used without added surface treatment in many industrial process drive turbines. Recent experience shows that untreated titanium may still exhibit significant erosion in some conditions, and titanium stages with high erosion potential may still require additional surface treatment. Titanium is compatible with several PVD and CVD coatings that can be used to improve erosion resistance.

As we noted in Sec. 3.4.3, there are also specialty alloys, such as Vasco M252, that are used in a limited number of low-pressure turbine stages because they exhibit good erosion resistance without additional surface treatment or overlays. To reiterate, Vasco M252 has the advantage of higher inherent strength than either titanium or 400 series stainless steel. It exhibits a high degree of inherent erosion resistance and has been used successfully in some last-stage blade applications without additional surface treatments.

3.7 Bladed Disk Assembly Processes

No matter how sophisticated the analysis and design of the individual components, and no matter how accurate and precise the manufacture of these components, in the end the processes, techniques, and inspection used in the final assembly of the bladed disk are critical to the successful and reliable operation of the turbine rotor. In this section, we are going to review some of the common processes used in the assembly of the bladed disks. Among these are the installation of the blades in or on the disk to produce the bladed disk assembly. Also covered is the installation of shrouds and auxiliary dampers, with particular emphasis on the installation and assembly of riveted shrouds, i.e., shrouds assembled using pain tends to hold the shroud in place. As these techniques are discussed, the most common forms of in-process inspection and typical acceptance criteria for acceptable assemblies will be reviewed. The processes to be covered are widely used and are common to most makes and models of industrial steam turbines and process drives.

3.7.1 Assembly of Bladed Disks with Circumferential Dovetail Roots

As noted previously, circumferential dovetail roots are by far the most common type in use in present-day process drive steam turbines, making it important for turbine design engineers and reliability

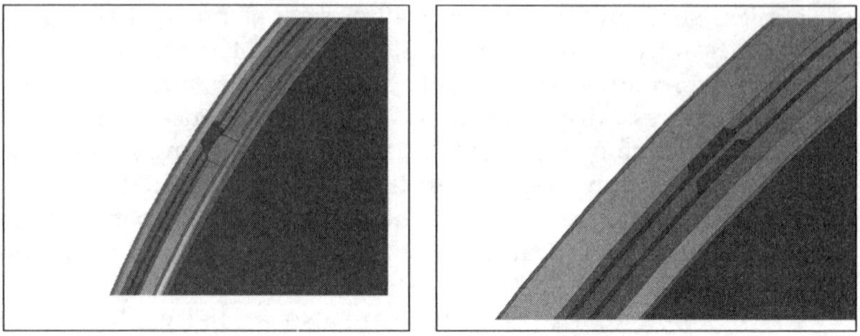

(a) (b)

Figure 3.30 Filling slot and filling gaps illustrated.

engineers alike to understand their assembly processes and their unique assembly requirements to produce a reliable turbine rotor. Dovetail roots are very simple features that involve contact between interlocking shapes on the blade root and the wheel groove. There are two main forms of dovetail roots previously discussed, the internal dovetail form with the wheel hooks on the inside of the wheel groove and the external or straddle root form where the wheel hooks are formed on the exposed lateral sides of the wheel. To install the blades in or on the wheel, there must be a discontinuity in the wheel grooves that allows installation of the blades. This discontinuity is commonly called the *filling slot*, which takes the form of a gap in the wheel grooves through which the blades are installed. Figures 3.30*a* and 3.30*b* illustrate the most common forms of filling slots for internal dovetails and external straddle roots, respectively. The assembler simply inserts the blades at the filling slot and then moves the blades around the wheel until the entire complement of blades has been installed.

Assembly of Dovetail Roots

Turbine blades with dovetail roots are installed in the wheel through a hole or a gap in the circular root form of the disk. For the internal dovetail configuration of Fig. 3.31*a*, the blades are installed via a filling slot, i.e., an opening milled into the wheel's root form that permits the blades to be inserted radially through the slot, then moved circumferentially around the wheel until the entire complement of blades has been installed. The filling slot must then be filled either with a special blade, usually referred to as a locking bucket, or with a blank spacer piece, which is retained by one or more pins through the wheel rim and the locking bucket root blank. A similar installation for straddle roots is shown in Fig. 3.31*b*, where the blades are installed through a gap in the circular teeth on the disk. This gap is then closed using a locking bucket that, like that for an internal root form, is retained by one or more pins through the root of the blade and the wheel rim.

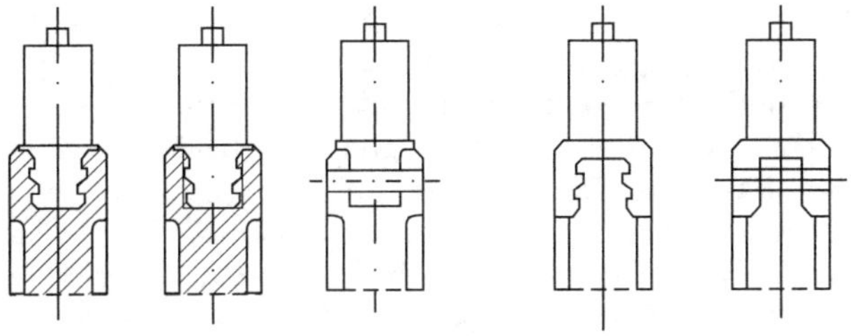

Internal dovetail root; blades are installed through filling slot (see inset), locking blade fixed by one or more locking pins through the wheel rim.

(a)

Straddle root; blades are installed via gap in wheel rim teeth, fixed in position by one or more locking pins through the wheel rim.

(b)

FIGURE 3.31 Filling slot and locking pins.

Even though the wheel has been designed for a specified integer number of blades, the tolerance stack up for a large number of blades, perhaps 100 or more in many wheels, rarely results in a bladed disk that closes exactly as designed. Most assembly specifications for wheel assembly call for the locking bucket to exhibit a slight circumferential interference with the adjacent blades when it is inserted in the filling slot. This interference tends to force the blades radially outward against the root locks. This effect reduces uncertainty and potential errors in the rotor balancing process due to loose blades. It also tends to create a stiffer blade support during the tenon peening process, where riveted shroud bands are used.

The assembler is shooting for an assembly that is "tight but not too tight." The tightness of the fit at the root of the blade may have a significant influence on the vibratory behavior of individual blades and the final bladed disk assembly. Turbine design engineers recognize the design advantage of the friction damping that results when adjacent blade roots move relative to one another. A final assembly with either too much or too little contact force between adjacent roots may not provide the desired dynamic behavior.

To achieve appropriate tightness or interference when the locking bucket is installed, several options are available. The simplest method utilizes shims cut in the shape of the bucket root form. During assembly of the wheel, the assembler checks the fit and inserts shims between as many blades as necessary to achieve the specified fit. Typically only one shim per blade is used, and it is desirable to space the shims periodically around the wheel to avoid concentrating shims in the last few blades. The assembler must measure and plan ahead, or measure for the last blade or blades, then remove sufficient blades and reinsert with shims to close the wheel. When closing the wheel using shims, it is important that the shims be spaced around the wheel and not concentrated in one space close to locking bucket, to minimize the possibility of introducing excessive unbalance in the final assembly.

Another method to ensure proper closure of the wheel involves the use of the blades that are deliberately produced either slightly oversized or slightly undersized, known, respectively as overs and unders, that are used in combination with standard-size blades to ensure that the wheel closes with the current number of blades. When using this method, the assembler installs a complete set of standard blades and measures the space remaining for the locking bucket. From this information, the number of oversize or undersize blades can be calculated. Next either the entire set or a significant portion of the set is removed from the wheel, then reinstalled with the appropriate number of overs or unders interspersed around the wheel.

An alternate method which is sometimes used is called *mill to close*. In this method the standard blades are made deliberately a slight amount oversize. Then based on the measurement of the space left for the locking bucket, some of or all the blades are milled to remove a small amount of material, ensuring that the row of blades closes correctly.

It sometimes happens, especially in bladed disks with a large number of small blades, that the row appears to close correctly with either one more or one fewer blade than is specified by the designer. If a review of the modal behavior of the bladed disk indicates that there is no significant impact on the vibratory behavior of the blades and disk, it is often preferable to accept the small discrepancy in the number of blades rather than use an excessive number of shims or oversize or undersize blades. However, the decision to accept a bladed disk assembly with a discrepancy in the number of blades can only be made after a thorough engineering review of the effects on the modal behavior of the bladed disk assembly.

Blade Row Closures for Circumferential Dovetail Roots

For internal roots, a locking bucket or locking piece is installed in the filling slot in the wheel groove through which the blades are inserted. The locking piece must nearly fill the slot, with close contact on the sides of the groove. Since the locking piece is held in place with one or more hardened steel pins, reamed and installed after the locker is in place, any gap that will result in unsupported pin length is undesirable.

For external or straddle roots, the locking bucket or locking piece fills the gap in the wheel locks that permits installation of the blades. Like the locking bucket for internal roots, the locking bucket or locking piece must have close contact with the sides of the wheel to avoid any unsupported length of the pins.

The design and installation of the locking blades and retaining pins are critical to a successful circumferential root design. Unlike a

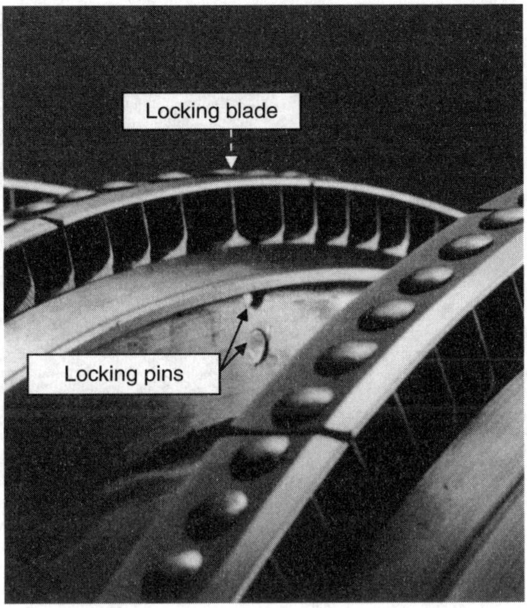

FIGURE 3.32 Locking pins. (*Photograph courtesy of Dresser-Rand Company.*)

standard blade with circumferential roots, where the load is carried by a relatively large contact patch and the load is shared by all the teeth in bending, the locking bucket is retained by only one or two pins whose diameter is only a fraction of the width of the root. The pins themselves are highly loaded in shear, and so must be produced from a very strong material, such as tool steel. The fit of the pin in its bore is also important. A tight fit or even a slight interference fit is desirable, as it minimizes bending deflection in the pins and helps to ensure the pins are loaded mainly in shear, with minimal bending stresses. Figure 3.32 illustrates a typical locking pin installation for a blade row with internal dovetail roots.

3.7.2 Bladed Disk Assembly—Axial Fir Tree Roots

As noted before, axial fir tree roots are most commonly used on high-speed and highly loaded turbine stages. The assembly process for axial fir tree roots is very straightforward, requiring little or none of the measurements and adjustments that are required to ensure proper closure of the stage with circumferential dovetail roots. However, there are still some precautions and required practices to ensure the best results in the bladed disk assembly. It is common practice, especially for larger blades, to measure and record the mass weight or the moment weight of each blade, from which installation sequence or order can be calculated to minimize the residual unbalance to

Figure 3.33 Axial roots retained with washers and rivets. (*Photograph courtesy of Dresser-Rand Company.*)

variations in blade mass. By their nature axial fir tree roots are easily installed and removed from the wheel, making it easy to exchange the positions of a few blades to fine-tune the residual unbalance of the stage.

When axial roots are used, any interlocking shrouds, such as Z-locks or center-milled integral shrouds, require the simultaneous insertion of all the blades in the stage. When Z-lock shrouds that are designed with significant preload between blades are installed, considerable force may be required to overcome friction between the blades and grooves. A dead blow hammer is often used to work each blade in by a small amount, working around the wheel multiple times until the blades are all in their desired axial engagement. Axial roots require some positive means to retain the blades in the correct axial location. Many axial roots are retained by the use of pins or bolts installed at the bottom of the wheel slot, as shown in Fig. 3.33, with washers on each end that overlap the wheel face and ends of the roots to prevent axial movement. An alternate design, shown in Fig. 3.34, uses bend-up tabs instead of washers. A sheet metal retainer with a dog bone shape is placed in the bottom of the wheel slot, with a larger tab on either side of the wheel. After the blade is installed above the retainer, the tabs are bent up to overlap the wheel face and end of the blade root, preventing axial movement of the blades relative to the wheel.

3.7.3 Bladed Disk Assembly—Pinned Roots

Because the root forms do not automatically determine the blade location, the assembly process must correctly locate each blade in relation

FIGURE 3.34 Axial roots with "dog bone" or bent-tab type of retainers. (*Photograph courtesy of Dresser-Rand Company.*)

to the turbine rotor at the correct radius to the turbine centerline and with the airfoil stacking axis in the correct angular relation, not only to the rotor but also to adjacent blades in the stage.

Depending on the tools available and the preference of the process engineer, blades and wheel rim may have pilot holes predrilled, or the drilling may all be done in its entirety after the blade is fixtured in its correct location in the wheel. In either case, the final holes are align-reamed through both blade and wheel rim simultaneously at very close tolerances, ensuring proper fit of the pins. To ensure close fit, in some cases the pins may be ground to match the size of individual bores. In many cases, pins that are sized for a slight interference fit must be cooled to facilitate assembly after align-reaming. After align-reaming, the disk, blades, and pins become, in effect, a matched assembly. In this case, if they are removed for any reason, the blades and pins must be reassembled in their original locations on the disk.

Pins must also be positively retained in their bores. Thermal transients during start-up or large load changes may result in looseness even in interference-fit assemblies. Coupled with vibration and pressure loads on the face of the wheels, pins will "walk" out of their engagement in the bladed wheel rim if they are not positively retained. A common means to retain the pins involves staking or upsetting the material adjacent to the end of the pins in several locations at each end to prevent axial movement in the wheel, but they may also be retained by mechanical fastenings of one sort or another. Figure 3.35 illustrates a typical staked pin and root assembly.

FIGURE 3.35 Pinned root installation. (*Photograph courtesy of Dresser-Rand Company.*)

3.7.4 Riveted Shroud Installation

One of the most common shroud configurations is the riveted shroud. In this configuration, a segmented shroud band is attached to the blades by tenons that extend through holes in the shroud and that are then peened to deform and expand the tenon to permanentaly attach the shroud to the blade tips. Correct assembly of the shroud band segments and peening of the tenons are critical to long-term reliability. The most critical process is peening the tenon, to deform it into the classic rivet shape. While the riveting process can be automated to some extent, it still often depends on operator skill for acceptable results. For some cases, such as kidney-shaped tenons, assembly involves manual operations, where good results depend entirely on the craftsmanship and skill of the operator.

Stage designers often use the shroud to create a specific blade packeting plan to control the modal and vibratory behavior of the bladed disk. Shroud segments are specified by the turbine design engineer to group a specific number of blades into a specific number of packets. During assembly, it is therefore important to ensure that the correct number of blades is installed through the use of shims or over- and undersized buckets.

The use of very long shroud segments can create difficulty in assembling the shroud to the blades. Due to the slight angular displacement between adjacent blades, the holes in the shroud must be slightly oversize to permit installation over the tenons. Very long segments exacerbate this problem. When long or continuous shroud bands are specified, a common solution is to utilize a layered shroud structure. This may involve a ship-lap feature at adjacent segments, or it may involve a double-layer shroud, such as where six-blade shroud segments are overlapped by three blades at each end to create an effective double-layer continuous shroud band.

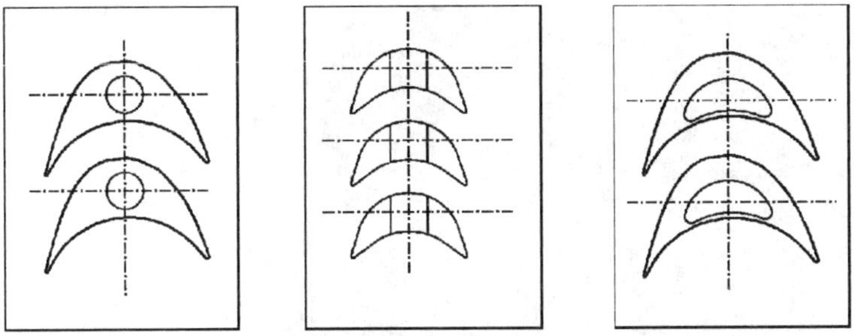

(a) Round tenon
- Most common form
- Turned or milled in separate operation
- Simple to install

(b) Sectional tenon
- Most easily machined
- Inexpensive
- Requires punched holes in shroud band
- Low stress applications

(c) "Kidney" tenon
- Most costly
- Large, "beefy" airfoil sections
- Heavy shroud band
- Difficult peening process
- Usually manually peened

Figure 3.36 Tenons for riveted shrouds can be round, sectional, or square "kidney" shapes.

As one might imagine, the design of the tenon has a major influence on the cost and on the difficulty of assembling and peening a riveted shroud assembly. The most common type of tenon is a simple round shape extending from the tip of the airfoil. Round tenons are the easiest to produce and install. Figure 3.36*a* shows the blade with a round tenon. Shroud band segments can be produced with either punched, drilled, or milled holes.

The least expensive tenons to produce during blade production are the so-called sectional tenons, as illustrated in Fig. 3.36*b*. Sectional tenons are square or rectangular and are very inexpensively formed by a simple milling operation during blade manufacture. While the tenons themselves are less expensive to produce, the matching holes in the shroud band must be produced by a punching process, which tends to limit its use to relatively thin shroud band material that can be easily punched. The rectangular tenon shape is evident in Fig. 3.37, showing a shroud with sectional tenons.

Kidney-shaped tenons, named for their shape and shown in Fig. 3.36*c*, are sometimes used when thick shroud band material must be attached to blades with large, robust airfoil sections. Since they can only be produced by a milling process, kidney-shaped tenons are the most expensive to produce and create the most difficulties during the peening process. In the past, kidney-shaped rivets were peened using manual processes. In most cases, this meant the assembler used a pneumatic hammer to expand and shape the end of the tenon into the desired rivet head shape. This process depended in large part on the craftsmanship and skill of the assembler. However, orbital riveting machines are now available that can be programmed to peen kidney-shaped rivets, resulting in more consistent results.

Figure 3.37 Sectional tenons after peening. Note the rectangular shape of the finished tenons. (*Photograph courtesy of Dresser-Rand Company.*)

Figure 3.38 Installing shroud segments prior to peening. (*Photograph courtesy of Dresser-Rand Company.*)

Shroud band segments must be properly prepared for installation, which is illustrated by Fig. 3.38. For very stiff blades, it is often necessary to mark the hole locations from the blades after installation. In this way, variations due to shims or oversize and undersize blades can be accounted for. On the other hand, if the blades are very flexible, such as drawn blades, the shroud may be drilled or punched based on the theoretical or design locations of the blades to ensure they are properly oriented in the wheel.

Shroud bands for sectional rivets are usually punched with a square or rectangular shape to match the tenon shape. Holes for round tenons may be either drilled or punched, depending on hole size and material thickness. In special cases, the holes may be produced by milling to minimize deformation of the segments or to achieve close hole diameter control. Kidney-shaped rivets require the use of shroud segments with holes that are either punched using specially shaped tools or milled to size and shape on a CNC milling center—both expensive processes.

In riveted shrouds, the critical operation is the process of upsetting and forming the tenon into the final "rivet" shape. Peening the tenons involves deliberate plastic deformation of the tenon, making it easy to understand the importance of high ductility in the blade material. Low ductility in the blade material may create serious problems during the peening process, including cracks and even fractures in the tenon.

In some cases, the tenon may be locally heated immediately prior to upsetting. To be effective, the material must be heated to a sufficient temperature to reduce the material's ultimate strength. Heating may be used for tenons with large cross-sectional area to reduce the total force and energy imparted to the blade. It may also be used to allow peened tenons to be used with materials having higher strength or lower ductility that would otherwise rule out the use of peened tenons. For example, riveted shrouds have been installed on titanium blades by heating the tenons prior to upsetting. While heating the tenons can be very effective, it must be approached with great caution. If the heating process is not carefully controlled, there is a danger that the blade or shroud material may attain high enough temperature to change its material properties.

The peening process can be carried out using one of several related processes, illustrated conceptually by Fig. 3.39. One common procedure uses an automatic hammer, like that shown by

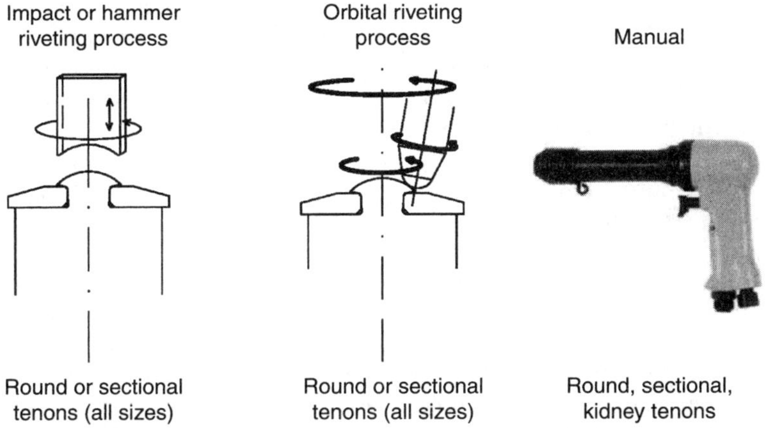

FIGURE 3.39 Peening processes.

Figure 3.40 Automatic hammer-type riveting machine. (*Photograph courtesy of Dresser-Rand Company.*)

Fig. 3.40, with its anvil shaped to produce the desired rivet shape. The hammer strikes the tenon up to 200 times per minute, while simultaneously rotating at about 15 rpm, with the speed controlled by the operator.

The orbital riveting process is more automated, with an orbital head forcing a hardened head against the tenon. The anvil rotates about its own axis, while the head simultaneously nutates the anvil's axis about the tenon's centerline. The anvil's axis is usually held at a slight angle relative to the tenon's axis.

For irregularly shaped rivets, such as Kidney-shaped tenons, the only option may be manual peening using an air-operated impact hammer. This process may, of course, be used on any tenon if automated equipment is not available. With the manual peening process, the results are almost entirely dependent on the skill, experience, and craftsmanship of the operator.

Figure 3.41 illustrates the characteristics of a well-executed peened tenon. Figure 3.42 is a photograph of an example of correctly formed tenons. The tenons in a riveted shroud cannot be easily inspected after assembly, so how do we ensure good-quality rivets?

As with any process that is not easily checked or inspected after completion, the only practical approach is to develop a detailed process specification, controlling as many of the critical process variables as possible. In addition to machine parameters such as hammer size and speed, process development must include qualification tests of both the process and the operators. Most power hammers are at least partly dependent on the skill and experience

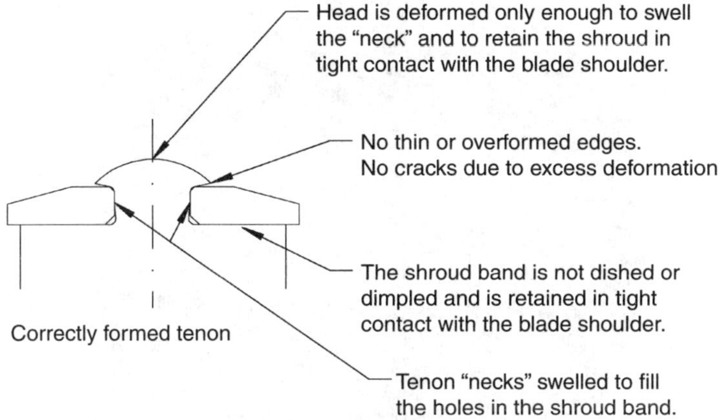

FIGURE 3.41 Cross section of correctly upset and formed tenon.

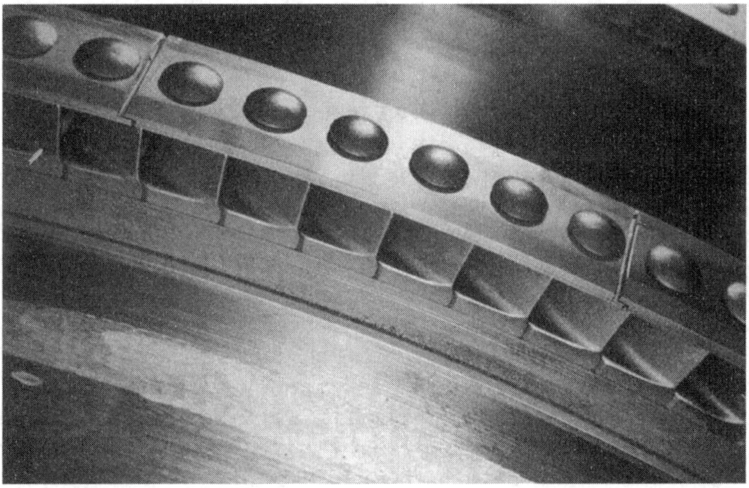

FIGURE 3.42 Example of correctly formed tenons. (*Photograph courtesy of Dresser-Rand Company.*)

of the operator. For manual peening with an air hammer, operator qualification is the single most critical step in the overall process qualification and documentation. Process and operator qualification testing relies on the production of sample riveted tenons on sample blades in either the production disk or a test dummy disk. The blades and shroud bands are assembled using the same process as the production disk. The sample blades are then removed, and the tenons are sectioned so that important parameters such as neck filling, head shape, and tight contact between the shroud band and

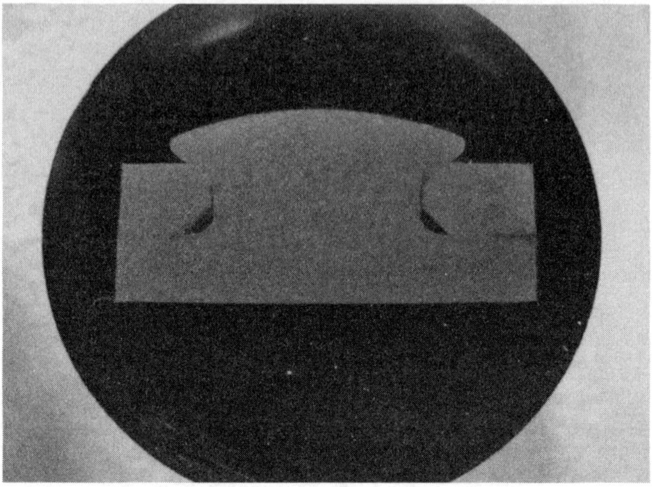

FIGURE 3.43 Qualification sample for peened tenon illustrating correctly formed head, good contact at neck, and negligible distortion in shroud band. (*Photograph courtesy of Dresser-Rand Company.*)

the shoulders of the blade tip can be checked. Figure 3.43 is a photograph of a qualification sample that illustrates all the attributes required for an acceptable peened tenon. In most cases, the OEM or service shop will perform qualification tests on a regular periodic basis, usually quarterly or semiannually, and will retain the results of the process and personnel qualifications in its records. For critical stages or units, or for unusual designs, it may be advisable to carry out qualification tests on the specific design being used to ensure results meet the acceptance criteria. Although the assembled shroud cannot be completely inspected, there are still some common external checks that can be carried out, including subjecting the peened tenons to a dye penetrant inspection process or, in limited cases, ultrasonic inspection to check for cracks in the heads.

Although it is difficult to inspect the finished tenons after the peening process is completed, some common problems that can affect reliability are easily identified by looking for their telltale signs.

A common error in manual processes results in a simple mushrooming of the edges of the tenon, without sufficiently upsetting the neck of the tenon to fill the hole in the shroud band, seen in Fig. 3.44 The edges of the head often are overstretched, resulting in paper-thin edges and possible cracking. With insufficient metal in contact to resist centrifugal loads on the shroud, the shroud becomes loose, losing contact with the blade shoulder. Loose holes and insufficient contact with the blade shoulder reduce the effectiveness of the shroud as both damper and structural stiffener.

Figure 3.46 Photo of friction band shroud wire installed. (*Photograph courtesy of Dresser-Rand Company.*)

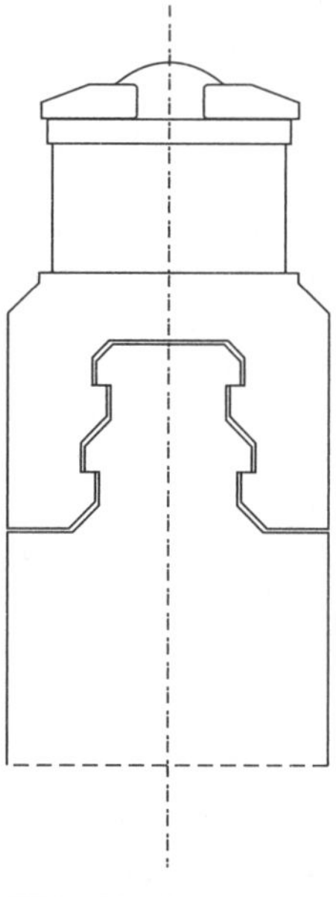

Figure 3.47 Double-covered shroud construction—conventional riveted shroud installed atop blades with integral shrouds.

Double-covered blades are expensive to produce, so they are typically only used in very highly loaded stages, such as the control stages in high-speed process drives and in some critical marine turbine applications.

Most integral shrouds are trimmed after installation to eliminate axial runout from the faces of the shroud, and to eliminate radial runout of the OD surfaces of the shroud. To facilitate this operation, blades are produced with extra stock on the integral shrouds just for this final trimming operation.

3.8 Inspection, Testing, and Quality Assurance

In the preceding discussions, we touched on many of the manufacturing processes as well as some of the critical inspection processes to ensure a reliable bladed disk assembly. However, given the importance of the blades themselves, as well as that of the complete bladed disk assembly, a discussion of the role of inspection, testing, and quality assurance activities associated with the production assembly and operation of steam turbine blades is warranted.

Many engineers mistakenly believe that quality assurance and quality control activities are the sole province of the inspectors and quality assurance department. There are very good reasons, however, that turbine design engineers and reliability engineers must be involved in the development and execution of a comprehensive quality plan that addresses all the critical manufacturing and assembly processes associated with the production of steam turbine blades and bladed disk assemblies. At the design stage, and especially during the design of individual components, virtually all the important design parameters (material properties, part dimensions, and assembly tolerances, among others) are only assumptions or specifications on the part of the designer. As the reader will learn in subsequent chapters, the calculated blade frequencies, blade loading limits, and overspeed limits are all sensitive to the final, as-built values of these parameters. To ensure that the final product performs in accordance with the designer's predictions, the actual values of key design parameters must meet or exceed the designer's assumed values. The vehicle that is used to make sure this occurs is the *Quality Plan*, also known as the *Inspection and Test Plan* or *ITP*.

The inspection and test plan has three major objectives. The first is to ensure that the minimum requirements for each component, as defined by the turbine design engineer, are met. It is also the vehicle to ensure that any additional or special requirements stemming from contract or customer specifications are defined and carried out during execution of component manufacture. And finally it provides a record of the actual production and assembly of the components and assemblies covered by the plan. If necessary, these records can be used in the future to diagnose and troubleshoot problems, and they form the basis for improved processes and results on future projects. A good-quality plan lays out an objective set of processes and acceptance criteria for the project, leaving as little as possible to later subjective decisions by manufacturing personnel, inspectors, and customer representatives. As a minimum, the inspection and test plan must detail the characteristics to be controlled, any controlling documents and standards that pertain to these characteristics, the acceptance criteria for these characteristics, the methods by which these characteristics are to be measured and documented, record-keeping requirements, and finally a list of any inspection points requiring observation by a customer representative.

Although the standard was specifically written to address repairs of existing rotating machinery, API Standard 687 contains some excellent examples of ITPs for a variety of turbomachinery rotor repairs, including steam turbines, which are covered by Section 4 in the standard. The proforma quality plans in this document provide an excellent reference for turbine design engineers, reliability specialists, inspectors, and end users. The inspection activities and documentation requirements are, of course, applicable to brand new steam turbines as well as repair and reblading activities for existing turbine rotors.

CHAPTER 4

System of Stress and Damage Mechanisms

Newton's laws of motion form the basis of analyzing any mechanical structure to evaluate response under a specified force. The definition of force and its effects on the structure are embedded in Newton's laws. It is appropriate and essential to start a chapter on mechanical design of a mechanical structure, e.g., steam turbine blades, with a discussion of these laws, which are taught and discussed in physics and mechanics. It is imperative to realize the importance, and to understand certain ingredients, of these laws as these are applied to assess the behavior of mechanical structures.

Newtonian mechanics is the system of mechanics based upon Newton's laws of motion in which mass and energy are considered to be separate from each other. In newtonian mechanics, mass and energy are considered to be conservative, in contrast to their treatment in relativistic mechanics.

First Law A body will stay in a state of rest or in a state of constant motion unless forced to change. In other words, in the presence of a resultant force, a body will acquire a velocity that will vary with time.

Second Law Force acting on a body is proportional to the rate of change of momentum of the body.

$$\text{Momentum } M = m(dx/dt)$$

$$\begin{aligned} \text{Force } F &\propto d\,[m(dx/dt)]/dt \\ &\propto (dm/dt)(dx/dt) + m(d^2x/dt^2) \\ &= Km(d^2x/dt^2) \end{aligned} \tag{4.1}$$

when the mass of the structure does not change and K is a constant.

Finally,

$$F = Kma \tag{4.2}$$

The expression in Eq. (4.2) is valid only when

1. The mass of the body does not change with time.
2. The units of force, mass, and acceleration are chosen to make $K = 1$.

Newton's second law is applicable to mechanical structures when the mass does not change. However, for dynamic systems there must be displacement that changes with time, leading to acceleration of the body. If the mass of the blade changes due to deposition, corrosion, and erosion, then the first term of Eq. (4.1) must be considered. Singh et al. (1994) reported a case of a steam turbine blade failure caused by chemical buildup that changed the natural frequency in the resonance region.

Third Law If two particles interact, the force exerted by the first particle on the second particle (called the *active force*) is equal in magnitude and opposite in direction to the force exerted by the second particle (called the *reaction force*). Simply stated, any action has an equal and opposite reaction. This is the governing law that allows one to write the force balance on structure that eventually leads to the equation of motion for dynamics as well as for steady-state stress evaluation.

The implications of Newton's laws of motion have also been expressed as follows:

1. If the force is as defined by Newton, then energy must conserve, or
2. If energy is conserved, the force must be as defined by Newton's laws of motion.

4.1 Stress-Strain Behavior of Metals

The basic response of a material under load is obtained by plotting displacement versus applied load. This is generally done through a test on a uniaxial polished specimen. A schematic view during deformation of a gage length of such a specimen is shown in Fig. 4.1.

The P_0, A_0, and L_0 are the original force, area, and length, respectively, of the gage portion of the specimen. As deformation occurs, the length of the specimen increases and the area is reduced to maintain the constant volume of the material. After the ultimate strength of the material, the cross section starts necking, and at final fracture the cross-sectional area becomes A_f and the final gage length is L_f, as measured by application of a controlled load on a specimen. By

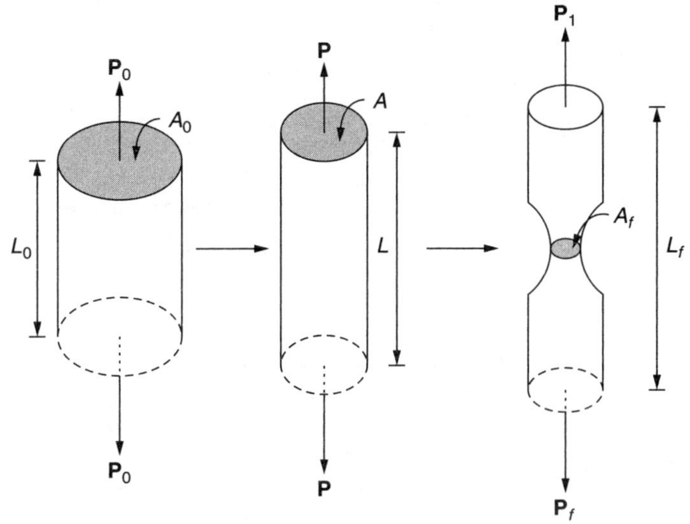

FIGURE 4.1 A schematic deformation of the gage length of test specimen.

monitoring the applied load and by measuring the elongation in a known length of the specimen, a curve can be plotted that describes the relationship between load and elongation. The resulting curve is usually converted into force per unit area (stress) and elongation per unit length (strain). It is called the *stress-strain curve* (Fig. 4.2).

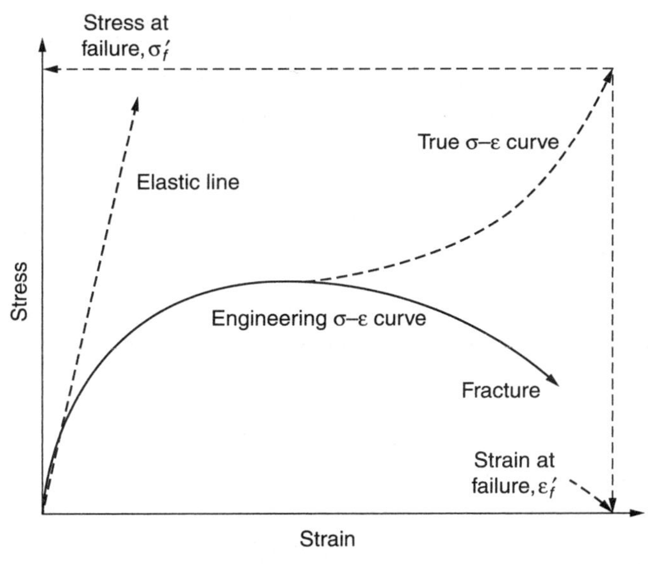

FIGURE 4.2 A typical stress-strain curve.

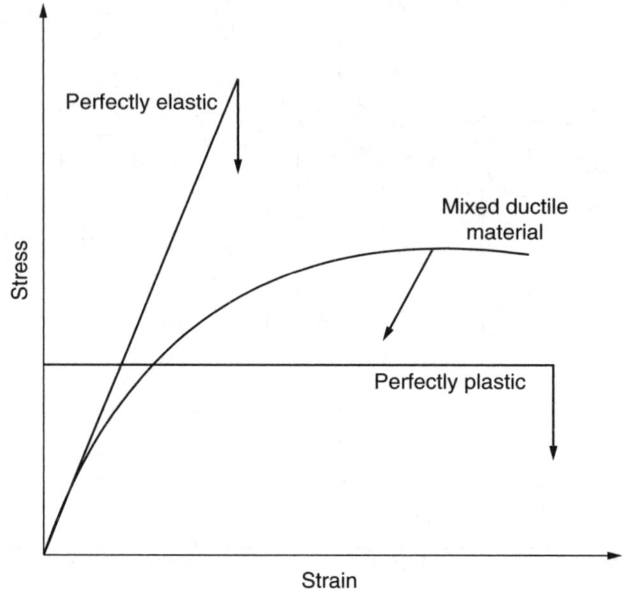

FIGURE 4.3 Typical behavior of perfectly elastic, perfectly plastic, and mixed-type materials.

Materials may display various behaviors as detailed below and shown in Fig. 4.3.

1. Material may behave in a perfectly elastic manner, which means when load is removed from a material, then it will acquire its original position before it breaks.
2. Other materials may behave in a perfectly plastic manner, which means there will be increasing displacement at a constant load until it breaks.
3. Most structural materials display a combination of elastic and plastic behavior.

Some of the terms used to estimate reliability of a mechanical structure in terms of stress and strain are discussed next (Mitchell, 1979).

Engineering stress S is defined as the ratio of P_0 to A_0.

$$S = P_0/A_0 \tag{4.3}$$

True stress σ is the ratio of P to A.

$$\sigma = P/A \tag{4.4}$$

Engineering strain e is defined as the change in the gage length divided by the initial gage length and is usually expressed as a percentage.

$$e = \Delta l / l_0 = (l - l_0)/l_0 \tag{4.5}$$

It is more reasonable to use the instantaneous gage length than the initial gage length in the definition of strain, and the resulting definition of strain is more convenient for many purposes. In devising such a definition, one considers only an infinitesimal amount of deformation, so that the length can be considered as remaining constant during this infinitesimal deformation. Thus, the infinitesimal strain $d\varepsilon$ will be the infinitesimal change in length dl divided by the characteristic length l prevailing during the infinitesimal deformation. Strain obtained from Eq. (4.6) is called *true strain.*

$$d\varepsilon = dl/l \tag{4.6}$$

The expression for a finite amount of true strain can be found by integrating Eq. (4.6). Because of the logarithmic nature of this expression, true strain is sometimes called *logarithmic strain* and is usually expressed as a decimal quantity.

$$\varepsilon = \ln(l/l_0) \tag{4.7}$$

The relationship between true strain and conventional strain can be stated as follows:

$$\varepsilon = \ln(l/l_0) = \ln(1 + e) \tag{4.8}$$

For consistency in volume, $A_0 l_0 = Al$, and when the diameters are D_0 and D,

$$\varepsilon = \ln(A_0/A) = 2\ln(D_0/D) \tag{4.9}$$

True tensile stress to cause fracture is represented by

$$\sigma_f = P_f / A_f \tag{4.10}$$

True fracture ductility is represented by

$$\varepsilon_f = \ln(A_0/A) = \ln[100/(100 - \%\text{RA})] \tag{4.11}$$

where RA is the reduction in area of the specimen.

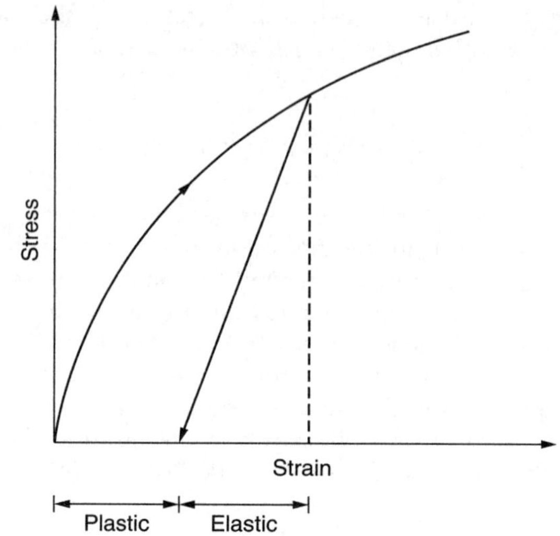

FIGURE 4.4 Elastic and plastic strains after unloading.

The relationship between true stress and engineering stress can be stated by using the consistency of the volume of the material during deformation.

$$\sigma = s\,(1 + e) \tag{4.12}$$

After a material has gone through a plastic deformation, the total strain is the summation of the elastic and plastic portions of strain. These strains are represented by Eq. (4.13) and shown in Fig. 4.4.

$$\text{Total strain } \varepsilon_t = \varepsilon_e + \varepsilon_p \tag{4.13}$$

The elastic portion of strain can be determined by using Hooke's law, given here:

$$\varepsilon_e = \sigma / E \tag{4.14}$$

The plot of the log of plastic strain versus the log of stress displays a straight-line behavior. The relationship between stress and strain can be written as follows:

$$\log\sigma = \log K + n \log \varepsilon_p$$

This relationship reduces to the following equation.

$$\sigma = K\,(\varepsilon_p)^n \tag{4.15}$$

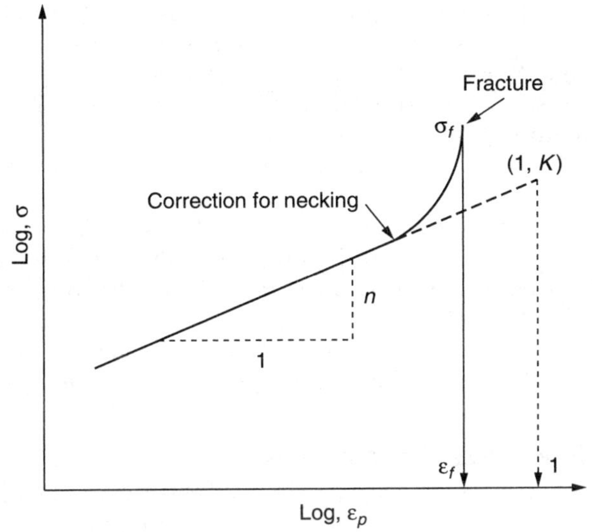

Figure 4.5 Log-log plot of plastic strain versus stress.

or

$$\varepsilon_p = (\sigma / K)^{1/n} \tag{4.16}$$

where n is the monotonic strain hardening exponent and K is the monotonic strength coefficient. Substitute Eq. (4.14) and Eq. (4.16) into Eq. (4.13) to get the following relationship.

$$\varepsilon_t = \sigma / E + (\sigma / K)^{1/n} \tag{4.17}$$

The above equation is known as the *Ramberg-Osgood relationship*. Also from Eq. (4.15)

$$\sigma_f = K\varepsilon_f^n \tag{4.18}$$

Then

$$\varepsilon = \sigma / E + \varepsilon_f (\sigma / \sigma_f)^{1/n} \tag{4.19}$$

Equation (4.19) is another form of the Ramberg-Osgood relationship.

4.1.1 Engineering Stress-Strain Properties

The modulus of elasticity E is the slope of the stress-strain curve below the elastic limit. It is also defined as stress ($\sigma = E$ when $\varepsilon = 1$) to cause an elastic strain of unity in Hooke's law, $\sigma = E\varepsilon$.

The tensile yield point S_{ty} is defined as the stress needed to cause sudden gross yielding and applies only if there is a "flat top" region of yielding on the stress-strain curve.

The tensile yield strength S_{ys} is the stress needed to cause a specified amount of inelastic strain, which is usually 0.2 percent and is determined by constructing an offset line with a slope of E.

The percent of reduction in area %RA [$= 100(A_0 - A_f)/A_0$] is defined as the reduction in cross-sectional area required to cause fracture, where A_f is measured at the resulting minimum cross section after fracture.

Poisson's ratio ν is the ratio of diametrical strain to longitudinal strain and is determined at stresses below the yield point.

4.1.2 True Stress-Strain Properties

The E, σ_{ty}, and σ_{ys} are the same as defined earlier. When measured at small strains, true and engineering values are nominally the same.

True fracture strength $\sigma_f (= P_f/A_f)$ is the true tensile stress needed to cause fracture and should be corrected for the influence of lateral stresses due to necking. The load at fracture divided by the area at fracture is higher (e.g., about 20 percent when %RA = 70 percent) than the corrected value and depends on the severity of the neck.

True fracture ductility $\varepsilon_f [= \ln(A_0/A_f)]$, is the true plastic strain required to cause fracture and is equal to ln(100/100 – %RA).

The strain hardening exponent n is the slope of the log ε_p – log σ plot. This is the power to which the plastic strain must be revised to be proportional to the stress and can be shown to be equal to the true strain at which necking occurs.

The strength coefficient K is the true stress to cause a true plastic strain of unity in the equation $\sigma = K\varepsilon_p{}^n$. If σ_f is corrected, then $K = \sigma_f/\varepsilon_f{}^n$. If $\varepsilon_f < 1$, the flow curve is extrapolated to a plastic strain of unity to obtain K.

Halford (1963) presented a graphical method to estimate the strain hardening exponent n when the σ-ε curve is available. This method is based on the energy principle and can be explained with the aid of Fig. 4.6.

The relationship between true stress and true plastic strain in the nonlinear region is given by

$$\sigma = K\varepsilon_p{}^n \tag{4.20}$$

The true plastic strain–true stress relationship ($\sigma = K\varepsilon_p{}^n$) is used for the nonlinear portion of the σ-ε curve.

The cross-hatched areas (Fig. 4.6) are designated as the plastic strain energy U_p and the complementary plastic strain energy U_c.

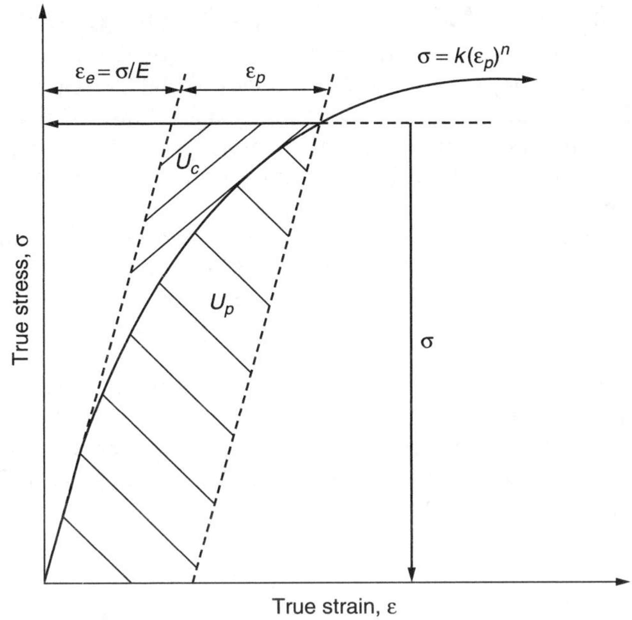

FIGURE 4.6 Typical true stress-strain curve.

If the true stress–true plastic strain curve obeys the relation represented by Eq. (4.20), then the ratio U_c/U_p is equal to n.

$$U_p + U_c = \sigma\varepsilon_p \tag{4.21}$$

Rewriting Eq. (4.21) yields the following:

$$U_c/U_p = \sigma\varepsilon_p/U_p - 1 \tag{4.22}$$

Integrating Eq. (4.20) with respect to ε_p, we find the area under the curve to be

$$U_p = \int \sigma d\varepsilon_p = [K/(n+1)](\varepsilon_p)^{n+1} \tag{4.23}$$

Substituting Eq. (4.23) into the right-hand side of Eq. (4.22) yields

$$U_c/U_p = (n+1) - 1 = n \tag{4.24}$$

where n can be evaluated as described only if Eq. (4.20) describes the stress-strain curve in the nonlinear regime.

4.1.3 Measure of Material's Engineering Energy Capacity

The work done is equal to the energy absorbed by the material. Most of this energy represents potential energy that is released upon unloading for deformation within the elastic limit. The potential energy at the proportional limit (σ_p) is thus the triangular area shown in Fig. 4.7. It is important to note that beyond the elastic limit a portion of the stored energy is not recovered.

In a tension test the amount of work done in deforming the specimen to a value of x_1 is represented by

$$U = \int P dx \tag{4.25}$$

In Eq. (4.25), P is the applied force, L_i is the gage length, and the displacement $\Delta L = x_1$.

The volume of material in the gage length is $A_i L_i$, and the work done per unit volume is obtained by dividing both sides of the equation by this volume.

$$u = U/(A_i L_i) = \int_0^{x_1} (P/A_i)\, d(x/L_i) = \int_0^{\varepsilon_1} \sigma\, d\varepsilon \tag{4.26}$$

Hence, u is the work done per unit volume of material to reach a strain ε', which is equal to the area under the stress-strain

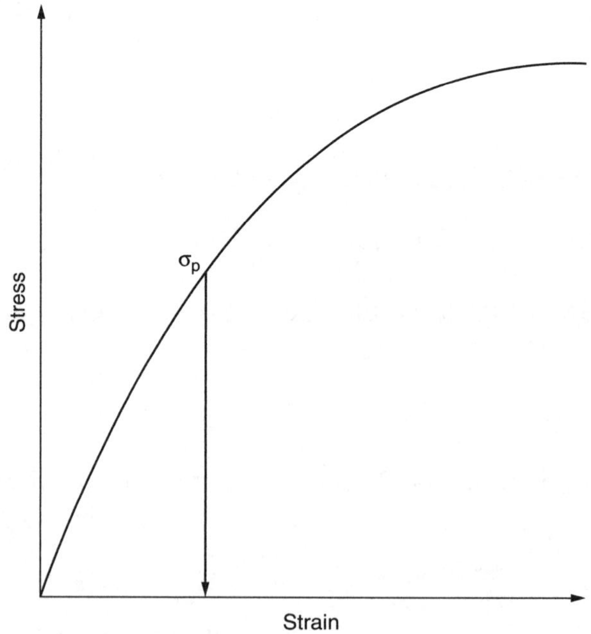

FIGURE 4.7 Definition of resilience.

curve up to the strain ε'. *Resilience* is defined by the following equation.

$$u_r = (\sigma_p^2)/(2E) \tag{4.27}$$

However, in view of the difficulty in defining σ_p, it is preferable to use this same equation, but with the offset yield strength σ_o replacing σ_p.

$$u_r = (\sigma_o^2)/(2E) \tag{4.28}$$

4.2 Stress Tensor and Strain Tensor

The magnitudes and directions of the stresses and strains in a component are needed for evaluating the maximum load that the component can satisfactorily carry. The stresses and the strains are also used in finding the optimum dimensions of the component to carry a specific load. That is, the stresses and strains are used as criteria for the possibility of failures within the material under the expected loading conditions. Different theories are propounded based on the type of loadings (static or dynamic) and on the type of materials (brittle or ductile) of the component.

A force F is applied to a beam in the left part of Fig. 4.8; the right part shows a free-body diagram of the right side of the beam separated from the left side. For equilibrium per Newton's third law, a balancing reaction force R is developed at the cutaway plane and will be equal to the applied load F.

The *average stress* at the cutaway plane is defined as

$$S_{av} = R/A \tag{4.29}$$

According to Newton's third law,

$$R = F \tag{4.30}$$

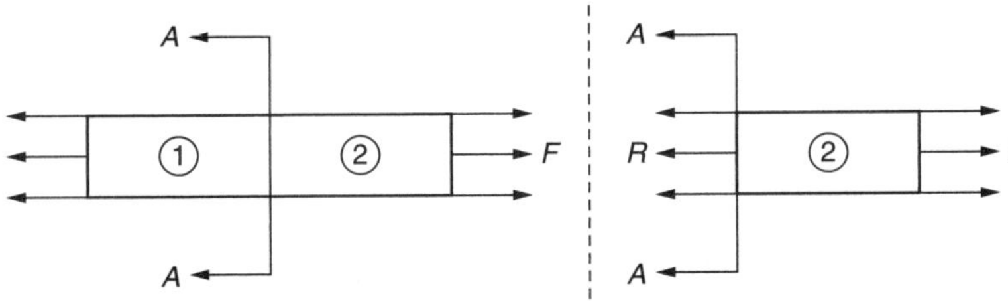

FIGURE 4.8 Free-body diagram and force equilibrium.

Thus,

$$S_{av} = F/A \tag{4.31}$$

Stress is defined as the reaction force R per unit area and generated by the body in order to resist the changes being imposed by the applied force F.

4.2.1 Stress at a Point and Stress Concentration

A quantity that can be described completely by its magnitude is called a *scalar quantity*, e.g., mass, density, and temperature. A quantity that needs direction in addition to magnitude for complete description is called a *vector quantity*, e.g., velocity, force, and acceleration. Stress and strain are tensor quantities and, to be completely defined, need magnitude, direction, and a point of application. This distinction is important because magnitude and direction alone are not sufficient to express stress and strain, but the location where these properties exist is also essential.

The average stress at a point, such as in Fig. 4.9, is stated as

$$S = \Delta P/\Delta A$$

The local stress at the point is

$$\sigma = Lt(\delta R/\delta A) = dR/dA \qquad \delta A \to 0 \tag{4.32}$$

There are situations in which the reaction forces needed to balance the applied load are not uniformly distributed, e.g., as in the presence of a notch such as shown in Fig. 4.10. This means that the stress at

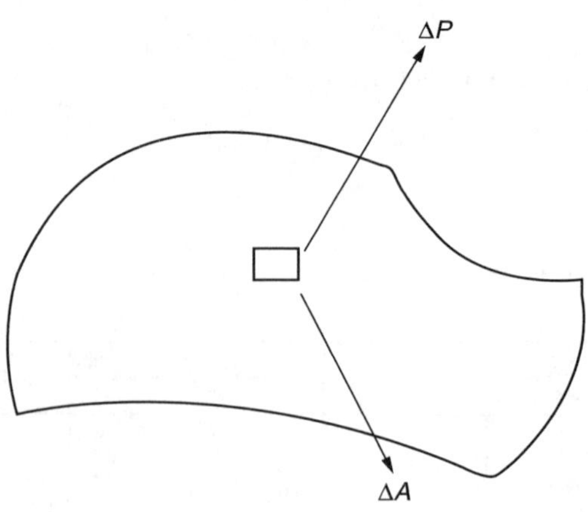

FIGURE 4.9 Force and stress on a small area.

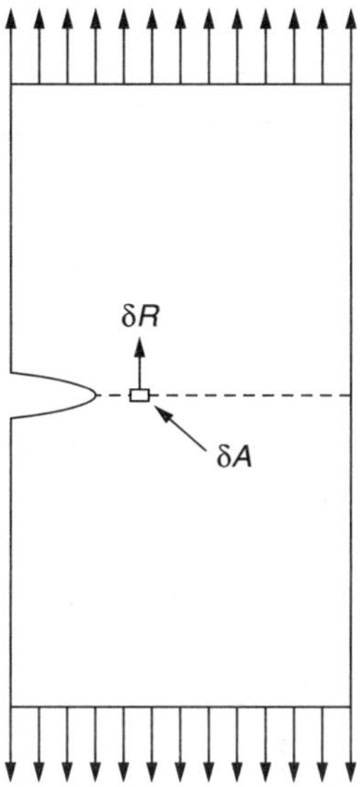

Figure 4.10 Definition of local stress.

each point across the cross section will be different and will not be uniformly distributed, as shown in Fig. 4.10. In such a situation the definition of stress at a particular point helps in the determination of local stresses.

Equation (4.32) offers a way to estimate stress from point to point in any structure. The estimation of stress and strain at each location is important because the damage to a structure starts from the location of maximum stress. Due to a notch or a discontinuity across the cross section, the distribution of force is not uniform, even though the average stress through the cross section might have a lower magnitude. This variation of force is exhibited in a stress variation through the cross section, as shown in Fig. 4.11. Damage to a structure starts from the location where stress is concentrated even though the average stress across a cross section might have an acceptable magnitude.

It is clear that Eq. (4.32) provides a way to estimate stress concentration if the stress concentration exists while average stress does not. Determination of localized stress is done by multiplying the average cross-sectional stress by a theoretical stress concentration factor. This method works very well when the localized stress is less than the yield strength of material, i.e., in the absence of any plastic deformation. Finite element analysis may be used to estimate actual localized stress, but it takes a lot of resources. Many times the stress is so localized that it can be estimated by some simple rules without spending valuable resources on finite element analysis. A popular method to estimate the peak stress is called the *Neuber rule*.

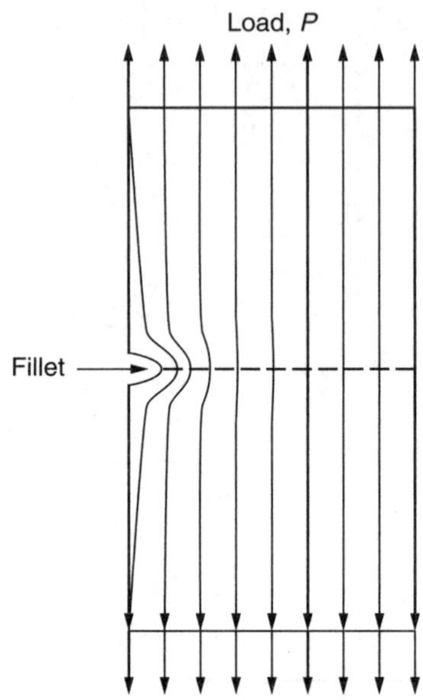

FIGURE 4.11 Stress contours showing concentration of stress near fillet.

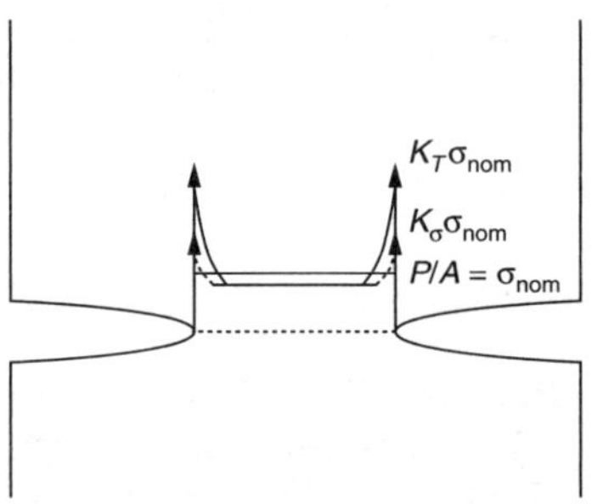

FIGURE 4.12 Localized stress in the vicinity of a fillet.

Many mechanical components of turbines have fillets and notches, as in Fig. 4.12. This feature is known to magnify the average stress in the vicinity of the fillets and notches, a phenomenon known as *stress concentration*. Predetermined tables and equations of many geometric and loading situations are available to estimate the stress concentration K_T. These are based on the assumption of elastic behavior of the material.

Therefore, the stress concentration K_T is known as the *elastic* stress concentration. The resulting peak stress estimated by multiplying the average stress to the elastic stress concentration is expected to lie on the straight line representing the hookean line, i.e., a straight line drawn from the origin of the stress-strain plot with a slope representing Young's modulus. If the magnitude of the peak stress is greater than the elastic limit, then localized plastic deformation is expected and the actual state of the material must be determined by its proper location on the stress-strain curve of the material.

Fatigue cracks initiate and propagate from these notches. It has been observed that special care must be exercised when the resulting stress is greater than the yield strength of the material. After yielding or just beyond the elastic limit, the correct state is determined by the stress-strain curve.

Figure 4.13 shows the hookean line. Clearly if the real stress-strain curve is considered, then the actual stress beyond the elastic limit will be smaller than estimated by the elastic stress concentration. However, the actual strain will be greater than estimated by elastic considerations. This means that the actual stress concentration K_σ will be less than K_T, but the actual strain concentration K_ε will be greater than K_T, as determined from the linear equation $E = \sigma/\varepsilon$. Within the elastic limit these three concentrations ($K_T = K_\sigma = K_\varepsilon$) are equal. After the yielding of material, a plastic analysis must be performed. Equations of the theory of elasticity can be analytically solved. It is rare, however, that the exact solution can be obtained for a practical case. Many finite element codes are capable of elastic-plastic analysis, which is expected to provide a better state of stress and strain at the location of peak

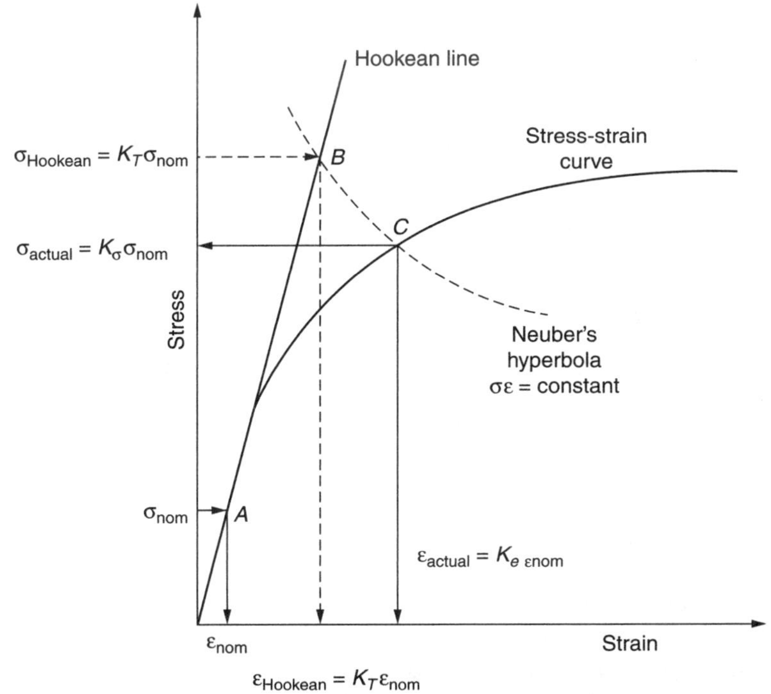

FIGURE 4.13 Neuber's hyperbola.

stress. It may be time-consuming or even costly to warrant this type of analysis. Depending on the situation an approximate solution might be desirable. A widely used approximate method is given by Heinz Neuber, and it has been determined that this method provides accuracy of about ±15 percent.

Neuber's method is explained through Fig. 4.13, which shows the situation of a fillet where the stress concentration occurs. Stress at point A is the nominal stress σ_{nom}, while stress at point B is magnified in the vicinity of the fillet. If the peak stress is large enough to yield the material, then it should follow the σ-ε curve. These points are shown in Fig. 4.13. The maximum elastic stress, as estimated by multiplying the nominal stress by the theoretical stress concentration K_T is represented by point C on the hookean line. Neuber's equation, given below, allows estimation of the location of point C by using the elastic peak stress of point B.

One way of expressing Neuber's rule is by the following equation:

$$K_\sigma K_\varepsilon = K_T^{\,2} \tag{4.33}$$

where $K_\sigma = \sigma_{\text{actual}}/\sigma_{\text{nom}} = \textit{actual}$ stress concentration (4.34)

$K_\varepsilon = \varepsilon_{\text{actual}}/\varepsilon_{\text{nom}} = \textit{actual}$ strain concentration (4.35)

$K_T = \sigma_H/\sigma_{\text{nom}} = \textit{elastic}$ stress concentration (4.36)

σ_H = stress from linear analysis

E = Young's modulus

Substituting Eqs. (4.34) through (4.36) into Eq. (4.33) yields the following:

$$(\sigma_{\text{actual}}/\sigma_{\text{nom}})(\varepsilon_{\text{actual}}/\varepsilon_{\text{nom}}) = (\sigma_{\text{H}}/\sigma_{\text{nom}})^2 \tag{4.37}$$

$$\sigma_{\text{actual}}\,\varepsilon_{\text{actual}} = (\sigma_H/\sigma_{\text{nom}})^2(\sigma_{\text{nom}}\,\varepsilon_{\text{nom}})$$

$$\sigma_{\text{actual}}\,\varepsilon_{\text{actual}} = (\sigma_H)^2(\varepsilon_{\text{nom}}/\sigma_{\text{nom}}) = (\sigma_H)^2/E = \text{constant} \tag{4.38}$$

The Ramberg-Osgood equation for the stress-strain curve is given by

$$\varepsilon_{\text{actual}} = \sigma_{\text{actual}}/E + (\sigma_{\text{actual}}/K)^{1/n} \tag{4.39}$$

The two unknown quantities $\varepsilon_{\text{actual}}$ and σ_{actual} can now be estimated by using Eqs. (4.38) and (4.39). Many numerical schemes have been propounded to solve these equations. It may be easier to solve these equations graphically by drawing Neuber's hyperbola as represented by Eq. (4.38). The intersection point of this curve and the σ-ε curve is the location of point C in Fig. 4.13. Actual stress and strain now can be read from the plot.

4.2.2 Three-Dimensional Expression for Stress at a Point

The stress system at a point in the three directions x, y, and z is shown in Fig. 4.14. It is easier to visualize the complete description of the stress system by enclosing the point in question by a cube of very small size. All the components of stress are shown on the different faces of the cube. Mathematically, it is easier to express the stress designation by two subscripts in place of just by σ (normal stress) and a τ (shear stress). The first subscript denotes the plane that the stress acts upon, and the second subscript denotes its direction in the double-subscript designation. The plane is defined by the direction of its normal direction.

For example, stress σ_{xx} means that it is acting on plane x and in the x direction, and σ_{xy} means the stress is acting on x plane, but in the y direction. It can be seen that if the subscript is repeated, then it is a normal stress; if subscripts are not repeated, then it is a shear stress. In most real analysis the chosen coordinate system x, y, z will not be the direction of maximum and minimum stresses. If the coordinate system 1, 2, 3 is assumed to be the direction of the principal stress (maximum magnitude) and it is at an angle from the x, y, z system, the following relationship is defined and will be used in the following equations. Directional cosines are the cosines of the angles between two coordinate systems. These are defined as follows:

$$\cos(1X) = \cos\alpha = l \qquad \cos(2Y) = \cos\beta = m \qquad \cos(3Z) = \cos\gamma = n$$

Stress Tensor

The six stress components σ_{xx}, σ_{yy}, σ_{zz}, σ_{xy}, σ_{yz}, and σ_{zx} completely define the state of stress at a general point of a continuous medium in

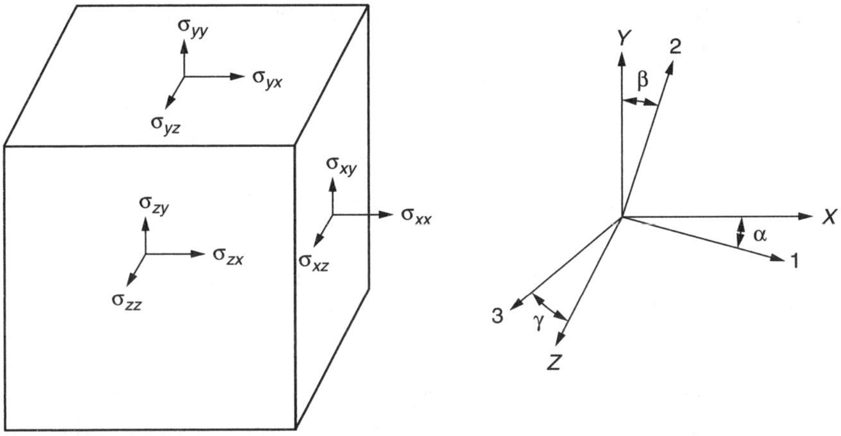

FIGURE 4.14 Three-dimensional stress system.

any orthogonal coordinate system x, y, z. The stress components are components of a stress tensor σ_{ij}.

Principal Normal Stresses

At any point, it is always possible to choose three orthogonal axes in such a way that the shearing stresses at the point vanish, so that

$$\sigma_{12} = \sigma_{23} = \sigma_{31} = 0$$

These orthogonal axes are called the *principal axes*. The normal stresses acting in the principal directions are called *principal stresses* and are denoted by σ_1, σ_2, and σ_3.

For a three-dimensional representation of stresses, let an infinitesimal cube surround point C. Each plane of the cube is designated by the direction of its normal vector. The stress will be designated by a double subscript, that is, σ_{ij}. The first subscript denotes the plane on which it acts. The second subscript defines the direction of the stress; that is, σ_{ij} is a stress acting on a plane i and is pointing in direction j. With this notation, it can be seen that stresses with a repeated subscript are normal stresses (for example, σ_{xx}, σ_{yy}, etc.) while the stresses with different subscripts (for example, σ_{xy}, σ_{zy}) are shear stresses.

4.2.3 Principal Stresses and Direction Cosines

Principal stresses are defined as maximum normal stresses, and for a three-dimensional system there will be three of these stresses acting on planes normal to each other. The shear stresses on the normal planes must be zero.

Let the coordinate system (1, 2, 3) represent the directions of the principal stresses S. Then, from the theory of elasticity, the following can be written:

$$(S - \sigma_{xx})\, l - \sigma_{xy}\, m - \sigma_{xz}\, n = 0$$
$$-\sigma_{xy}\, l - (S - \sigma_{yy})m - \sigma_{yz}\, n = 0$$
$$-\sigma_{xz}\, l - \sigma_{yz}\, m - (S - \sigma_{xz})\, n = 0$$

where l, m, and n are direction cosines of the principal stresses.

The above equation can be rewritten as follows:

$$\begin{bmatrix} (S-\sigma_{xx}) & -\sigma_{xy} & -\sigma_{xz} \\ -\sigma_{xy} & -(S-\sigma_{yy}) & -\sigma_{yz} \\ -\sigma_{xz} & -\sigma_{yz} & -(S-\sigma_{xz}) \end{bmatrix} \begin{Bmatrix} l \\ m \\ n \end{Bmatrix} = 0 \qquad (4.40)$$

For a nontrivial solution, the determinant

$$\begin{vmatrix} (S-\sigma_{xx}) & -\sigma_{xy} & -\sigma_{xz} \\ -\sigma_{xy} & -(S-\sigma_{yy}) & -\sigma_{yz} \\ -\sigma_{xz} & -\sigma_{yz} & -(S-\sigma_{xz}) \end{vmatrix} = 0 \qquad (4.41)$$

The solution of Eq. (4.41) will give the values of the three principal stresses (σ_1, σ_2, and σ_3).

$$S^3-(\sigma_{xx}+\sigma_{yy}+\sigma_{zz})\,S^2+(\sigma_{xx}\sigma_{yy}+\sigma_{yy}\sigma_{zz}+\sigma_{zz}\,\sigma_{xx}-\sigma_{yz}^{\;2}-\sigma_{xz}^{\;2}-\sigma_{xy}^{\;2})S$$
$$-(\sigma_{xx}\sigma_{yy}\sigma_{zz}+2\sigma_{yz}\sigma_{xz}\sigma_{xy}-\sigma_{xx}\sigma_{yz}^{\;2}-\sigma_{yy}\sigma_{xz}^{\;2}-\sigma_{zz}\sigma_{xy}^{\;2})=0$$

The three roots of the above equation give the values of the three principal stresses σ_1, σ_2, and σ_3. As the values of the principal stresses are unique, the coefficients and constants in the above equation must be constant; i.e., these are independent of the coordinate system. These can be expressed as follows:

$$\begin{aligned} \sigma_{xx}+\sigma_{yy}+\sigma_{zz} &= I_1 \\ \sigma_{xx}\sigma_{yy}+\sigma_{yy}\sigma_{zz}+\sigma_{zz}\,\sigma_{xx}-\sigma_{yz}^{\;2}-\sigma_{xz}^{\;2}-\sigma_{xy}^{\;2} &= I_2 \\ \sigma_{xx}\sigma_{yy}\sigma_{zz}+2\sigma_{yz}\sigma_{xz}\sigma_{xy}-\sigma_{xx}\sigma_{yz}^{\;2}-\sigma_{yy}\sigma_{xz}^{\;2}-\sigma_{zz}\sigma_{xy}^{\;2} &= I_3 \end{aligned} \qquad (4.42)$$

Constants I_1, I_2, and I_3 in Eq. (4.42) are called *stress invariants*. If the principal axes are chosen as the coordinate system, then Eq. (4.42) can be written as follows:

$$\begin{aligned} \sigma_{xx}+\sigma_{yy}+\sigma_{zz} &= I_1=\sigma_1+\sigma_2+\sigma_3 \\ \sigma_{xx}\sigma_{yy}+\sigma_{yy}\sigma_{zz}+\sigma_{zz}\,\sigma_{xx}-\sigma_{yz}^{\;2}-\sigma_{xz}^{\;2}-\sigma_{xy}^{\;2} &= I_2=\sigma_1\sigma_2+\sigma_2\sigma_3+\sigma_3\sigma_1 \\ \sigma_{xx}\sigma_{yy}\sigma_{zz}+2\sigma_{yz}\,\sigma_{xz}\sigma_{xy}-\sigma_{xx}\sigma_{yz}^{\;2}-\sigma_{yy}\sigma_{xz}^{\;2}-\sigma_{zz}\sigma_{xy}^{\;2} &= I_3=\sigma_1\sigma_2\sigma_3 \end{aligned}$$

Principal Stress Directions

Substituting principal stress values in Eq. (4.40) will determine the direction cosines after equation $l^2+m^2+n^2=1$.

Substitute either of the principal stresses in Eq. (4.40) for S, for example, σ_1:

$$\begin{bmatrix} \sigma_1-\sigma_{xx} & -\sigma_{xy} & -\sigma_{xz} \\ -\sigma_{xy} & -(\sigma_1-\sigma_{yy}) & -\sigma_{yz} \\ -\sigma_{xz} & -\sigma_{yz} & -(\sigma_1-\sigma_{xz}) \end{bmatrix} \begin{Bmatrix} l_1 \\ m_1 \\ n_1 \end{Bmatrix} = 0$$

If l_1, m_1, and n_1 are direction cosines for coordinate 1, the above equations can be rewritten as the following, which is known as the eigenvalue problem.

$$\begin{bmatrix} \sigma_{xx} & \sigma_{xy} & \sigma_{xz} \\ \sigma_{xy} & \sigma_{yy} & \sigma_{yz} \\ \sigma_{xz} & \sigma_{yz} & \sigma_{xz} \end{bmatrix} \begin{Bmatrix} l_1 \\ m_1 \\ n_1 \end{Bmatrix} = \sigma_1 \begin{bmatrix} \mathbf{1} & \mathbf{0} & \mathbf{0} \\ \mathbf{0} & \mathbf{1} & \mathbf{0} \\ \mathbf{0} & \mathbf{0} & \mathbf{1} \end{bmatrix} \begin{Bmatrix} l_1 \\ m_1 \\ n_1 \end{Bmatrix} \tag{4.43}$$

Definitions of Stresses

Stress tensor: Six stress components σ_{xx}, σ_{yy}, σ_{zz}, σ_{xy}, σ_{yz}, σ_{zx} completely define the state of stress at a general point of a continuous medium, in any orthogonal coordinate system x, y, z. The stress components are components of a stress tensor σ_{ij}.

Principal normal stresses: At any point, it is always possible to choose three orthogonal axes in such a way that the shearing stresses at the point vanish, so that

$$\sigma_{12} = \sigma_{23} = \sigma_{31} = 0 \tag{4.44}$$

These axes are called principal axes, and the normal stresses acting in the principal directions are called principal stresses, denoted by σ_1, σ_2, and σ_3.

Mean normal stress: This is invariant with respect to the coordinate system and is defined as

$$\sigma_{\text{mean}} = 1/3(\sigma_{xx} + \sigma_{yy} + \sigma_{zz}) = 1/3(\sigma_1 + \sigma_2 + \sigma_3) = \tfrac{1}{3} I_1 \tag{4.45}$$

Octahedral shearing stress: This is given by

$$\tau_0 = \tfrac{1}{3}[(\sigma_1 - \sigma_2)^2 + (\sigma_2 - \sigma_3)^2 + (\sigma_3 - \sigma_1)^2]^{1/2} \tag{4.46}$$

The octahedral shearing stress measures the intensity of stress, which is responsible for bringing a solid substance into the plastic state. In terms of the components of the stress tensor, the octahedral shearing stress may be expressed by

$$\tau_0 = \tfrac{1}{3}[(\sigma_{xx} - \sigma_{yy})^2 + (\sigma_{yy} - \sigma_{zz})^2 + (\sigma_{zz} - \sigma_{xx})^2 + 6\,(\sigma_{xy}{}^2 + \sigma_{yz}{}^2 + \sigma_{zx}{}^2)]^{1/2} \tag{4.47}$$

Equivalent and effective stresses: Equivalent total stress for the von Mises theory is defined as

$$\sigma_e = \tfrac{1}{2}^{1/2}\,[(\sigma_{xx} - \sigma_{yy})^2 + (\sigma_{yy} - \sigma_{zz})^2 + (\sigma_{zz} - \sigma_{xx})^2 + 6\,(\sigma_{xy}{}^2 + \sigma_{yz}{}^2 + \sigma_{zx}{}^2)]^{1/2} \tag{4.48}$$

Strains

Similar to the stress tensor, strain tensors are defined by the following statements.

Elastic strain is the instantaneous strain upon stress application, and it is instantaneously recovered on unloading. The elastic strain is linearly related to the stress divided by the elastic modulus, and all other strains will be described *as inelastic*. The six strain components at a point are ε_x, ε_y, ε_z, $\frac{1}{2}\gamma_{xy}$, $\frac{1}{2}\gamma_{yz}$ and $\frac{1}{2}\gamma_{zx}$, where x, y, z is an orthogonal coordinate system.

Mean normal strain is invariant and is defined as

$$\begin{aligned} e &= \tfrac{1}{3}(\varepsilon_x + \varepsilon_y + \varepsilon_z) \\ &= \tfrac{1}{3}(\varepsilon_1 + \varepsilon_2 + \varepsilon_3) \\ &= \tfrac{1}{3}\, \mathrm{I}\varepsilon \end{aligned} \tag{4.49}$$

Strain Invariants

$$\mathrm{I}\varepsilon = \varepsilon_1 + \varepsilon_2 + \varepsilon_3 = \varepsilon_{ij} \tag{4.50}$$

$$\mathrm{II}\varepsilon = (1/2!)\,\delta_{pq}^{\ \ ii}\,\varepsilon_{pi}\,\varepsilon_{qj} \tag{4.51}$$

$$\mathrm{III}\varepsilon = (1/3!)\,\delta_{pqr}^{\ \ ijk}\,\varepsilon_{pi}\,\varepsilon_{qj}\,\varepsilon_{rk} \tag{4.52}$$

and $\delta_{pq}^{\ \ ij}$ is called the *Kronecker delta*.

Octahedral unit shear is a measure of the intensity of the plastic distortion suffered by the material. The strains ε_1, ε_2 and ε_3 are the principal strains in the following expression.

$$\gamma_0 = \tfrac{2}{3}\,[(\varepsilon_1 - \varepsilon_2)^2 + (\varepsilon_2 - \varepsilon_3)^2 + (\varepsilon_3 - \varepsilon_1)^2\,]^{1/2} \tag{4.53}$$

In the cartesian coordinates, the octahedral shear is represented by

$$\gamma_0 = \tfrac{2}{3}[(\varepsilon_x - \varepsilon_y)^2 + (\varepsilon_y - \varepsilon_z)^2 + (\varepsilon_z - \varepsilon_x)^2 + (3/2)(\gamma_{xy}^{\ 2} + \gamma_{yz}^{\ 2} + \gamma_{zx}^{\ 2})]^{1/2} \tag{4.54}$$

Equivalent and Effective Strains Equivalent total strain for the von Mises theory is defined as

$$\begin{aligned} \varepsilon_e &= \tfrac{1}{2}[(\varepsilon_i - \varepsilon_j)(\varepsilon_i - \varepsilon_j) + (3/2)\,\gamma_{ij}\,\gamma_{ij}]^{1/2} \\ &= \tfrac{2}{3}[(\varepsilon_x - \varepsilon_y)^2 + (\varepsilon_y - \varepsilon_z)^2 + (\varepsilon_z - \varepsilon_x)^2 \\ &\quad + (3/2)(\gamma_{xy}^{\ 2} + \gamma_{yz}^{\ 2} + \gamma_{zx}^{\ 2})]^{1/2} \end{aligned} \tag{4.55}$$

Similarly, the plastic equivalent strain is defined as

$$\begin{aligned}\varepsilon_e'' &= \tfrac{1}{2}[(\varepsilon_i'' - \varepsilon_j'')(\varepsilon_i'' - \varepsilon_j'') + (3/2)\,\gamma_{ij}''\gamma_{ij}'']^{1/2} \\ &= \tfrac{2}{3}[(\varepsilon_x'' - \varepsilon_y'')^2 + (\varepsilon_y'' - \varepsilon_z'')^2 + (\varepsilon_z'' - \varepsilon_x'')^2 \\ &\quad + (3/2)\,(\gamma_{xy}''\gamma_{xy}'' + \gamma_{yz}''\gamma_{yz}'' + \gamma_{zx}''\gamma_{zx}'')^{1/2}\end{aligned} \tag{4.56}$$

Anelastic strain is the transient strain for a stress change. Part of this strain may not be recovered on stress reversal and becomes a permanent strain.

Plastic strain is the time-independent permanent strain that occurs on loading above the elastic limit and is not recoverable on unloading.

Transient creep strain is the transient strain at a constant load that occurs after a load change. Part of this strain may not be recovered on load reversal and becomes a permanent strain.

Creep strain is the permanent strain that changes with time at a creep rate $d\varepsilon/dt$, and it is a function of the stress, the strain, and their histories at a given temperature.

4.3 Deformation and Fracture Damage

There are many reasons for the degradation and damage to a material. Some of, but not all, the most common reasons are described in Fig. 4.15. Deformation damage or failure occurs when a change in the physical dimensions or shape of a component is sufficient for its function to be lost or impaired. When a component is cracked to the extent that a component is separated into two or more pieces, it is termed *fracture*. Wear occurs when surface material is removed due to abrasion or sticking between solid surfaces that touch and have relative

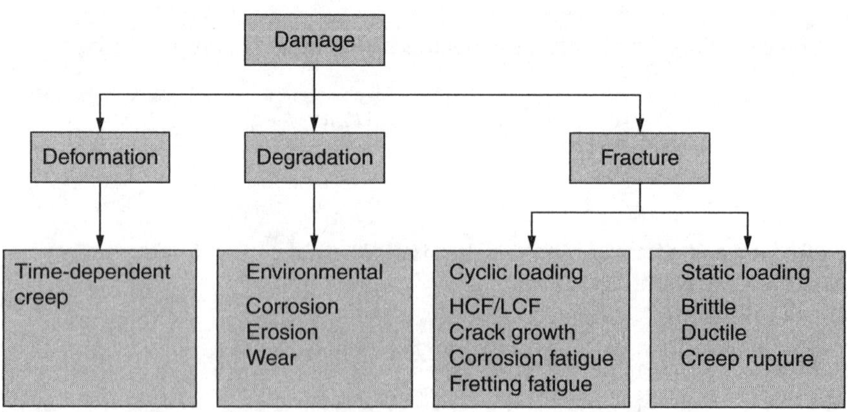

FIGURE 4.15 Basic definition of damage.

motion. Erosion is wear that is caused by a fluid (gas or liquid). Loss of material due to chemical action is called corrosion.

4.3.1 Theories of Failure under Static Loads

To estimate the maximum load that may be applied to a component without causing failure and to avoid the experimental determinations for an infinite number of combinations of stresses, several theories have been advanced.

Maximum Normal Stress Theory (Rankine Theory)

Material will fail under any condition of loading when the *maximum normal stress* at any point reaches the limiting value. The limiting value is determined by an axial pull test in tension or in compression regardless of what the stresses may be on other planes at that point. The Rankine theory is good for brittle materials and inadequate for ductile materials.

Mathematically, this can be expressed as

$$\sigma_u = S_m \tag{4.57}$$

where σ_u = maximum principal stress at critical point
S_m = limiting stress as determined from axial test

Maximum Normal Strain Theory (Saint Venant's Theory)

Material will fail at a point under any condition of loading when the *maximum normal strain* (De Saint Venant, 1833) at that point reaches a critical value as determined by an axial test in tension or compression, regardless of what the stresses or strains may be on any other planes in the component.

Mathematically,

$$\varepsilon_u = \varepsilon_m \tag{4.58}$$

where ε_u = maximum principal strain at the point
ε_m = critical strain determined in an axial test

Maximum Shearing Stress Theory (Special Case of Coulomb's Theory Proposed by Guest)

Under any condition of loading, the material will fail when the maximum shear stress reaches the value of the limiting shear stress as experimentally determined under pure shear.

Mathematically,

$$\sigma_{ij} = S_{mp} \tag{4.59}$$

where σ_{ij} = maximum shear stress in material
S_{mp} = limiting shear stress determined from test

Internal Friction Theory and Mohr's Theory (Formulated by Coulomb and Generalized by Mohr)

Failure in a material is due to a sliding action within the material. The resistance to sliding comes from shearing stress and frictional resistance. Failure is assumed to occur when the maximum shear stress is greater than the combined resistance to sliding. Mohr assumed that the limiting shear stress on the plane along which failure occurs is a function of the normal stress acting on that plane. This theory does not establish a relationship between stresses, but is left on an experimental determination. Several combinations of maximum and minimum principal stresses resulting in failure are determined experimentally, and the Mohr circles subsequently can be constructed as in Fig. 4.16. The critical combination of normal and shear stresses may be determined by drawing the envelope of the circles.

Maximum Strain Energy Theory (Developed by Beltrami, Huber, and Haigh)

The material will fail, regardless of the combination of stress and strain at the point, when the value of strain energy per unit volume at the point in the material reaches the maximum value of strain energy per unit volume that the material is capable of absorbing under an axial loading condition. See Fig. 4.17.

Strain energy per unit volume is equal to the area under the stress-strain curve. Below the proportional limit, the strain energy is given as follows:

$$U_m = \tfrac{1}{2}\,\sigma_{av}\,\varepsilon_{max}$$

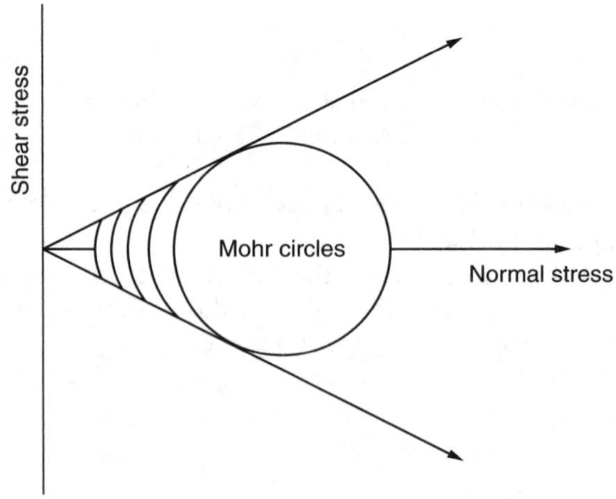

FIGURE 4.16 Internal friction theory.

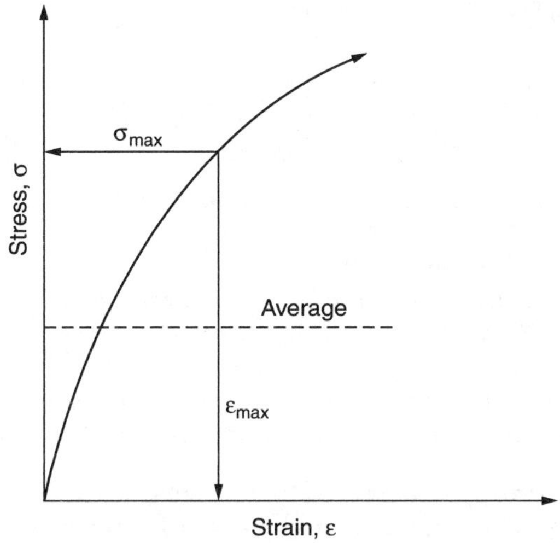

FIGURE 4.17 Strain energy theory.

Hooke's law is valid, then

$$U_m = S_m^{\ 2}/2E$$
$$= E\,\varepsilon_m^{\ 2}/2 \qquad (4.60)$$

The strain energy in a biaxial stress system per unit volume is defined as

$$U = \tfrac{1}{2}\,S_u\varepsilon_u + \tfrac{1}{2}\,S_v\varepsilon_v \qquad (u \text{ and } v \text{ are two normal directions})$$
$$= \tfrac{1}{2}(S_u/E)(S_u - \nu S_v) + \tfrac{1}{2}\,(S_v/E)(S_v - \nu S_u) \qquad (4.61)$$

(where ν is Poisson's ratio)

or

$$U = (\tfrac{1}{2}E)\,(S_u^{\ 2} + S_v^{\ 2} - 2\nu\,S_uS_v) \qquad (4.62)$$

For the failure

$$U = U_m$$

or

$$(S_u^{\ 2} + S_v^{\ 2} - 2\nu\,S_uS_v) = S_m^{\ 2} \qquad (4.63)$$

Hencky–von Mises Theory (Developed by Hencky and von Mises Independently)

$$(S_u - S_v)^2 + (S_v - S_w)^2 + (S_w - S_u)^2 = 2S_m^{\ 2} \qquad \text{(a 3D representation)} \qquad (4.64)$$

For biaxial stress, $S_w = 0$

$$(S_u - S_v)^2 + (S_v)^2 + (-S_u)^2 = 2S_m^{\ 2}$$

$$(S_u^{\ 2} + S_v^{\ 2} - S_u S_v) = S_m^{\ 2} \tag{4.65}$$

This is sometimes called shear distortion or shear energy theory.

Comparison of Theories

Comparison of test data from a large number of experiments provides information about the validity or usefulness of the different theories of failure.

In a uniaxial tension test, when the specimen starts to yield, the following six quantities are reached simultaneously.

1. The principal stress ($\sigma = P/_a$) reaches the tensile elastic strength (elastic limit or yield point) of the material.
2. The maximum shearing stress ($\tau = \frac{1}{2} P/a$) reaches the shearing elastic limit or yield point τ_{yp}, of the material $\tau_{yp} = \frac{1}{2}\sigma_{yp}$.
3. The tensile strain ε reaches the value ε_e.
4. The total strain energy W absorbed by the material per unit volume reaches the value $W_e = \frac{1}{2}(\sigma_e^{\ 2}/E)$.
5. The strain energy of distortion W_d (energy accompanying change in shape) absorbed by the material per unit volume reaches a value $W_{de} = [(1+\mu)/3E]\,\sigma_e^{\ 2}$.
6. The octahedral shearing stress reaches the value

$$\tau_{Ge} = [(2)^{0.5}/3]\,\sigma_e = 0.47\sigma_e \tag{4.66}$$

The six limiting values described above occur simultaneously in a tensile specimen in which the state of stress is uniaxial. If, however, the state of stress is biaxial or triaxial, the foregoing six quantities will not occur simultaneously and it is a matter of considerable importance in design as to which one of the quantities is assumed to limit the loads that can be applied to a member without causing inelastic action.

The data indicate that the maximum normal stress theory is in close agreement with the test data for brittle cast iron, while the Hencky–von Mises theory is in reasonable agreement with the test data for ductile materials. The maximum shear stress theory gives conservative results for the ductile metals and ultraconservative results for the brittle materials subjected to stress of the opposite sign. The maximum unit strain theory is unsafe for ductile materials. The trend in design practice is toward the use of the maximum normal stress theory for brittle materials and the Hencky–von Mises theory for ductile materials.

4.3.2 Creep

Creep is a time-dependent permanent deformation caused by stress. A permanent deformation may occur due to a sustained loading maintained for an extended time. Deformation may get accelerated due to high temperature even at a relatively low stress. A crack may also initiate in the vicinity of the localized stress due to creep in a shorter time. Creep damage is traditionally considered in the casing for steam turbines. It also becomes important for blades, e.g., in the fillet area, at the base of the airfoil, or in the root area where high localized stress may exist. Sometimes the temperature of a stage may rise due to flow conditions or change of pressure of the stage. Some common terms used in considering creep damage are given below.

Creep limit is defined as the maximum stress a given material can withstand in a given time without exceeding a specified quantity of creep.

Creep rupture strength is the stress that, at a given temperature, will cause a material rupture in a given time.

Creep strength is the stress that, at a given temperature, will result in a creep rate of 1 percent deformation in 100,000 h.

Creep deformation accumulates with time. This deformation may become so large that the component fails to perform its intended function, or it might eventually rupture if the duration is large. Virtually any material will creep upon approaching the melting temperature.

Creep properties are determined by performing a test under a constant load or stress on a specimen. The resulting strain is recorded as a function of time, as shown in Fig. 4.18. After an instantaneous strain, a decelerating strain stage called *primary creep* is reached, which is followed by a steady minimum creep rate called *secondary creep*. Finally, an accelerating stage, *tertiary creep*, is reached, ending in fracture.

This is essentially a thermally activated process. The laws of physics state that the rate of this activation process may be governed by the Arrhenius rate equation, which is

$$d\varepsilon/dt = Ae^{-Q/RT} \tag{4.67}$$

The physical constant Q is called *activation energy* and is the barrier that must be overcome for the molecular motion to occur.

Many methods to deal with creep have been advanced. The two methods most widely used in mechanical design are briefly described next.

Sherby-Dorn (S-D) Parameter

It is assumed that the activation energy is constant. The coefficient A in the rate equation is a function of stress σ. The equation can be rewritten as follows:

$$d\varepsilon/dt = A(\sigma)e^{-Q/RT} \tag{4.68}$$

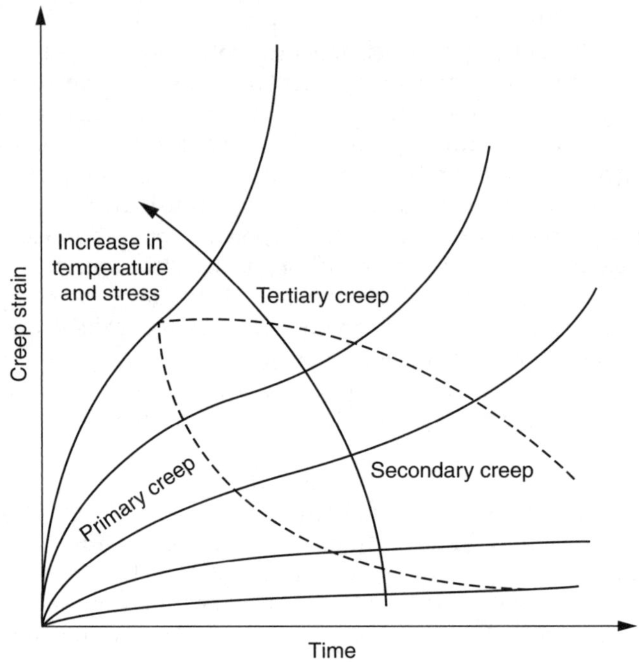

FIGURE 4.18 Schematic basic creep behavior of a material.

or

$$d\varepsilon = A(\sigma)e^{-Q/RT}\,dt \tag{4.69}$$

Equation (4.69) is integrated from time 0 to t. For a steady-state creep, the constant of integration is neglected.

$$\varepsilon = A(\sigma)e^{-Q/RT}t \tag{4.70}$$

For a given value of θ ($= e^{-Q/RT}t$), creep strain versus stress will be a unique curve and the term θ ($= e^{-Q/RT}t$) is called *temperature-compensated time*. It has been observed that the creep strain at rupture is almost constant for a given temperature-compensated time to rupture θ_r for a rupture time denoted by t_r.

Taking the log of the expression for temperature-compensated time $\theta = e^{-Q/RT}t$ to base 10 yields the following relationship, using the value of $\theta = 0$ when $t = t_r$:

$$\log \theta_r = \log t_r - (Q/RT)\log e \tag{4.71}$$

or

$$\log \theta_r = \log t_r - 0.217(Q/T) \tag{4.72}$$

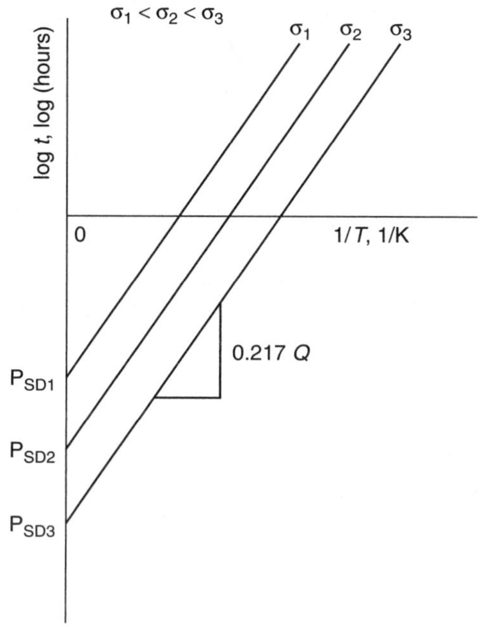

FIGURE 4.19 Value of constant for SD parameter.

when $R = 2$ cal/kmol.

The Sherby-Dorn (S-D) parameter is defined as follows:

$$P_{SD} = \log t_r - 0.217(Q/T) \tag{4.73}$$

The typical value of Q for various steels, including stainless steels, is 90,000 cal/mol. For pure aluminum, the value of Q is 36,000 cal/mol. Once Q is estimated or known, a plot of P_{SD} versus $1/T$ can be made as shown in Fig. 4.19. The data for all stresses and temperatures should form a single curve; i.e., these collapse to a single curve. Using such a plot in conjunction with Eq. (4.73), the rupture times t_r can be estimated for a given combination of stress and temperature. The required data to obtain the P_{SD} versus stress plot involves shorter rupture times than the design life of the component.

The advantage of using this correlation is that the data of a shorter t_r at a higher temperature can be used to assess creep rupture life for a longer t_r at a lower temperature.

Larson-Miller (L-M) Parameter (LMP)

The method described by Larson and Miller (1952) is widely used in design and is called the Larson-Miller parameter (LMP). The LMP curves for many materials are available in the literature.

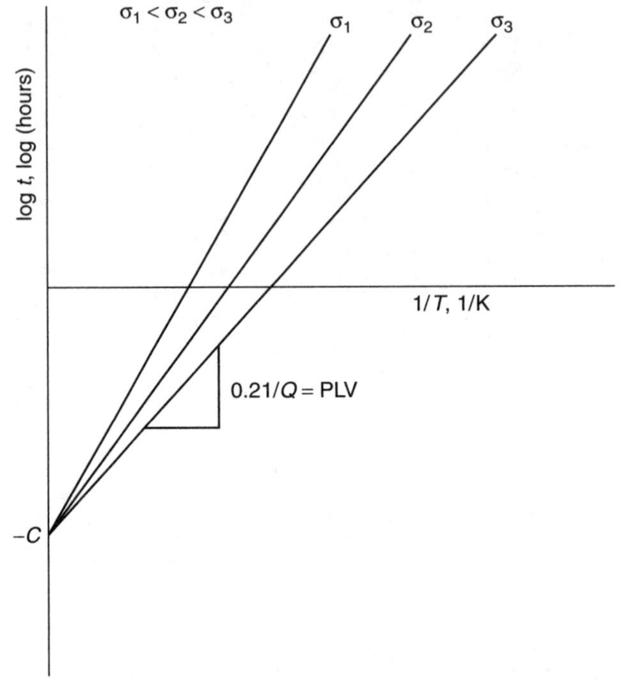

FIGURE 4.20 Value of constant for LMP.

Equation (4.72) may be rearranged as follows:

$$0.217Q = T(\log t_r + C)$$

where $C = -\log \theta_r$.

$$\text{LMP} = (T + 273)(\log t_r + C) \tag{4.74}$$

The value of C is the intercept on a plot of $\log t_r$ versus $1/T$ for a given stress, as depicted in Fig. 4.20. Values of the constant C for rupture of various steels are often near 20. This value of C may be used as an estimate if more specific information is not available.

For a given material, a plot of stress versus LMP results in a single curve. A typical representative LMP curve for AISI 403 steel is shown in Fig. 4.21.

Causes of High Internal Temperatures in Steam Turbines

In some instances the temperature of a stage may rise beyond expected values. Generally, there are three basic categories that can increase internal temperatures in steam turbines: abnormal steam conditions, abnormal operation, and internal obstruction. Occasionally, internal

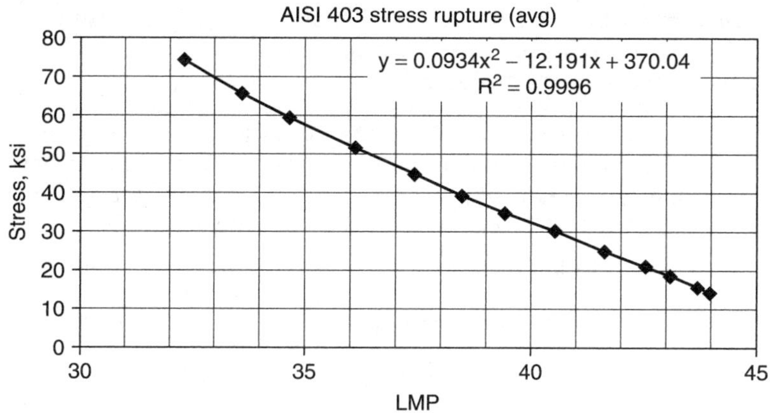

Figure 4.21 A typical LMP representation.

rub may also increase internal temperatures, but that is not the object of concern here.

The causes of high internal temperatures are described in the general terms of a simple, straight-through turbine. In this context, induction and extraction machines can be considered as separate machines in series. The rules and circumstances for overheating can be applied to each section separately; thus, one section can be abused while other sections are operating normally.

Abnormal Steam Conditions during High Inlet Temperature or High Exit Pressure High exit pressure will cause the "back end" of the flow path to run at higher than normal temperatures. Under these conditions the exit temperature cannot realistically exceed the inlet temperature.

Abnormal Operation If a turbine, or section of a turbine, is run at normal speed with abnormally low flow, temperatures will be high. If it is an unloaded machine and driving itself, temperatures cannot realistically exceed inlet temperatures. If the turbine has abnormally low flow and is driven externally, by a motor, generator, another turbine, or a section thereof, temperatures could be very high.

There are arguably circumstances surrounding an unloaded turbine, driving itself, with significant leakage or uncontrolled extraction flows, where the exit temperature could exceed the inlet temperature.

Internal Obstruction If the flow path is obstructed by deposits or damage, temperatures ahead of the obstruction will be abnormally high. There could also be a smaller effect downstream of the obstruction, due to inefficiency. In such circumstances the temperature of turbines would not realistically exceed the inlet temperature.

Evaluation of the creep life of the blades becomes important either during the design phase or when a failure occurs. The following examples will explain this point.

1. A mechanical structure made of alloy steel is subjected to creep under simple tension at a stress of 20 ksi at 700°F, and the measured rupture life is 30 yr. Estimate time for rupture if the structure is operated at a temperature of 750°F.

   Larson-Miller method:

$$
\begin{aligned}
P_{\text{LM}} &= (T+460)\,(\log t_r + C) \\
&= (700+460)(\log 262{,}800 + 20) \\
&= 29{,}486.77 \\
\log t_r &= P_{\text{LM}}/(T+460) - C \\
&= 29{,}486.77/(750+460) - 20 \\
&= 4.37 \\
t_r &= 10^{4.37} = 2.67 \text{ yr}
\end{aligned}
$$

2. Estimate for how long the structure can be safely run at 20 ksi at 750°F after the structure has been in operation for 15 years at 700°F. Use the Larson-Miller method.

   The expression for Miner's rule is given below.

$$
\begin{aligned}
\sum(\Delta t_i/t_{ri}) &= 1 \\
15/30 + \Delta t_{750}/2.67 &= 1 \\
\Delta t_{530} = (1 - 0.05)\ 2.67 &= 1.34 \text{ yr}
\end{aligned}
$$

Estimation of Creep Rupture and Damage

The creep rupture life in the case of constant stress across a cross section can be estimated by using an LMP of the material. The horizontal line is drawn from the stress value on the vertical axis to touch the LMP curve. From this point the corresponding value of LMP can be determined (Fig. 4.22).

Once the value of LMP is determined, Eq. (4.74) is used to estimate the time to rupture at the given temperature T of the component.

$$\text{LMP} = (T + 273)(\log t_r + C)$$

Cumulative Creep Rupture or the accumulation of creep strain in the case of a multiple application of stress for a given time (Fig. 4.23) can be estimated by utilizing either P_{SD} or LMP.

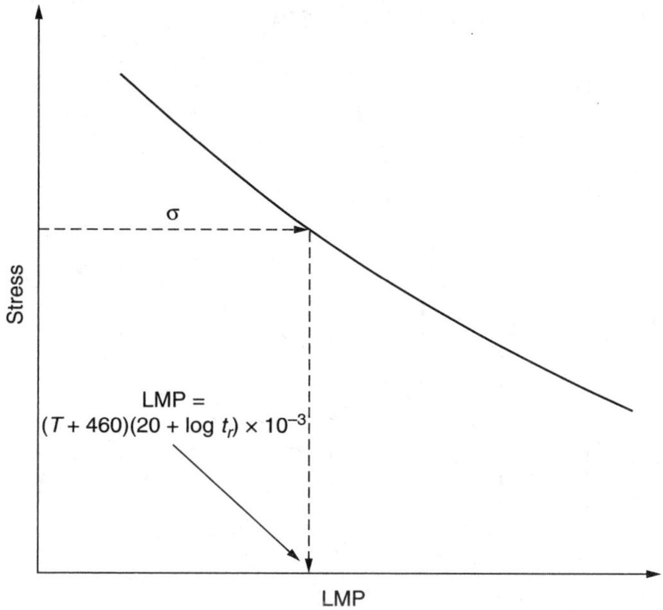

Figure 4.22 Rupture in the case of constant stress across the cross section.

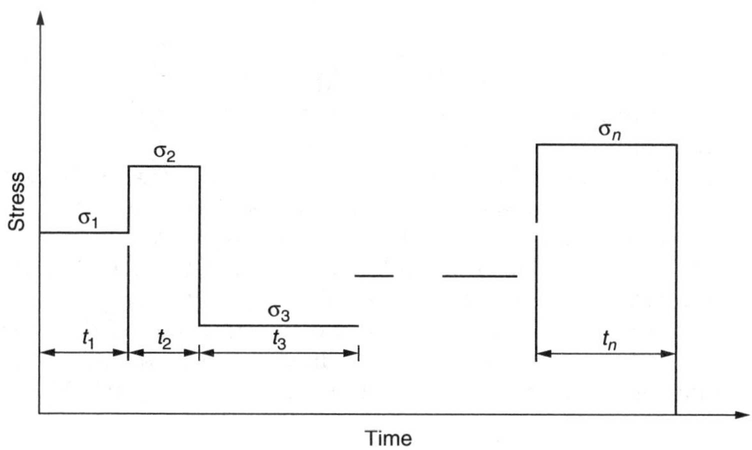

Figure 4.23 Multiple stress levels at different time duration.

During multiple applications of changing stress, creep strain accumulates. One wonders what the strain-time path will be, and an answer is not clear; but many engineering rules to estimate accumulative damage have been used to attempt to ascertain one.

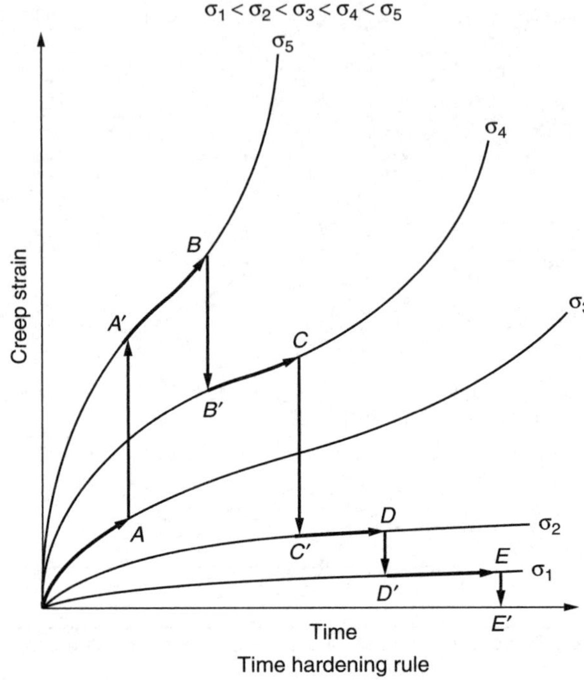

FIGURE 4.24 Time hardening rule shown schematically.

The time hardening rule states that the principal factor that governs the creep rate is the length of time at the particular temperature involved, independent of stress history. Figure 4.24 schematically depicts this rule.

The strain hardening rule states that the principal factor that governs the creep rate is the strain regardless of the stress history required to produce the strain. Such a behavior is shown in Fig. 4.25.

The life fraction rule is a combination of the time hardening rule and the strain hardening rule. This is shown graphically in Fig. 4.26.

If stresses change, LMP curves and the equation defining LMP can be used to estimate creep rupture life. A rough estimate of creep rupture life is made by applying a time fraction rule similar to Miner's rule used in cyclic loading:

$$\sum(\Delta t_i / t_{ri}) = D \tag{4.75}$$

where $i = 1, 2, 3, \ldots, n$.

Here D represents the magnitude of damage done due to a combination of all stresses during their applied time. A value of D equal to unity represents the rupture of the material, and a value of D less than unity represents the amount of damage before rupture.

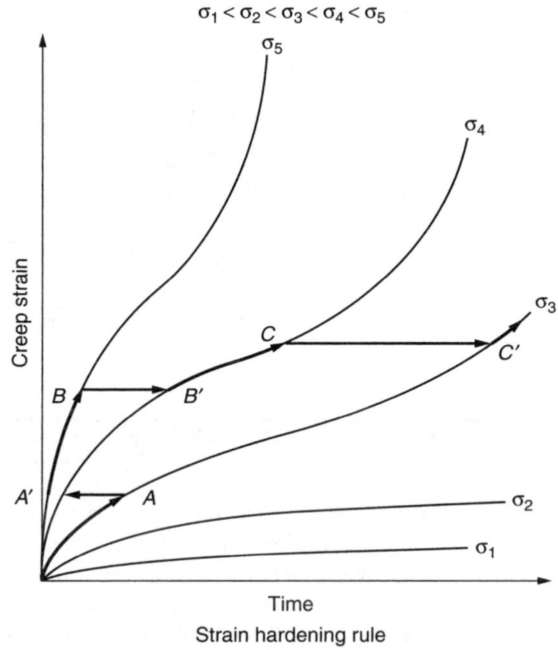

FIGURE 4.25 Strain hardening rule shown schematically.

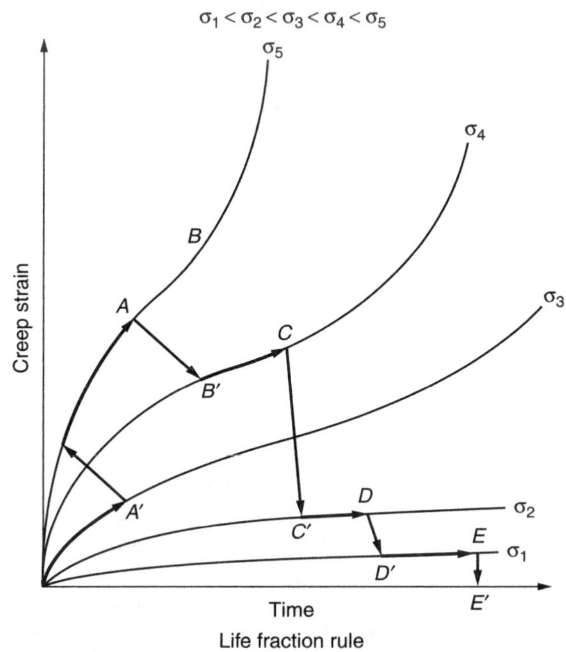

FIGURE 4.26 Life fraction rule shown schematically.

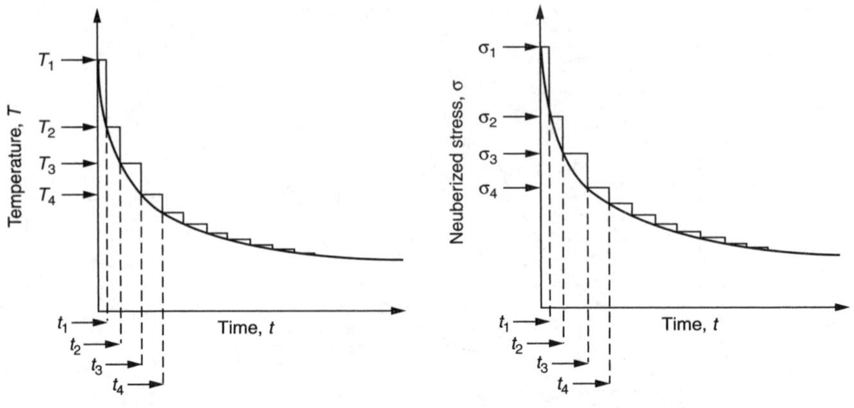

FIGURE 4.27 Case of localized stress and temperature.

The above method may also be useful in the case of stress changing with time. Figure 4.27 shows the case where stress and temperature change continuously from a high to a low value. One way to estimate damage that can be done is given in Eq. (4.74). First, divide the curves in small time duration; then use the peak stress and higher temperature to estimate the rupture time for each segment. After that use Miner's damage rule [Eq. (4.70)] to estimate the amount of damage done to the material.

$$\begin{array}{l} \sigma_1, T_1, t_1, \cdots, \text{LMP}, \cdots, t_{r1} \\ \sigma_2, T_2, t_2, \cdots, \text{LMP}, \cdots, t_{r2} \\ \sigma_3, T_3, t_3, \cdots, \text{LMP}, \cdots, t_{r3} \\ \quad\vdots \qquad \vdots \qquad \vdots \\ \sigma_n, T_n, t_n, \cdots, \text{LMP}, \cdots, \text{t}_{rn} \end{array} \tag{4.76}$$

Applicable Miner's damage rule:

$$\text{Damage } D = \sum{}_1 (t_i / t_{ri}) \tag{4.77}$$

where $i = 1, 2, 3, \cdots, n$.

Creep Rupture in Case of Localized Stress

To estimate rupture life, a magnitude of actual stress should be utilized. In the vicinity of a stress riser, stress varies from a high value to a lower value through thickness. Peak stress is estimated by multiplying stress concentration by average stress. Many times the estimated peak stress is greater than the yield strength of the material in question, and the actual stress is lower than estimated. Therefore, it is prudent to use and estimate the actual stress. Finite element analysis provides a better estimate of the peak stress. Many times the stress is so localized that it

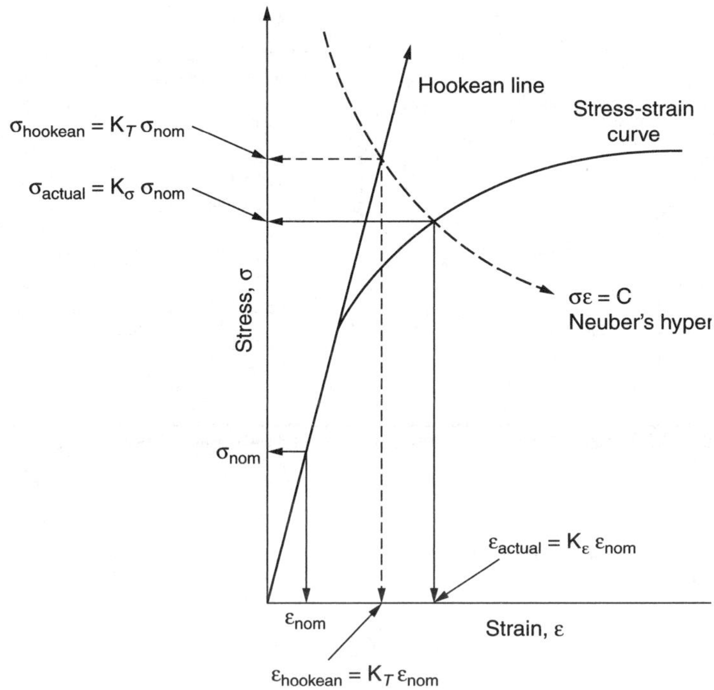

FIGURE 4.28 Graphical demonstration of Neuber's rule.

can be estimated by some simple rules without spending valuable resources on finite element analysis. A popular method to estimate the peak stress is called *Neuber's rule* and was described in detail earlier. Figure 4.28 schematically shows Neuber's method.

An important case of creep damage occurs when the localized stress is large enough to initiate a crack in a very small time through a very small depth. Once the crack is initiated, it may grow under the varying stress, which is known as a phenomenon called *creep crack growth*. Let us assume that the actual stress (stress from an elastic-type solution has been Neuberized) and temperature variation through the thickness are as shown in Fig. 4.29.

The depth under consideration is divided into small segments as shown in Fig. 4.29. Stress and temperature are estimated from the curve at these locations as shown below.

$$\sigma_1, \sigma_2, \sigma_3, \cdots, \sigma_n$$

$$T_1, T_2, T_3, \cdots, T_n$$

$$a_1, a_2, a_3, \cdots, a_n$$

Next calculate rupture time t_{r1} for the set of σ_1 and T_1 by using the LMP. Consider this as the crack initiation of crack length a_1.

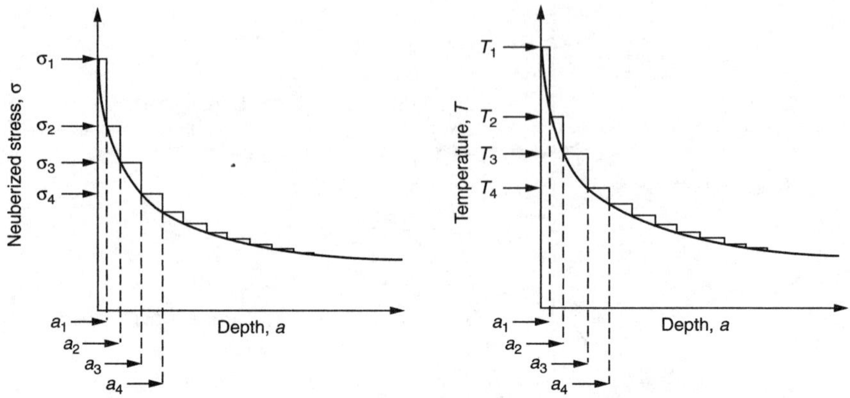

FIGURE 4.29 Neuberized stress and temperature through thickness.

For each set of σ and T, the t_r values are estimated using LMP as listed below.

$$\sigma_1, T_1, \cdots, t_{r1}$$
$$\sigma_2, T_2, \cdots, t_{r2}$$
$$\sigma_3, T_3, \cdots, t_{r3}$$
$$\vdots$$
$$\sigma_n, T_n, \cdots, t_{rn}$$

Once the t_{r1} values are estimated, the crack growth is calculated by use of *linear elastic fracture mechanics* (LEFM). The theory of LEFM calculates a quantity called the *stress intensity factor* K_I, which has been shown to be a function of stress, crack length, geometry, and type of loading. To demonstrate this calculation method, let us consider a root attachment of an airfoil, as shown in Fig. 4.30.

The expression of K_I for the above case can be approximated for blade root and is

$$K_I = \sigma(\pi a)^{1/2} F(a/b) \tag{4.78}$$

$$F(a/b) = [A_0 + A_1(a/b) + A_2(a/b)^2 + A_3(a/b)^3 + A_4(a/b)^4]/(1 - a/b)^{1/2} \tag{4.79}$$

A_0 through A_4 are constants.

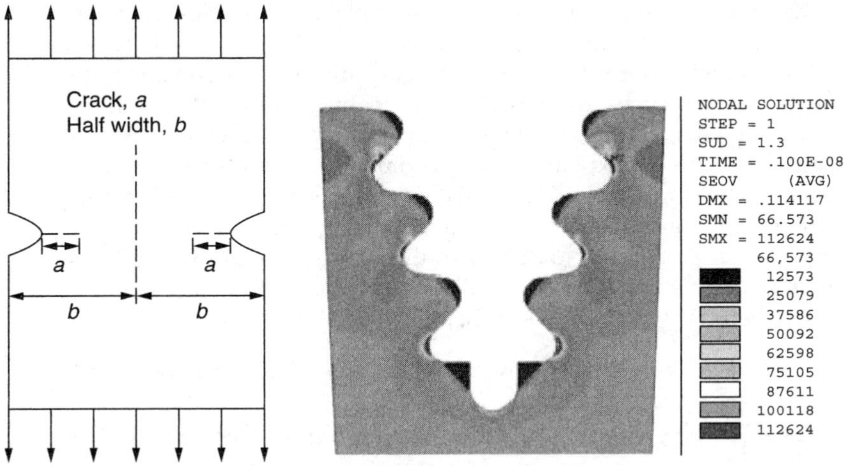

FIGURE 4.30 Stress localization in a root attachment.

Using Eq. (4.78), the values of K_I corresponding to each combination of σ and a are estimated starting from σ_2 and a_2 as shown below.

$$\begin{gathered}\sigma_2, a_2, \cdots, K_{I2}\\ \sigma_3, a_3, \cdots, K_{I3}\\ \sigma_4, a_4, \cdots, K_{I4}\\ \vdots\\ \sigma_n, a_n, \cdots, K_{In}\end{gathered}$$

To estimate the crack growth with time, the following expression for da_c/dt (refer to BS 7910:1999) is used.

$$da_c/dt = 0.014[(K_I)^2/(\sigma t_r)]^{0.85} \tag{4.80}$$

Finally, the total time t to final fracture is calculated from

$$t = t_{r1} + \sum_i (a_i - a_{i-1})\{0.014[(K_{Ii})^2/(\sigma_i t_{ri})]^{0.85}\} \tag{4.81}$$

where $i = 2, 3, \cdots, n$.

4.3.3 Damage due to Cyclic Loading

A Historical Perspective

The industrial revolution of the 1800s saw the advent of rotating and reciprocating machinery. The failure of metals due to repeated loads became a recognized problem; specifically, ductile metals failed in a brittle manner. Failure occurred at loads that were considered to be "safe," and the vast majority of component failures in ground, air, and sea vehicles were attributed to fatigue failure. Thus, considerable efforts and resources are expended in dealing with and avoiding/alleviating the fatigue problems. Many different viewpoints have been advanced to explain this phenomenon of material distress. These studies have ranged from dislocation mechanisms to phenomenological material behavior to full-scale structure. It is beyond the scope of this discussion to deal with the above in detail. The following is a list of some of, but not all, the key contributions to explain fatigue behavior in materials, as listed by Mitchell (1979).

1829	Albert in Germany: failure because of repeated loads first documented.
1839	Poncelet in France: first introduced the term fatigue.
1849	Institution of Mechanical Engineers (IMechE) in United Kingdom: crystallization theory of metal fatigue debated.
1864	Fairbairn: first experiments on effects of repeated loads.
1871	Wohler: first systematic investigation of fatigue behavior of railroad axles, rotating bending test: *S–N* curve: concept of endurance limit.
1886	Baushinger: notes change in "elastic limit" caused by cycling; stress-strain hysteresis loop.
1903	Ewing and Humfrey: microscopic study disproves old crystallization theory; fatigue deformation takes place by slip similar to monotonic deformation.
1910	Bairstow: investigates changes in stress-strain response during cycling; hysteresis loop measured; multiple-step tests; concept of cyclic hardening and cyclic softening.
1955	Coffin (GE) and Manson (NASA Lewis) (working independently): thermal cycling; low-cycle fatigue, plastic strain consideration.

Cyclic Stress-Strain Curve

Generally, metals are metastable under repeated plastic straining. The properties can be very different from monotonic properties. Depending upon the initial state of the material and test conditions, a metal may exhibit different behavior, namely, cyclic softening, cyclic hardening, cyclic stability, and a combination of the above behaviors. Figure 4.31 depicts these various behaviors.

Cyclic Softening under Controlled Strain Amplitude Cycling

The top of Fig. 4.32 shows strain versus time loading history and is called a *strain-controlled test*. In such a test, applied strain is kept to a constant value. The bottom of Fig. 4.32 shows that to maintain the same level of strain during cyclic loading, stress has to be reduced

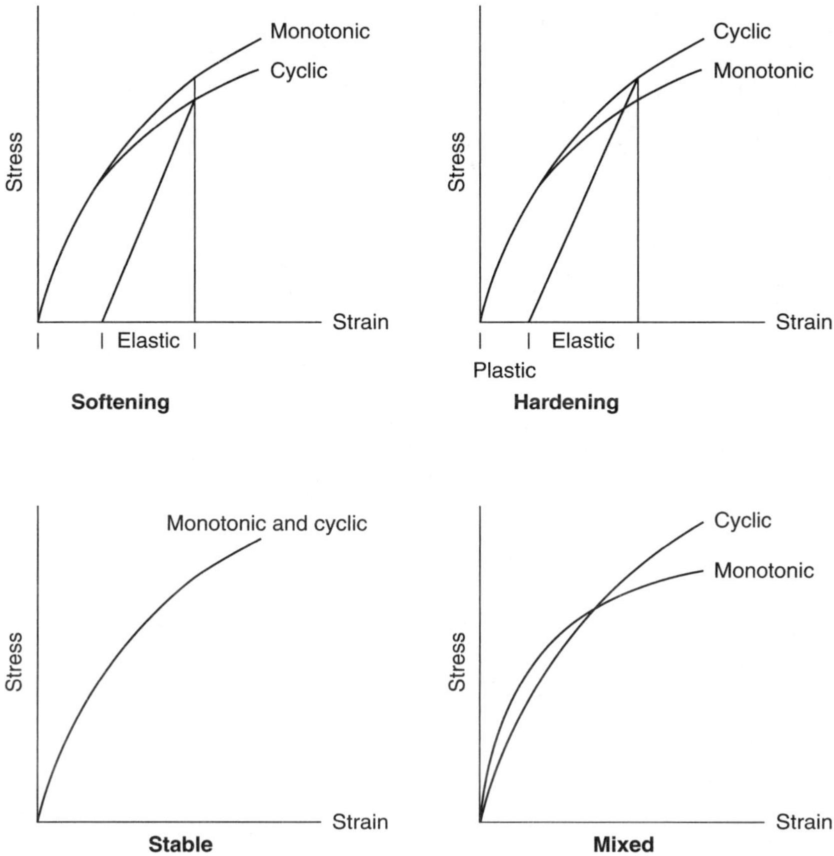

FIGURE 4.31 Representative stress-strain curves in monotonic and cyclic loadings.

because material softens. The middle of the figure shows the hysteresis loop, and during each loading step, stress is reduced. The example of materials exhibiting this behavior is cold worked pure metal and many steels.

Cyclic Hardening under Controlled-Strain Amplitude Cycling

The behavior of the cyclic hardening material is exactly opposite that of the cyclic softening material. The top of Fig. 4.33 shows a strain versus time loading history and is called a *strain-controlled test*. In such a test, applied strain is kept to a constant value. The bottom of Fig. 4.33 shows that to maintain the same level of strain during cyclic loading, stress has to be increased because material hardens. The middle of the figure shows the hysteresis loop, and during each loading step, stress is raised. Some of the cyclic hardening materials are annealed pure metal, aluminum alloys, and quenched steels.

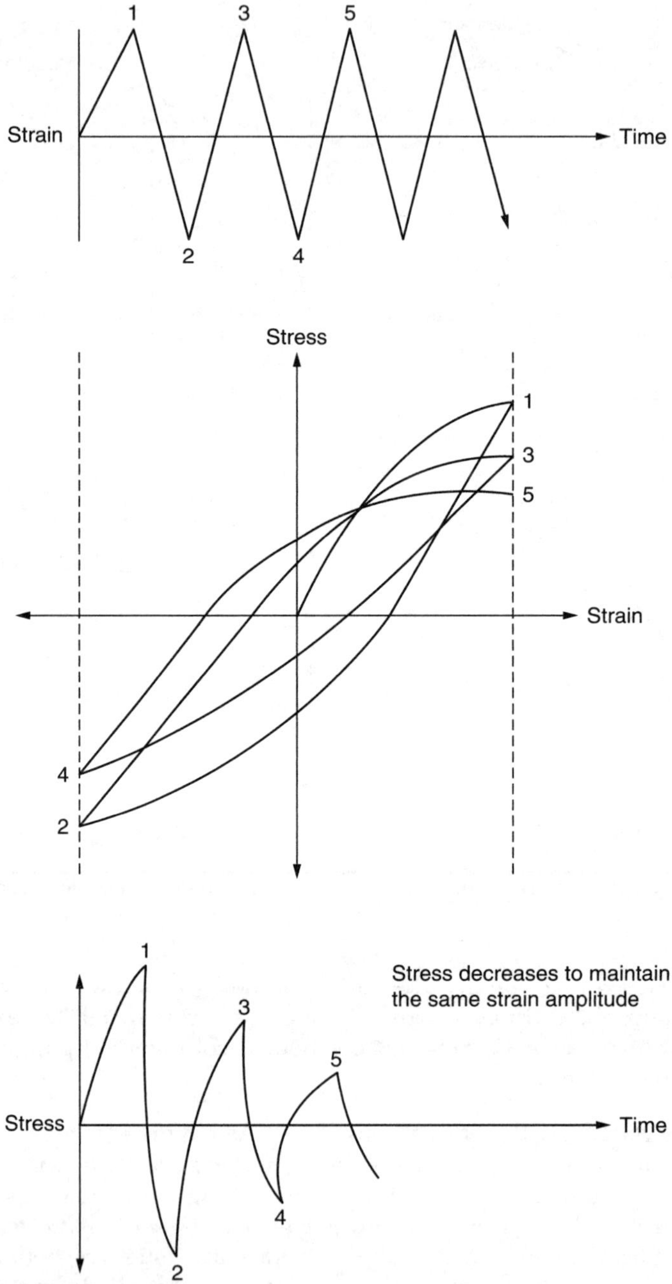

FIGURE 4.32 Cyclic softening, strain-controlled.

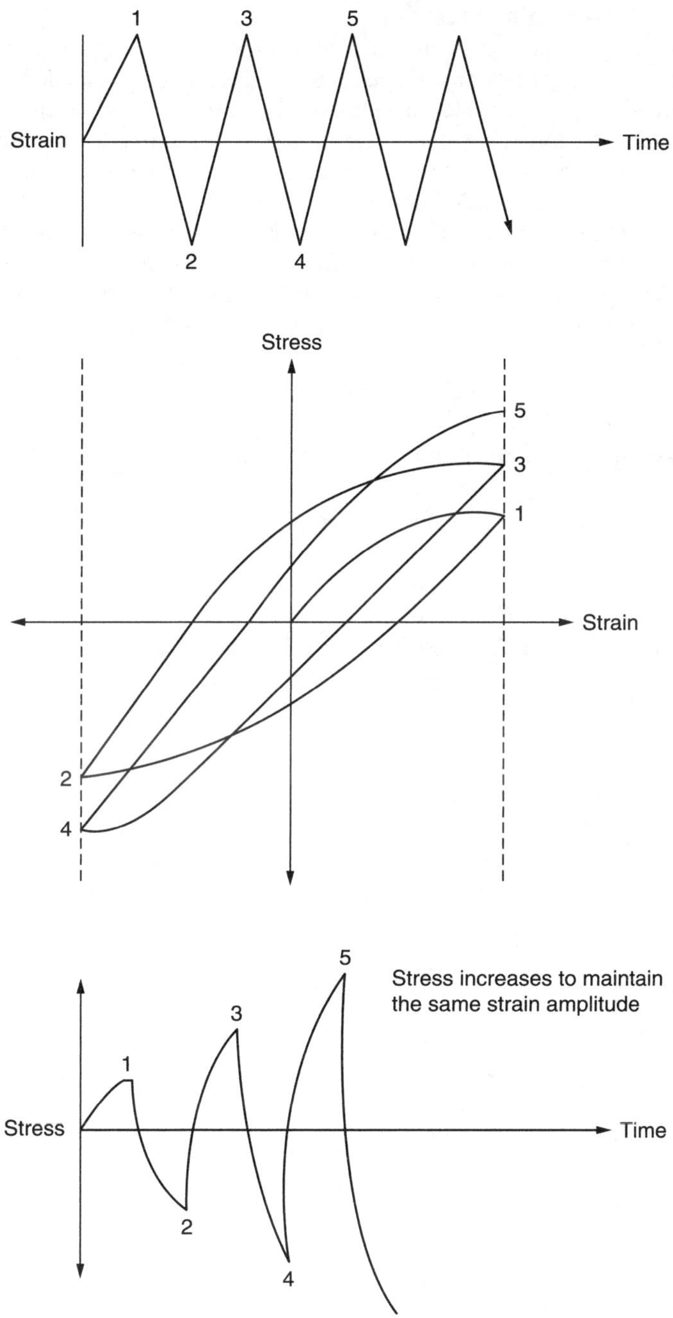

FIGURE 4.33 Cyclic hardening, strain-controlled.

Cyclic Stress-Strain Properties

Cyclic stress-strain properties are determined by testing smooth polished specimens under axial cyclic strain control. The locus of a single loading-unloading and loading history beyond the yield point results in a hysteresis loop (Fig. 4.34). The cyclic stress-strain curve is defined as the locus of tips of stable "true" stress-strain hysteresis loops obtained from companion specimens (Fig. 4.35).

The height of the loop from tip to tip is called the *stress range* $\Delta\sigma$. One-half of the width from tip to tip is defined as the *strain amplitude* $\Delta\varepsilon/2$. The plastic strain amplitude is found by subtracting the elastic strain amplitude $\Delta\varepsilon_e/2$ from the total strain amplitude.

$$\Delta\varepsilon_p/2 = \Delta\varepsilon/2 - \Delta\varepsilon_e/2 \tag{4.82}$$

Hooke's law gives the value of the elastic strain:

$$\Delta\varepsilon_e/2 = \Delta\sigma/2E \tag{4.83}$$

where E is the modulus of elasticity.

The plot of the log of stress amplitude versus plastic strain amplitude displays a straight-line behavior (Fig. 4.36) similar to monotonic loading.

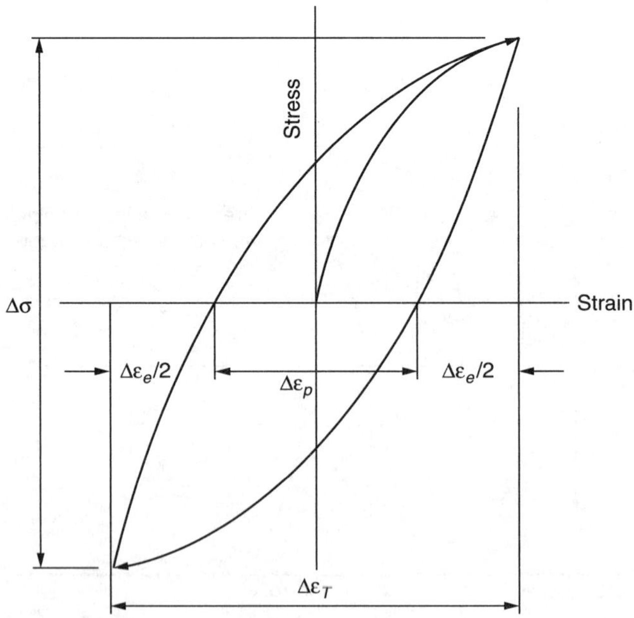

FIGURE 4.34 Hysteresis loop to define strain ranges.

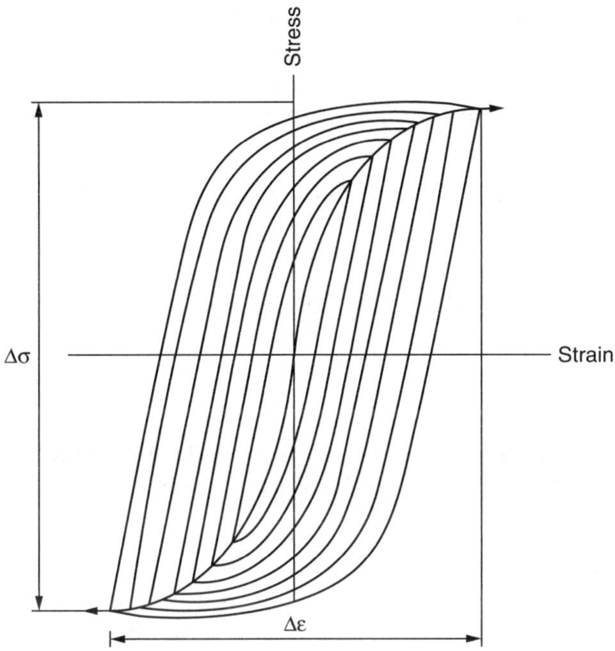

Figure 4.35 Multiple stable hysteresis loops for increasing loads.

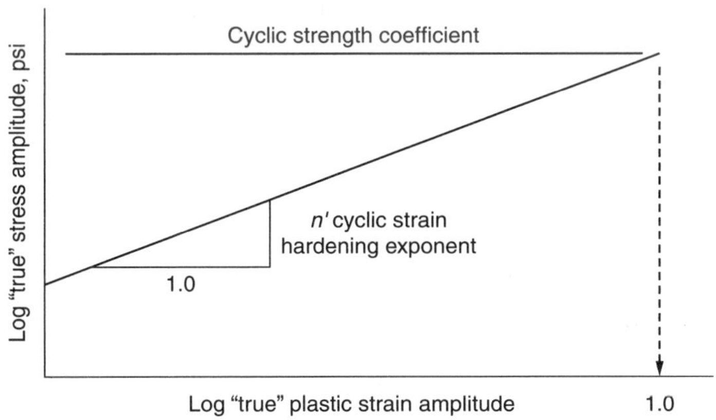

Figure 4.36 True stress versus true plastic strain.

The *cyclic yield strength* (0.2 percent σ_{ys}) is the stress to cause 0.2 percent inelastic strain as measured on a cyclic stress-strain curve.

The *cyclic yield hardening exponent n'* is the power to which the plastic strain amplitude must be raised to be proportional to stress amplitude:

$$\Delta\sigma/2 = K'(\Delta\varepsilon_p/2)^{n'} \tag{4.84}$$

where $\Delta\varepsilon_p/2$ = plastic strain amplitude.

Cyclic strength coefficient K' is the true stress to cause a plastic strain of unity.

It may be desirable to convert measured strains to stress to estimate the cyclic life.

$$\begin{aligned}\Delta\varepsilon/2 &= \Delta\varepsilon_e/2 + \Delta\varepsilon_p/2 \\ &= \Delta\sigma/2E + \Delta\varepsilon_p/2\end{aligned} \tag{4.85}$$

Equation (4.86) is the Ramberg-Osgood equation for cyclic σ-ε curve and is obtained by substituting expressions for $\Delta\varepsilon_p/2$ in Eq. (4.85).

$$\Delta\varepsilon/2 = \Delta\sigma/2E + (\Delta\sigma/2K')^{1/n'} \tag{4.86}$$

The knowledge of measured $\Delta\varepsilon/2$ can be transferred to stress amplitude by the above formula, using a numerical technique.

Fatigue Properties

Generally, fatigue resistance is described in terms of the number of constant-amplitude stress or strain reversals required to cause failure. Conventionally, for life estimation, a number of samples are tested at different stress levels until failure. The results are then plotted with stresses on the ordinate and the corresponding cycles to failure on the abscissa. The curve representing these data points is called the *S-N* curve (Fig. 4.37) and for a given material helps to estimate cycles to failure under an applied stress level.

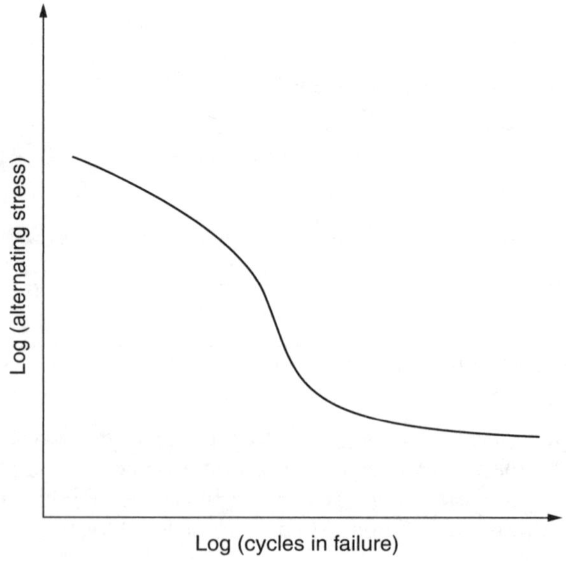

Figure 4.37 *S-N* curve.

The area under the stress-strain curve is denoted by the energy stored in the material. As the component is unloaded, some energy is released and some energy is stored in the material. The released energy is the elastic energy, as discussed earlier. The unreleased energy is due to the plastic deformation of the component. For life estimation it becomes easier if strain is considered in place of stress. A description of the strain-based life estimation method is presented next.

The plot of the log of "true" stress versus amplitude of cycles to failure displays a straight-line behavior (Fig. 4.38).

The relationship between stress and cycles can be written as follows:

$$\log(\Delta\sigma/2) = \log(\sigma_f'/2) + b\log(2N_f) \tag{4.87}$$

This reduces to

$$\Delta\sigma/2 = (\sigma_f'/2)\,(2N_f)^b \tag{4.88}$$

In terms of strain amplitude, the equation reduces to

$$\Delta\varepsilon_e/2 = (\sigma_f'/2E)\,(2N_f)^b \tag{4.89}$$

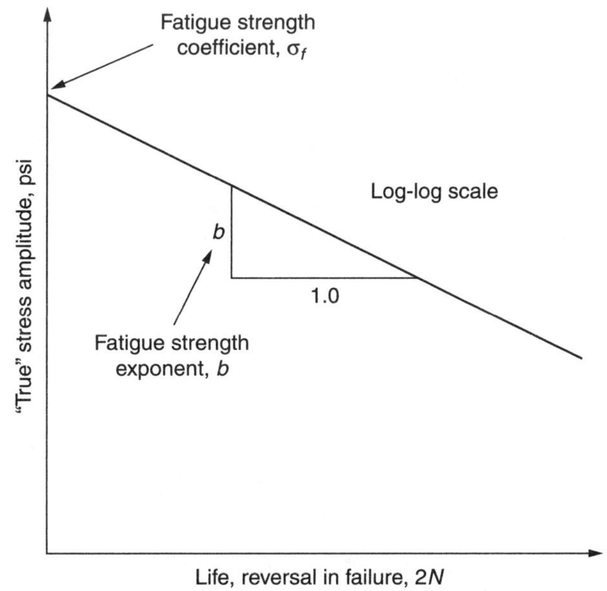

FIGURE 4.38 Stress amplitude versus cycles to failure.

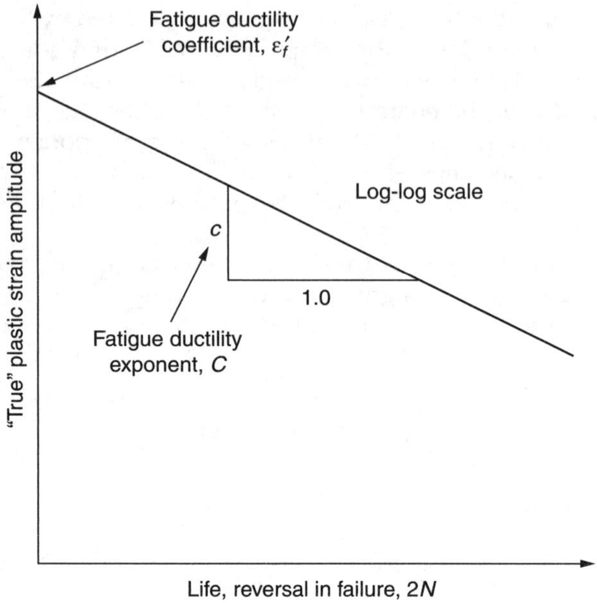

FIGURE 4.39 Plastic strain amplitude versus cycles to failure.

The plot of the log of true plastic strain amplitude versus amplitude of cycles to failure displays a straight-line behavior (Fig. 4.39). The relationship between strain and cycle can be written as follows:

$$\log(\Delta\varepsilon_p/2) = \log(\varepsilon_f') + c\log(2N_f) \tag{4.90}$$

$$\Delta\varepsilon_p/2 = \varepsilon_f'(2N_f)^c \tag{4.91}$$

Finally, the total strain amplitude is the summation of the elastic strain amplitude and the plastic strain amplitude.

$$\begin{aligned}\Delta\varepsilon/2 &= \Delta\varepsilon_e/2 + \Delta\varepsilon_p/2 \\ &= (\sigma_f'/2E)(2N_f)^b + \varepsilon_f'(2N_f)^c\end{aligned} \tag{4.92}$$

Equation (4.92) is known as the *Coffin-Manson equation for fatigue.*

Constants used in the above equations are material properties and are obtained through tests on the material of construction, which are defined below.

The *fatigue ductility exponent* c is the power to which the life in reversals must be raised to be proportional to the true strain amplitude and is the slope of the log $(\Delta\varepsilon_p/2)$ versus log $(2N_f)$ plot.

The *fatigue ductility coefficient* ε_f' is the true strain required to cause failure in one reversal and is the intercept of the log $(\Delta\varepsilon_p/2)$ versus $\log(2N_f)$ plot at $2N_f = 1$.

The *fatigue strength exponent b* is the power to which life in reversals must be raised to be proportional to true stress amplitude and is the slope of log $(\Delta\sigma/2)$ versus log $(2N_f)$ plot.

The *fatigue strength coefficient* σ_f' is the true stress required to cause failure in one reversal and is the intercept of the log $(\Delta\sigma/2)$ versus log $(2N_f)$ plot at $2N_f = 1$.

The *transition fatigue life* $2N_t$ is the life where the elastic and plastic components of the total strain are equal and is the life at which the elastic and plastic strain life lines cross.

The expressions of Eqs. (4.89), (4.91), and (4.92) are shown graphically in Fig. 4.40. On a log-log plot of strain amplitude versus cycles to failure the elastic and plastic portions will be straight lines. The total strain will be a summation of these strains at any given number of cycles to failure. The intercepts for the elastic and plastic lines are also indicated in Fig. 4.40.

The fatigue property of a material depends not only on the alternating stress, but also on the mean stress. Morrow (in Graham, 1968) suggested the following equation to include the effect of mean stress σ_0 by observing test data on many materials.

$$\Delta\varepsilon/2 = [(\sigma_f' - \sigma_0)/2E]\,(2N_f)^b + \varepsilon_f'\,(2N_f)^c \tag{4.93}$$

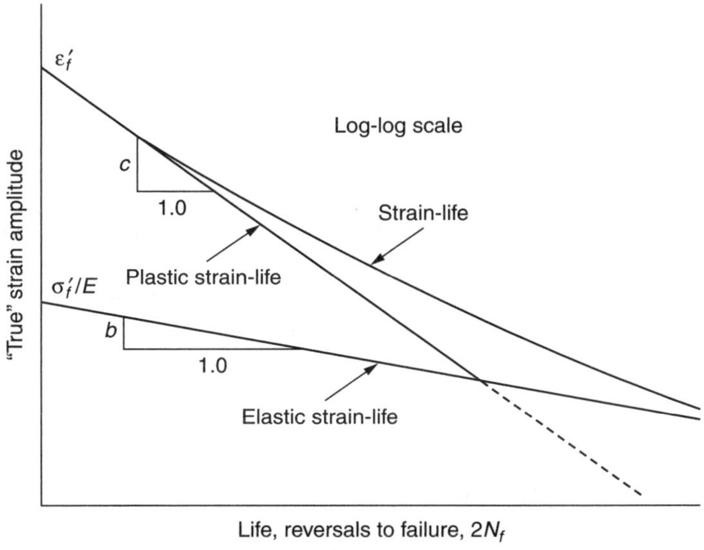

FIGURE 4.40 Strain amplitude versus cycles to failure.

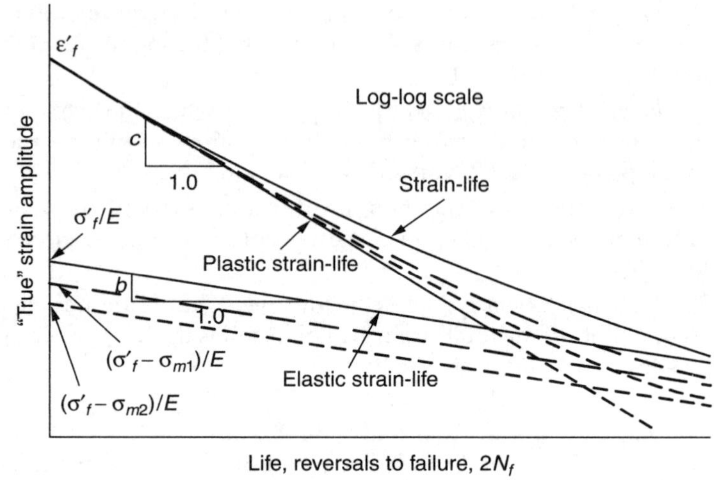

FIGURE 4.41 Effect of mean stress on fatigue life.

Figure 4.41 graphically demonstrates the implication of Eq. (4.93) for tensile mean stress. Researchers have found some relationships among properties that may be useful when some of them have to be approximated.

$$K' = \sigma_f' / \varepsilon_f^{n'} \tag{4.94}$$

$$n' = b/c \tag{4.95}$$

$$b = n'/(1 + 5n') \tag{4.96}$$

Universal Slope Method Material coefficients required to evaluate Eq. (4.92) are determined through fatigue tests, which are costly and time-consuming. Manson (1965) proposed a method [Eq. (4.97)] called the *universal slope method* that utilizes properties gathered from a mechanical test. It is a useful formulation to initially estimate curve when fatigue properties are not available.

$$\Delta\varepsilon = (3.5\sigma_u/E)\,(2N_f)^{-0.12} + D^{0.60}\,(2N_f)^{-0.60} \tag{4.97}$$

The clear implication of this equation is that fatigue strength in the elastic region is governed by the tensile strength, and it is assumed that $\sigma_f' = 3.5\sigma_u$. In the case of large $\Delta\varepsilon$ (small N_f), the fatigue properties or life is governed by ductility.

For example, let us use the SS 422 material to create $\Delta\varepsilon$ versus cycles to failure curve.

$$\sigma_u \text{ at room temperature} = 120{,}000 \text{ psi}$$
$$E \text{ at room temperature} = 29 \times 10^6 \text{ psi}$$
$$\% \text{ RA} = 35.0$$
$$\sigma_{\text{yield}} \text{ at room temperature} = 85{,}000 \text{ psi}$$
$$D = \text{Ln}[100/(100 - \%\text{RA})] = \text{Ln}[100/(100 - 35)] = 0.431$$

Substituting the appropriate values in Eq. (4.97), the expression for SS 422 [a stainless steel designation by American Society for Testing and Materials (ASTM)] is obtained.

$$\Delta\varepsilon = (1.45 \times 10^{-2})\,(2N_f)^{-0.12} + 0.431^{0.60}\,(2N_f)^{-0.60} \tag{4.98}$$

Figure 4.42 is the plot of Eq. (4.98) and includes the elastic and plastic portions of strain.

Nomenclature and Definitions The height of the hysteresis loop from tip to tip is called the *stress range* $\Delta\sigma$. The width from tip to tip of the

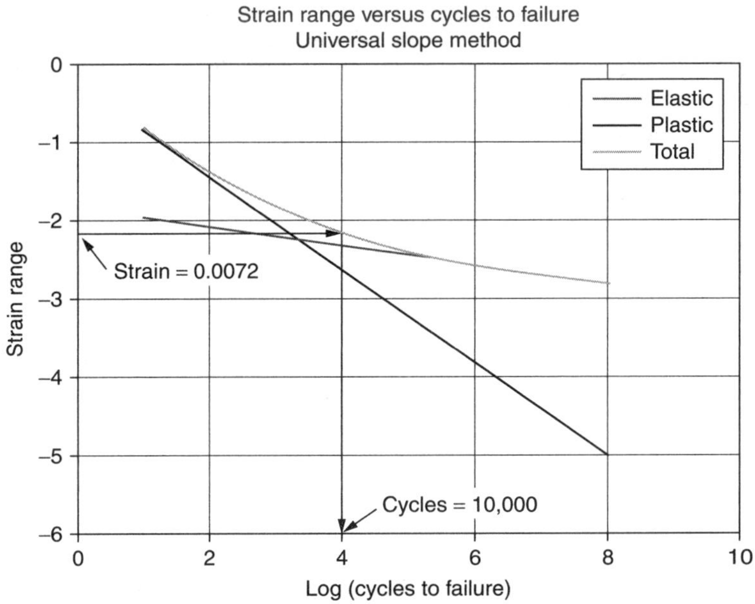

Figure 4.42 Strain versus life curve for AISI 422 estimated from the universal slope method.

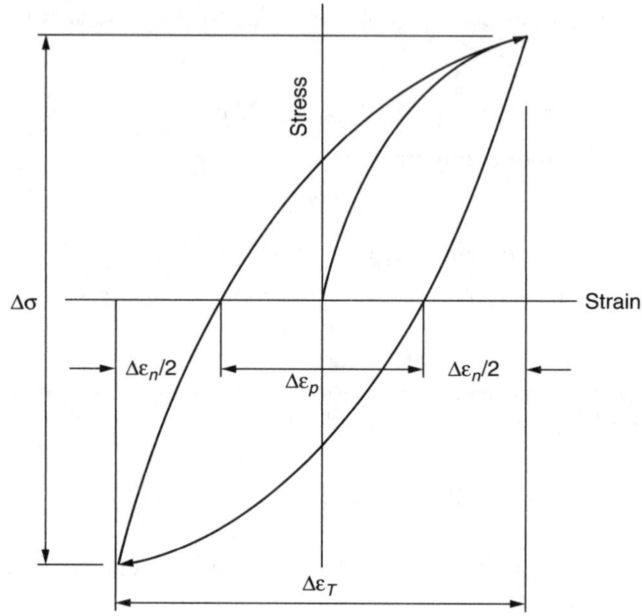

FIGURE 4.43 Hysteresis loop to define strain range.

loop is called *strain range* (Fig. 4.43), and one-half of the width from tip to tip is defined as *strain amplitude* $\Delta\varepsilon/2$. For complete reversed testing, one-half of the stress range is generally equal to the stress amplitude.

Mean stress is the average between the maximum and minimum stress of the fatigue loading cycle.

Alternating stress is the amplitude of the cyclic stress, or one-half the difference between the maximum and minimum stress of the cycle (Fig. 4.44).

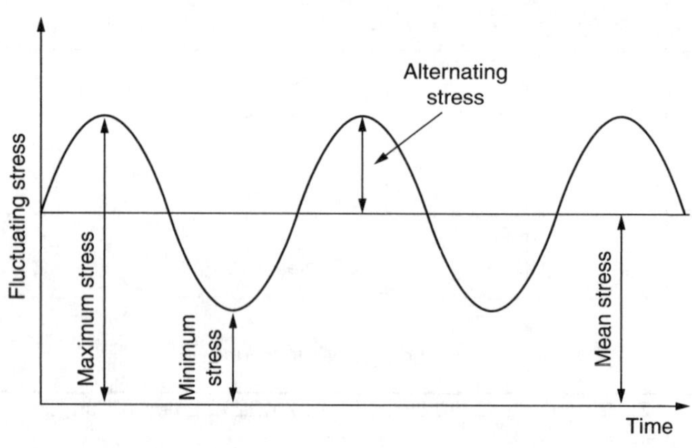

FIGURE 4.44 Alternating stress cycle.

Repeated Stress (Minimum Stress = 0)

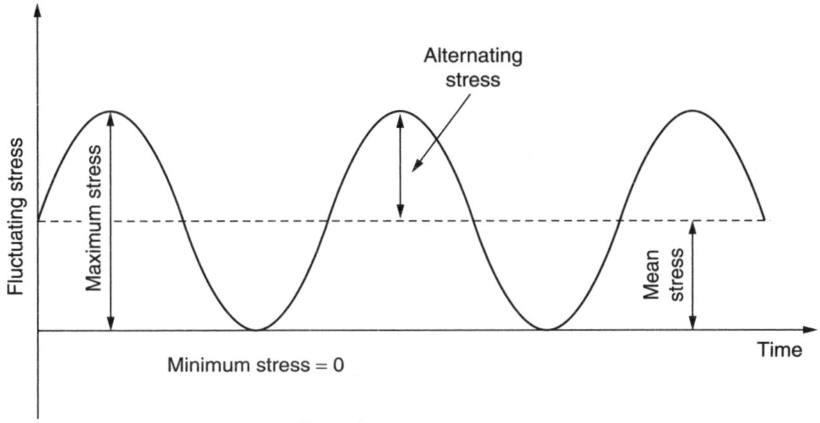

FIGURE 4.45 Repeated stress cycle 1.

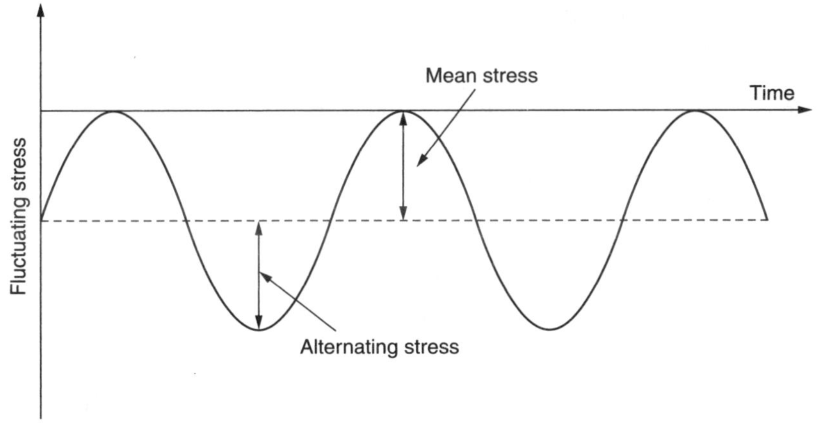

FIGURE 4.46 Repeated stress cycle 2.

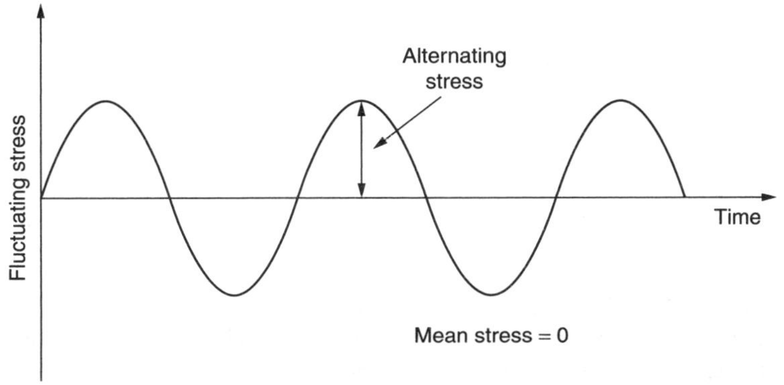

FIGURE 4.47 Reversed stress cycle.

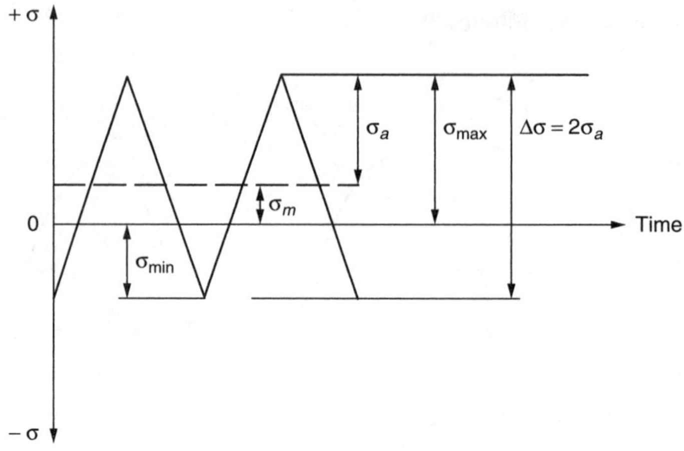

FIGURE 4.48 Definition of stress during cycle.

Traditionally, the effect of mean stress is determined by the ratio of minimum stress to maximum stress. There are many different ways of expressing these ratios, and Fig. 4.48 shows the definition of different stresses that have been used to define ratios.

The following are the most common ratios used to estimate the effect of mean stress on fatigue life.

R ratio:

$$R = \sigma_{min}/\sigma_{max} \tag{4.99}$$

$$R = (\sigma_m - \sigma_a)/(\sigma_m + \sigma_a) \tag{4.100}$$

A ratio:

$$A = \sigma_a/\sigma_m \tag{4.101}$$

V ratio:

$$V = \sigma_m/\sigma_a \tag{4.102}$$

Following are the conversions from one to another.

$$R = (1 - A)/(1 + A) = (V - 1)/(V + 1) \tag{4.103}$$

$$A = (1 - R)/(1 + R) = 1/V \tag{4.104}$$

$$V = (1 + R)/(1 - R) = 1/A \tag{4.105}$$

High Cycle Fatigue (HCF) For steel, the transition fatigue life is around 10^4 cycles. Traditionally, in terms of life, the life above 10^6 cycles is considered to be high cycle fatigue. Mechanistically, damage under

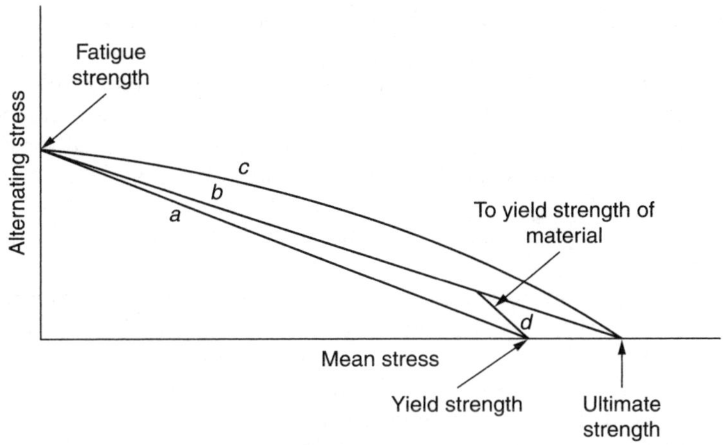

FIGURE 4.49 Effect of mean stress on fatigue strength.

negligible plastic strain can be considered high cycle fatigue. Thus, the fatigue strength depends on the mean strength of the material.

Mean stress data are generally shown on a constant life diagram (Fig. 4.49), which contains plots of combinations of alternating and mean stresses resulting in the same finite life to failure.

The Soderberg line is represented by line *a* and the equation of this line is given below.

$$S_a/\sigma_e + S_m/\sigma_y = 1 \tag{4.106}$$

The Goodman line is represented by line *b*, and the equation of this line is

$$S_a/\sigma_e + S_m/\sigma_{\text{ult}} = 1 \tag{4.107}$$

The Gerber line is represented by line *c*, and the equation of this line is

$$S_a/\sigma_e + (S_m/\sigma_{\text{ult}})^2 = 1 \tag{4.108}$$

Line *d* and the Past of Line *b* (Modified Goodman) Line *b* as representative of the Goodman line is drawn to the ultimate strength beyond the yield strength of the material. In fact, it allows small localized yielding until it reaches the ultimate strength. The modified Goodman line, a combination of lines *b* and *d*, ensures that at no time will the sum of alternating and mean stress be greater than the yield strength of the material. It might be proper to consider strain in place of stress for criteria even in the case of high cycle fatigue if yielding is at a very low level.

Adjustment of Fatigue Strength for Mechanical Components Fatigue properties are evaluated by tests on specimens machined from the material to be used in the mechanical components. To be accurate, these properties should be characterized using materials machined from actual manufactured parts. However, it is not practical for two reasons. First, at the design stage parts may not be available, and it may be very costly for some parts to be destroyed for this test. Second, it may not be possible to conduct the test in the actual operating environment.

There is an approximate method used to adjust the properties estimated on specimen testing in order for parts to be analyzed and used in service. For many applications this has proved to be adequate and is accomplished by using modifying factors to the specimen's properties. If the fatigue strength is estimated from test sample σ_e, then the fatigue strength σ_e' of the part can be approximated by

$$\sigma_e' = K_1 K_2 K_3, \cdots, K_n \sigma_e \tag{4.109}$$

The K_ns values depend on the influences of parameters being considered including, but not limited to,

Type of loading
Surface finish
Reliability factor
Strength reduction due to notch
Size effect
Surface treatment
Environment

The σ_e is considered to be the endurance strength of the material. When alternating stress is less than σ_e, the high cycle fatigue life is taken to be infinite. Many times test data are not available to support this assertion. However, assume that data may be available up to 10^{10} cycles. If the frequency of the alternating stress is 2000 Hz (blade frequency), then the structure will fail in about 16 days. Care is taken by using a factor in a Goodman-type diagram called the *factor of safety*. This does not provide any idea about life, which is a topic discussed in detail in a later section. A more appropriate consideration of the endurance limit can be obtained by considering Morrow's modified life relationship mentioned earlier. The endurance limit is also a function of mean stress, and the effect of mean stress is considered by using the Goodman or similar diagrams. The life relationship was described earlier in Eq. (4.93).

$$\Delta\varepsilon/2 = [(\sigma_f' - \sigma_m)/2E]\,(2N_f)^b + \varepsilon_f'\,(2N_f)^c$$

High cycle fatigue is considered where stress is low and life is very large. At these stress levels, the plastic deformation in the material should be considered negligible. The second part of the equation is concerned with the effect of plastic strain, and it can be neglected. Then the equation reduces to the following:

$$\Delta\varepsilon/2 = [(\sigma_f' - \sigma_m)/2E]\,(2N_f)^b \tag{4.110}$$

After substituting for alternating stress amplitude, it can be written as follows:

$$2\sigma_a/2\mathrm{E} = [(\sigma_f' - \sigma_m)/2E]\,(2N_f)^b \tag{4.111}$$

$$\sigma_a = (\sigma_f' - \sigma_m)\,(2N_f)^b \tag{4.112}$$

The magnitude of b is always negative. Let us substitute $b = -d$; then

$$2N_f = [(\sigma_f' - \sigma_m)/\sigma_a]^{1/d} \tag{4.113}$$

It can be seen that life can only approach infinity if $\sigma_a = 0$. In the case of failure in one reversal of load application, $2N_f = 1$, or $N_f = \frac{1}{2}$. This is the case for monotonic loading when the load is applied to failure. Substituting $2N_f = 1$ in Eq. (4.105), the following results:

$$1 = [(\sigma_f' - \sigma_m)/\sigma_a]^{1/d}$$

$$(\sigma_f' - \sigma_m)/\sigma_a = 1$$

$$\sigma_f' = \sigma_m + \sigma_a \tag{4.114}$$

In the case when the mean stress σ_m is zero, the failure occurs at one reversal or one-half of the loading cycle when $\sigma_a = \sigma_f'$. This is consistent with the monotonic stress-strain curve for the material.

Theoretical Stress Concentration Factor K_t It is the ratio of the maximum stress in the notch to the nominal stress, assuming that the material follows a linear stress-strain relationship and it behaves as the hypothetical isotropic material that forms the basis of the classical theory of elasticity.

Notch sensitivity index q is

$$q = (K_a - 1)/(K_t - 1) \tag{4.115}$$

where K_a is the alternating strength reduction factor

$$K_a = \sigma_a/\sigma_{an} \tag{4.116}$$

The nominal alternating stress of the fatigue loading cycle of a notched specimen is given by σ_{an}.

Low Cycle Fatigue (LCF) In the low-stress case, the resulting strain is almost elastic and life can be expected to be long. However, in the large-stress case, the resulting strain is very high and life is controlled by the plastic strain. The Coffin-Manson relationship deals with the situation when the elastic strain is negligible compared to the plastic strain. In reality in order to be accurate, the total strain [Eq. (4.93)] should be used for the life evaluation.

In steam turbines, a situation may arise near localized stress regions (peak fillet stress, base of blade or root section, etc.) where the plastic strain may be very high. A combination of plastic strain, temperature, and operating cycles may dictate the need to analyze life for LCF.

At the base of an airfoil let us assume that the total strain = 0.72 percent and the plastic strain is estimated to be 0.24 percent. Using the plot (Fig. 4.42) described earlier, the component has about 1000 life cycles only.

CHAPTER 5

Review of Fundamentals of Vibration

The last five decades have seen phenomenal theoretical development toward the understanding of the vibrational behavior of bladed disk systems. A very significant enabler has been the growth in the availability of ever more powerful computers.

Newtonian mechanics is the system of mechanics based upon Newton's laws of motion in which mass and energy are considered to be separate from each other. Mass and energy are considered to be conservative, in contrast to their treatment in relativistic mechanics. In structural analysis newtonian mechanics is used—this was briefly described earlier.

The intent of this chapter is to highlight some concepts that will be useful later in the interpretation of the vibration of blades and bladed disk systems. A mechanical structure such as a blade or bladed disk has infinite degrees of freedom (DOF). The number of natural modes of vibration of a mechanical structure equals the number of DOF. Use of the concept of generalized coordinates facilitates the solution of a multi-DOF system and decouples each mode of vibration. Each isolated natural mode of such a structure by itself (in a generalized coordinate system) can be considered as a single spring-mass-damper system. Therefore, the study of the dynamics of a single spring-mass-damper system will be helpful in understanding the behavior of a complex system, such as a bladed disk.

A force is applied to a spring, a viscous damper, and a mass, and their responses are depicted graphically in Fig. 5.1. The mathematical expressions for their responses under the action of a force are also given. A mechanical structure is thus considered to be composed of numerous springs (stiffness), dampers, and masses. It is important to realize that a continuum will exhibit a combination of the above responses.

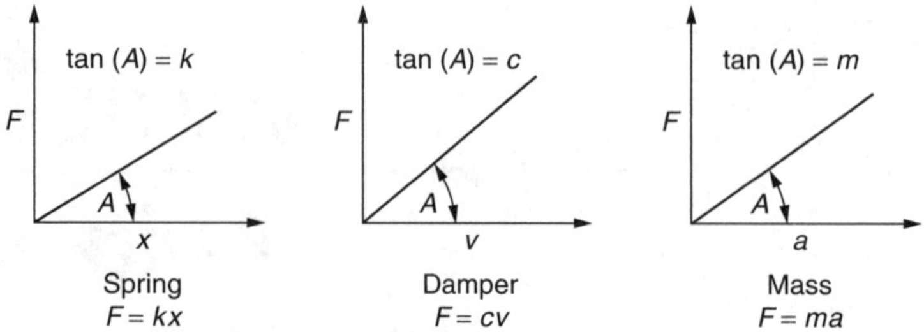

FIGURE 5.1 Behavior of spring, viscous damper, and mass under applied load.

5.1 Discrete Systems

In the earlier days, a spring-mass lumped model was used to study the vibrational behavior of a blade. This early approximation and method led to the concept of resonance. It showed that the coincidence of frequencies (forcing frequency and natural frequency) produced the largest amplitude of vibrational response. At the next level of complexity, the behavior of a single cantilevered beam was studied by considering it as a continuum rather than lumped. Theories of bending and the vibration of a beam were utilized. Later researchers realized that, in the case of banded (shrouded) blades, there existed more natural modes and natural frequencies than those estimated from a single beam model. This was due to the existence of coupling among the blades of a packet through the shroud band. The study (Weaver and Prohl, 1956) of this phenomenon identified the existence of many modes and frequencies in a packet of blades not detected by a single blade analysis. Packets of blades are mounted on a disk, and packets are coupled to one another through the disk. Analytical results as well as test data show that many modes and frequencies are influenced by the stiffness of the disk. These additional modes are not calculable through the analysis of a single packet of cantilevered blades and require the next level of complexity.

5.1.1 Single-Degree-of-Freedom (SDOF) System

The analysis of a spring-mass system (Fig. 5.2) provides insight in the understanding of the vibration characteristics of a mechanical system. The application of Newton's third law of motion yields the equation of motion for the above case.

$$ma + cv + kx = F(t)\cos A \tag{5.1}$$

where m = mass
a = acceleration of mass
c = viscous damping
v = velocity of mass

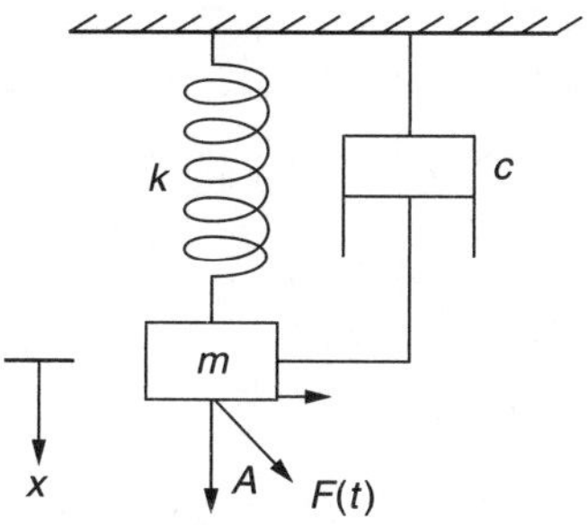

FIGURE 5.2 Single spring-mass-damper system.

k = stiffness
x = vertical displacement
$F(t) = F_0 \sin(\omega t)$ is time-dependent force
A = angle that defines the direction of force from vertical direction

Assume that a harmonic solution for the displacement $x(t)$ is the following:

$$x(t) = C_0 \sin(\omega t + \phi) \tag{5.2}$$

The solution of the differential Eq. (5.1) provides the absolute magnitude of the response magnifier and is expressed as

$$\text{Abs}\,(C_0k/F_0) = \cos A/[(1 - r^2)^2 + (2\zeta r)^2]^{1/2} \tag{5.3}$$

where $r = \omega/\omega_n$
$\omega_n = (k/m)^{1/2}$
$\zeta = c/[2(\text{km})^{1/2}]$
C_0 = amplitude of vibration
F_0/k = displacement if magnitude of applied steady-state force is F_0

Singh and others (Singh and Schiffer, 1982; Singh et al., 1988; Singh and Vargo, 1989) discussed the resonance of a circular symmetric structure and pointed out that resonance can be viewed as a state in which the energy built up in the system due to applied forces attains a maximum value. Thus, in the state of resonance, the magnitude of the response (stress, displacement) of the system also attains its maximum value, and the system's resistance to the exciting forces is minimum (Figs. 5.3 and 5.4). Here ζ represents the percent of critical damping. At the frequency ratio of $r = 1$, the exciting frequency equals the natural frequency and the magnifier is maximized.

Figure 5.5 shows the variation of the peak response of the mass as a function of the direction of the force. The degree of freedom of the

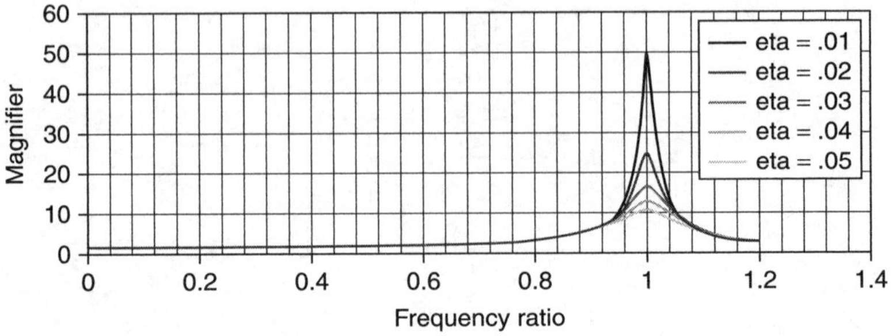

FIGURE 5.3 Absolute value of magnifier vs. frequency ratio.

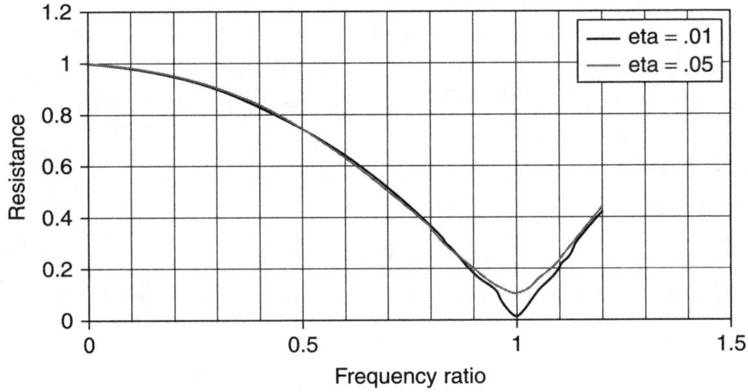

FIGURE 5.4 Resistance vs. frequency ratio.

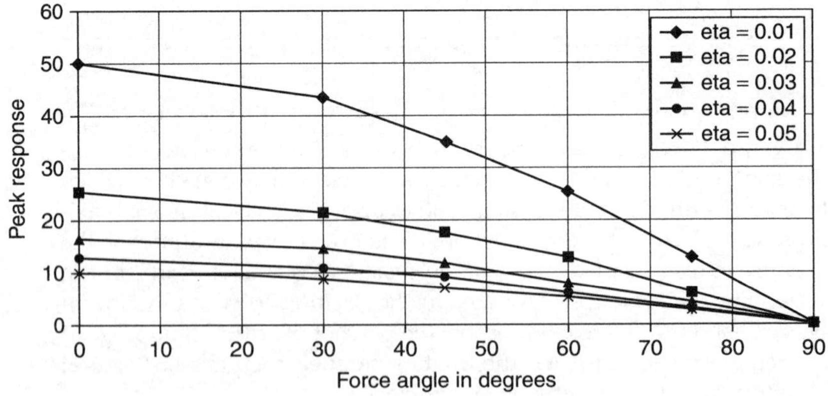

FIGURE 5.5 Peak response vs. angle of force.

spring-mass-damper system is the displacement in the vertical direction; i.e., the mass can move only up and down. The system will respond only to the vertical component of the force F. The response is greatest when the force acts in the vertical direction (angle A equals zero), and the response is zero when the vertical component of the force is zero (angle A equal to 90°). The mode shape, or motion pattern, in the vertical direction matches only with the vertical component of the force.

5.1.2 Multiple-Degree-of-Freedom (MDOF) System

The DOF of a system is equal to the number of independent coordinates necessary to describe the motion of the system. Free vibration will occur at the natural frequencies of the system. The system will go through a simple harmonic motion corresponding to one of the natural frequencies and that is called the principal mode of vibration or the normal mode of vibration. The number of modes will be equal to the DOF of the system. To explain this point, the solution of a 2-DOF system will be discussed here. The mathematics is not too complicated but shows the basic elements for using a multi-DOF system, which is shown in Fig. 5.6, where x_1 and x_2 are the independent coordinates and represent the DOF of the system.

The equations of motion of the system are written with the help of Newton's laws of motion as follows:

$$\begin{aligned} m_1(d^2x_1/dt^2) &= -k_1x_1 + k_c(x_2 - x_1) + F_1(t) \\ m_2(d^2x_2/dt^2) &= -k_2x_2 - k_c(x_2 - x_1) + F_2(t) \end{aligned} \tag{5.4}$$

The equations can be rewritten as follows after rearranging and expressing them in a concise (matrix) form:

$$\begin{bmatrix} m_1 & 0 \\ 0 & m_2 \end{bmatrix} \begin{bmatrix} d^2x_1/dt^2 \\ d^2x_2/dt^2 \end{bmatrix} + \begin{bmatrix} k_1 + k_c & -k_c \\ -k_c & k_2 + k_c \end{bmatrix} \begin{bmatrix} x_1 \\ x_2 \end{bmatrix} = \begin{bmatrix} F_1(t) \\ F_2(t) \end{bmatrix} \tag{5.5}$$

The first matrix is called the *mass matrix*, and the second one is called the *stiffness matrix*.

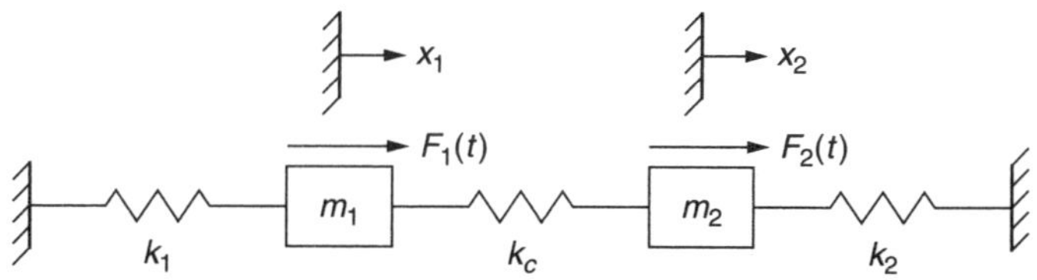

FIGURE 5.6 A 2-DOF system.

The solution to Eq. (5.5) is obtained by assuming a simple harmonic motion for the masses.

$$x_1 = A_1 \sin(\omega t)$$
$$x_2 = A_2 \sin(\omega t) \tag{5.6}$$

After substituting the values of the displacements for two masses, we obtain the following equations for free vibration

$$(k_1 + k_c - m_1\omega^2)A_1 - k_cA_2 = 0$$
$$-k_cA_1 + (k_2 + k_c - m_2\omega^2)A_2 = 0 \tag{5.7}$$

The nontrivial solution is obtained when the determinant of the left side is set to zero.

$$\begin{vmatrix} k_1+k_c - m_1\omega_2 & -k_c \\ -k_c & k_2 + k_c - m_2\omega^2 \end{vmatrix} = 0 \tag{5.8}$$

Equation (5.9) is the result of expanding the determinant.

$$\omega^4 - \omega^2[(k_1 + k_c)/m_1 + (k_2 + k_c)/m_2] + [k_1k_2 + (k_1 + k_2)k_c]/m_1m_2 = 0 \tag{5.9}$$

There are two positive solutions and two negative solutions for ω.

The amplitude ratios for the two positive solutions are given below.

$$A_1/A_2 = (k_2 + k_c - m_2\omega^2)/k_c$$
$$(A_1/A_2)^1 = (k_2 + k_c - m_2\omega_1{}^2)/k_c \tag{5.10}$$
$$(A_1/A_2)^2 = (k_2 + k_c - m_2\omega_2{}^2)/k_c$$

To appreciate and to gain insight into the meaning of the results, the problem is simplified by setting all masses equal to one another and setting all the stiffnesses of the springs to be equal to one another (Fig. 5.7).

$$m_1 = m_2 = m \qquad \text{and} \qquad k_1 = k_2 = k_c = k$$

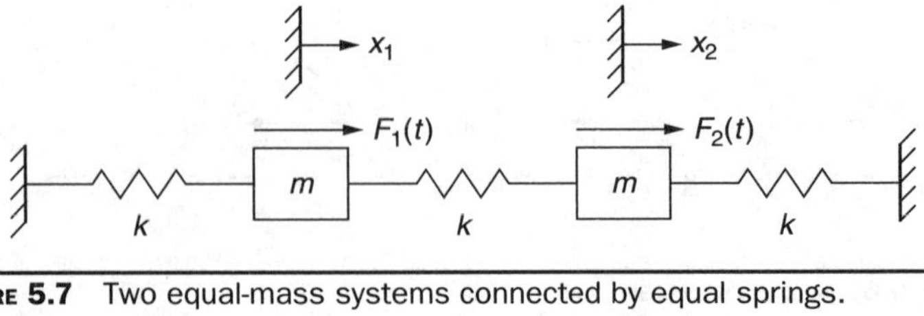

FIGURE 5.7 Two equal-mass systems connected by equal springs.

Equation (5.9) reduces to the following:

$$\omega^4 - \omega^2[(4k)/m] + (3k^2)/m^2 = 0$$

The roots of the equation are

$$\omega_1^2 = 2k/m + [4(k/m)^2 - 3k^2/m^2]^{1/2} = (k/m)(2-1) = k/m \quad (5.11)$$

and

$$\omega_2^2 = 2k/m - [4(k/m)^2 - 3k^2/m^2]^{1/2} = (k/m)(2+1) = 3k/m \quad (5.12)$$

Neglecting the negative values of ω, two values of the natural frequencies are given by the following expressions.

$$\omega_1 = (k/m)^{1/2} \quad (5.13)$$

$$\omega_2 = (3k/m)^{1/2} \quad (5.14)$$

The amplitude ratios for the two frequencies are described below. These are the two natural mode shapes associated with the natural frequencies.

The first mode is obtained by substituting the value of ω_1 in Eq. (5.10).

$$(A_1/A_2)_1 = [2k - m(k/m)]/k = 1 \quad (5.15)$$

The amplitudes of the displacements of the masses are equal, and the displacements are in phase as shown in Fig. 5.8.

The second mode is obtained by substituting the value of ω_2 into Eq. (5.10). For this mode shape the amplitudes are equal, but they are out of phase (Fig. 5.9).

$$(A_1/A_2)_2 = [2k - m(3k/m)]/k = -1 \quad (5.16)$$

It is important to note that the solution yielded two natural modes and two natural frequencies equal to the degrees of freedom (two

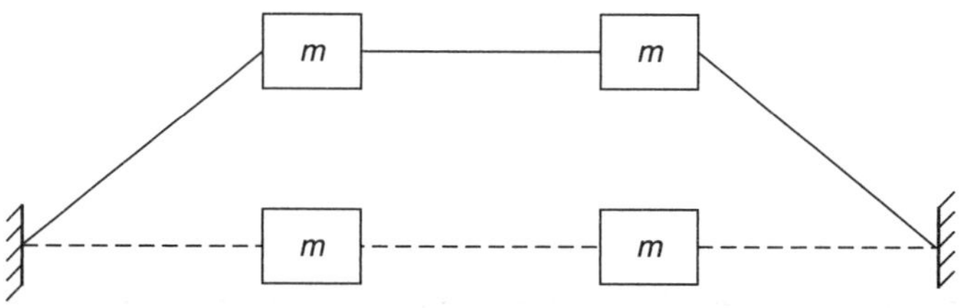

FIGURE 5.8 First mode shape.

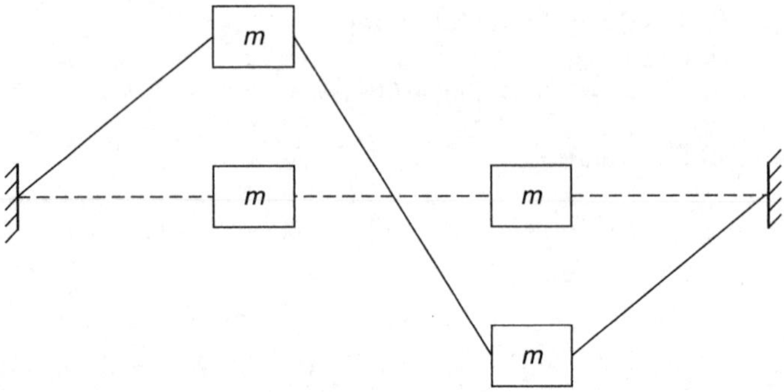

FIGURE 5.9 Second mode shape.

masses). In the first mode, masses move in the same direction, and in the second mode, they move in opposite directions. In other words, in the first mode masses move in phase with one another, but in the second mode they are out of phase with one another.

5.1.3 System with Equal Frequencies

A simple, but important system with equal frequencies is described by Johnson and Bishop (1956) and is shown in Fig. 5.10. It consists of a single mass M to which n equal masses m have been attached through springs of equal magnitude k. Each mass has one degree of freedom, which is in the vertical direction. This system resembles a bladed disk system as used in turbines, which will help us to understand its behavior. Mass M is that of a disk with infinite stiffness and grounded by spring stiffness x_0. Small mass m represents the mass of

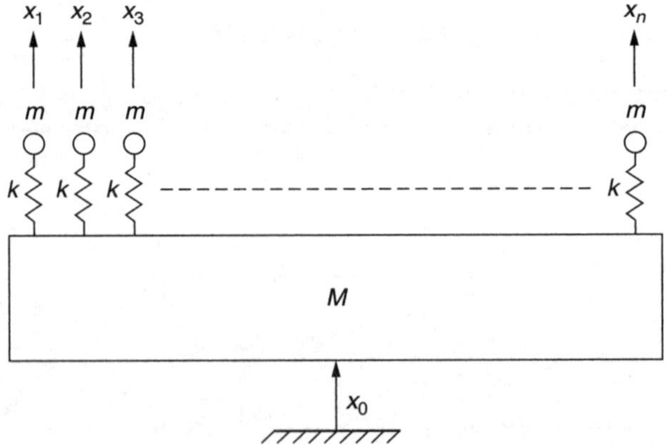

FIGURE 5.10 System with equal mass and stiffness.

the blade, and the blade's stiffness is k. The results of this analysis will show that there are a number of equal natural frequencies in this type of system, but the natural mode shapes will be different. The total number of degrees of freedom for this system is equal to the number of masses and is equal to $n + 1$. Therefore, there will also be $n + 1$ natural frequencies.

It is advantageous to use Lagrange's method to analyze this type of discrete system with a large number of degrees of freedom. The kinetic energy of the system is given by

$$2T = M(dx_0/dt)^2 + m(dx_1/dt)^2 + m(dx_2/dt)^2 + \cdots + m(dx_n/dt)^2 \quad (5.17)$$

The potential energy is given by

$$\begin{aligned} 2U &= k(x_0 - x_1)^2 + k(x_0 - x_2)^2 + \cdots + k(x_0 - x_n)^2 \\ &= nkx_0^2 + kx_1^2 + kx_2^2 + \cdots + kx_n^2 - 2kx_0 x_1 - 2kx_0x_2 - \cdots - 2kx_0x_n \end{aligned} \quad (5.18)$$

Lagrange's method is used to develop a set of differential equations representing the above system.

Let $v_i = dx_i/dt$. Then for a conservative system Lagrange's equation will be

$$d(dL/dv_i)/dt - dL/dv_i = 0 \qquad \text{when} \qquad L = T - U \quad (5.19)$$

Then the resulting equations of motion are

$$\begin{aligned} M(d^2x_0/dt^2) + nkx_0 - kx_1 - kx_2 - \cdots - kx_n &= 0 \\ m(d^2x_1/dt^2) + kx_1 - kx_0 &= 0 \\ m(d^2x_2/dt^2) + kx - kx_0 &= 0 \\ m(d^2x_n/dt^2) + kx_n - kx_0 &= 0 \end{aligned}$$

The characteristic determinant is given by

$$\begin{vmatrix} nk - M\omega^2 & -k & -k & -k & \cdots & -k \\ -k & k - m\omega^2 & 0 & 0 & \cdots & 0 \\ -k & 0 & k - m\omega^2 & 0 & \cdots & 0 \\ \cdots & \cdots & \cdots & \cdots & \cdots & \cdots \\ -k & 0 & 0 & 0 & \cdots & k - m\omega^2 \end{vmatrix} = 0 \quad (5.20)$$

For a nontrivial solution, the determinant is set to zero.The resulting characteristic equation is given by

$$\omega^2(k - m\omega^2)^{n-1}[Mm\omega^2 - (M + nm)k] = 0 \quad (5.21)$$

The natural frequencies are

$$\begin{aligned}
&\omega_0^2 = 0 \\
&\omega_1^2 = \omega_2^2 = \omega_3^2 = \omega_4^2 = \cdots = \omega_{n-1}^2 = k/m \\
&\omega_n^2 = k(M + nm)/Mm
\end{aligned} \tag{5.22}$$

The principal mode corresponding to ω_0 is a rigid body motion, and the other calculated mode shapes are

$$\begin{aligned}
&x_0 : x_1 : x_2 : x_3 : \cdots x_n :: 1{:}1{:}1{:}1{:}\cdots{:}1 \\
&x_0 : x_1 : x_2 : x_3 : \cdots x_n :: (-nm/M){:}1{:}1{:}1{:}\cdots{:}1 \\
&x_0 : x_1 : x_2 : x_3 : \cdots x_n :: \sin 2\pi/n : \sin 4\pi/n : \cdots \sin 2\pi \\
&x_0 : x_1 : x_2 : x_3 : \cdots x_n :: \cos 2\pi/n : \cos 4\pi/n : \cdots \cos 2\pi \\
&x_0 : x_1 : x_2 : x_3 : \cdots x_n :: \sin 4\pi/n : \sin 8\pi/n : \cdots \sin 4\pi \\
&x_0 : x_1 : x_2 : x_3 : \cdots x_n :: \cos 4\pi/n : \cos 8\pi/n : \cdots \cos 4\pi
\end{aligned} \tag{5.23}$$

5.2 Continuous Systems

Some of the general assumptions that are made for solving a continuous system are (1) continuously distributed mass, (2) continuously distributed elasticity (stiffness), (3) homogeneous and isotropic material, (4) material that obeys Hooke's law within the elastic limit, (5) infinite degrees of freedom, (6) particle performance of simple harmonic motion at natural frequency, and (7) each particle passing simultaneously through its respective equilibrium position.

5.2.1 Cantilever Beam

The governing differential equation for lateral vibration of a beam is obtained by considering forces and moments on an element of the beam, as shown in Fig. 5.11.

The summation of forces in the vertical direction is represented by

$$dV - p(x)\,dx = 0 \tag{5.24}$$

The moment about a point on the right face is represented by

$$dM - V\,dx - \tfrac{1}{2}\,p(x)(dx)^2 = 0 \tag{5.25}$$

and rewriting the above equations yields

$$dV/dx = p(x) \qquad \text{and} \qquad dM/dx = V + \tfrac{1}{2}\,p(x)\,dx$$

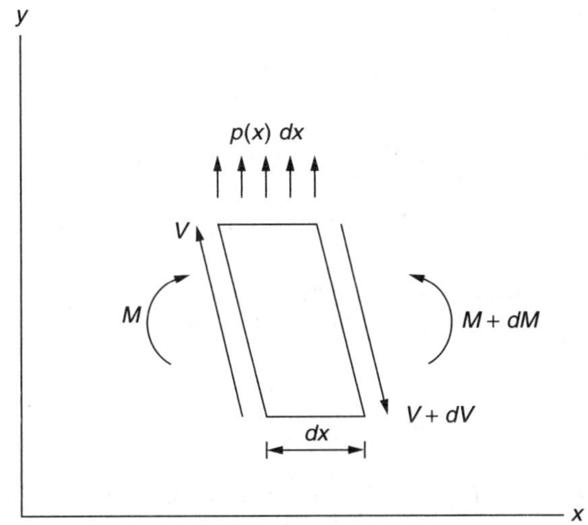

FIGURE 5.11 Forces on an element of a bent beam.

In the limiting case when dx is almost zero, the equation becomes

$$dM/dx = V$$

or

$$d^2M/dx^2 = dV/dx = p(x)$$

From the bending theory of the beam, the curvature of the bent beam is given by

$$M = EI(d^2y/dx^2) \tag{5.26}$$

Then

$$d[EI(d^2y/dx^2)]/dx^2 = p(x) \tag{5.27}$$

For a beam under its own weight the value of $p(x)$ is given by

$$p(x) = (w/g)\omega^2 y = m\omega^2 y \tag{5.28}$$

where m is the mass per unit length.

$$d^2[EI\,(d^2y/dx^2)]/dx^2 = p(x) = m\omega^2 y \tag{5.29}$$

With EI constant the equation becomes

$$\begin{aligned} EI\, d^2(d^2y/dx^2)/dx^2 - m\omega^2 y &= 0 \\ d^2(d^2y/dx^2)/dx^2 - m\omega^2 y/EI &= 0 \end{aligned} \tag{5.30}$$

Assume $n^4 = m\,\omega^2/EI$.

$$d^2(d^2y/dx^2)/dx^2 - n^4y = 0 \tag{5.31}$$

The solution of the above differential equation is given by

$$y = A\cosh nx + B\sinh nx + C\cos nx + D\sin nx \tag{5.32}$$

This defines the mode shapes corresponding to each ω, and the coefficients are evaluated when the initial boundary conditions are known.

$$\omega_n = n^2(EI/m)^{1/2} \tag{5.33}$$

It can be seen that EI is the stiffness of the beam and m is the mass.

The value of n depends on the boundary condition of the beam. For a cantilever beam, the following are the first through third modes.

$$\begin{aligned} (n_1l)^2 &= 3.52 && \text{first mode} \\ (n_2l)^2 &= 22.4 && \text{second mode} \\ (n_3l)^2 &= 61.7 && \text{third mode} \end{aligned} \tag{5.34}$$

The first three modes of a cantilever beam are shown in Fig. 5.12.

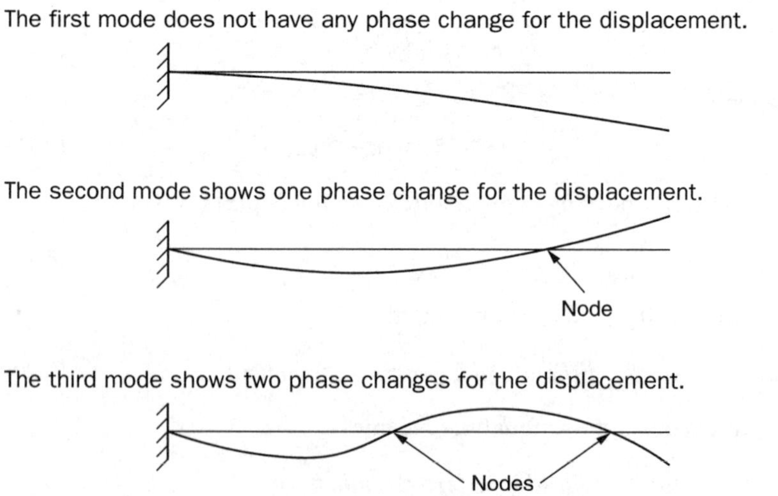

FIGURE 5.12 First three bending mode shapes of a beam.

5.2.2 Circular Plate

Figure 5.13 shows a circular plate of constant thickness of radius a. It is convenient in such a case to set up equations in the polar coordinates (r, θ) as follows.

Define an operator in polar coordinate as

$$G = [d^2/dr^2 + (1/r)d/dr + (1/r^2)\, d^2/d\theta^2]$$
$$[d^2/dr^2 + (1/r)\, d/dr + (1/r^2)\, d^2/d\theta^2] \tag{5.35}$$

The equation can be written as

$$G[W(r, \theta)] - \beta^4\, W(r, \theta) = 0 \tag{5.36}$$

where

$$\beta^4 = \omega^2\, \rho/D_E \tag{5.37}$$

$$D_E = Eh^3/[12(1 - \nu^2)] \tag{5.38}$$

and where h is the thickness of the plate and ν is the Poisson's ratio of the material

The mathematics involved to solve the problem gets complex, and it is not our intent to go through the complete solution. If the solution is desired, please consult a reference for a complete explanation. Only the final solution will be stated here, and the key points of the result will be discussed. These discussions are important to gain insight into the bladed disk behavior.

As there are two coordinates (r, θ), the equation for the frequency can be obtained from the definition of β, if it is known.

$$\omega = \beta^2(D_E/r)^{1/2} \tag{5.39}$$

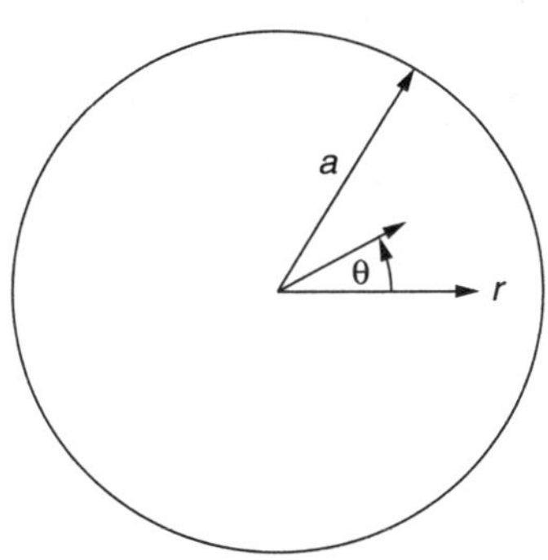

FIGURE 5.13 A constant-thickness circular plate.

The solution for displacement is obtained in terms of Bessel's functions, depends on boundary conditions, and is written as follows:

$$\begin{aligned} W_{mn}(\rho,\theta) = {} & [A_{1m} J_m(\beta r) + A_{3m} Y_m(\beta r) + B_{1m} I_m(\beta r) \\ & + B_{3m} J_m(\beta r)] \sin m\theta + [A_{2m} J_m(\beta r) + A_{4m} Y_m(\beta r) \\ & + B_{2m} I_m(\beta r) + B_{4m} J_m(\beta r)] \cos m\theta \end{aligned} \tag{5.40}$$

where $m = 0, 1, 2, 3, \cdots$

For a clamped plate, after much mathematical manipulation, the characteristic equation is written as

$$I_m(\beta a)\, J_{m-1}(\beta a) - J_m(\beta a)\, I_{m-1}(\beta a) = 0 \qquad m = 0, 1, 2, 3, \cdots \tag{5.41}$$

For a given m, Eq. (5.41) is solved numerically to obtain the values for β_{mn}. This in turn provides the values for the corresponding natural frequencies.

It is important to note that for each frequency ω_{mn} there are two corresponding natural modes, except for $m = 0$. Other modes for which m is not equal to zero are degenerate.

The analyses discussed here and for SMD, beam, collection of beam, and circular plates helps in the understanding of vibrational characteristics of a bladed disk. The reason for this argument is that a bladed disk is composed of these items and the mode demonstrated by the bladed disk is expected to have a combination of behaviors exhibited by individual components.

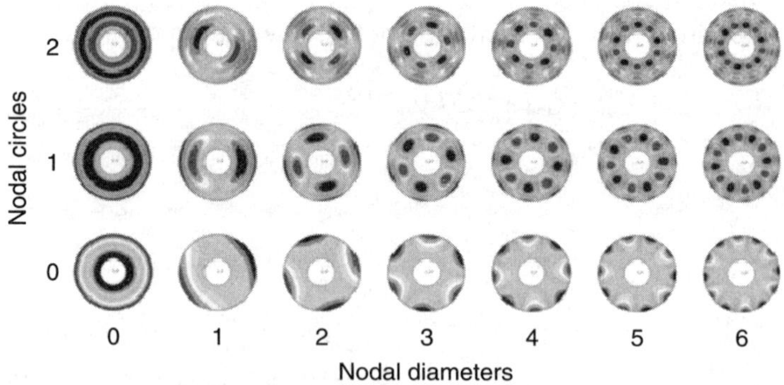

Figure 5.14 Some mode shapes of a circular disk.

CHAPTER 6
Damping Concepts

Energy dissipation in materials, termed *damping*, is caused by many physical mechanisms that depend on the material, temperature, applied stress, and frequency of the cyclic loading. Adequate damping can minimize the dynamic stresses of the disk system. The system damping is the sum of mechanical and aerodynamic damping. Materials under cyclic loading absorb energy, some of which may be stored as potential energy, but most of which is dissipated as heat. The amount of this energy may be small, and it may be very difficult to measure.

Damping behavior is of practical importance as the amount of damping affects the response under vibratory loading. Higher damping reduces the stresses in forced vibration near resonance and is also responsible for the decay of free vibration. Lazan (1968) described the damping properties of structural materials. A brief description is included here. Refer to Lazan (1968) for detailed study.

6.1 Rheological Model

The rheological model exhibits the behavior of low-stress damping (Fig. 6.1).

$$\sigma = \sigma_a \sin \omega t \tag{6.1}$$

where σ_a is the stress amplitude and ω is the frequency. The strain (displacement) response is sinusoidal with a phase shift (angle δ) relative to the stress.

$$\varepsilon = \varepsilon_a \sin (\omega t - \delta) \tag{6.2}$$

The stress-strain response forms an elliptical hysteresis loop, and the area of this loop is the dissipated energy, Δu, in each cycle per unit volume.

$$\Delta u = \pi \sigma_a \varepsilon_a \sin \delta \tag{6.3}$$

Pure sinusoidal force frequently used in the analysis of a vibrating system is uncommon in practice. Many harmonics of force are present that distort the form of the force.

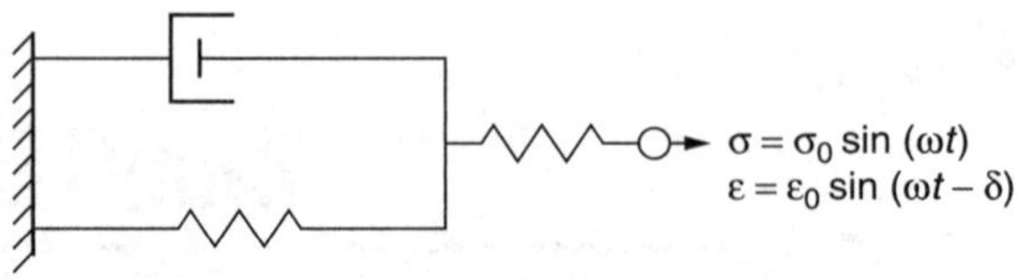

FIGURE 6.1 Rheological model for damping.

The characteristics of the forcing function may significantly affect the damping properties of materials during periodic vibrations.

6.2 Factors Affecting Damping

Some of, but not all, the factors that may affect damping properties are listed here:

1. Form of the force
2. Frequency
3. Amplitude of the force
4. Ratio of alternating to mean stress
5. State of stress—uniaxial or multiaxial
6. Prior stress history
7. Temperature
8. Radiation
9. Environment

The absorbed damping energy is generally dissipated as heat, and a small portion of the damping energy may be absorbed by structural change that raises the energy level of the system. The discussion here is limited to "mechanical damping," i.e., the energy dissipation within a material under cyclic loading. This also includes energy loss at an interface that may occur during relative motion.

For a complete system $D_s = F\,dx$ integrated over complete cycle

For a material $D = \sigma\,d\varepsilon$ integrated over complete cycle

where D_s is the total energy absorbed per cycle of loading by the entire structure (in lb/cycle) and D is the unit energy absorbed by a macroscopically uniform material per unit volume per cycle of loading (in lb/in^3 cycle).

Another important relationship between energy dissipation and stress amplitude when, all other conditions are equal, is stated below.

$$D = J\sigma^n \tag{6.4}$$

Here J and n are material constants; J is the damping constant, and n is the damping exponent.

The low-stress amplitude leads to two conditions worth noting that characterize linear viscosity (dashpot) and are known collectively as *linear damping*.

- The damping exponent $n = 2$.
- The hysteretic loop is elliptical in form.

6.3 Viscous Damping

Previous treatment of a spring-mass model system used the viscous damping model to solve the vibration behavior, and the basic differential equation was written as

$$ma + cv + kx = 0 \qquad \text{for free vibration}$$

The solution for frequency is given by

$$\omega_1 = -c/(2m) + [(c/2m)^2 - k/m]^{1/2}$$

$$\omega_2 = -c/(2m) - [(c/2m)^2 - k/m]^{1/2}$$

6.4 Critical Damping

The value of c that reduces the radical to zero in the above equation is called the *critical damping* c_c.

$$c_c/2m = (k/m)^{1/2} = \omega_n \tag{6.5}$$

$$c_c = 2m\omega_n \tag{6.6}$$

The damping ratio defined earlier is

$$\zeta = c/c_c \tag{6.7}$$

6.5 Proportional Damping

Sometimes for convenience of analysis, a particular type of damping concept is used called *proportional damping*. The mathematics for analyzing the example with viscous damping stated earlier in Chap. 5 was rather cumbersome. In this case the damping ratio c is assumed to be proportional to the stiffness and mass.

$$c = \beta k + \gamma m \tag{6.8}$$

where β and γ are damping ratios.

Many ways of defining damping have been used for mathematical formulations of vibration problems. Some are based on test data and others on concepts developed based on physical principles. The relationship between them is summarized below for convenience. If $\Delta\omega$ is the difference in frequency at the one-half power point, δ = log decrement, ζ = ratio of damping coefficient to critical damping, and Q = amplitude factor, then

$$\eta = \Delta\omega/\omega = \delta/\pi = 2\zeta = 1/Q \tag{6.9}$$

6.6 Frictional Damping and Z-Lock Shroud

Damping to reduce the vibration amplitude is induced by rubbing actions at the mating surfaces in many mechanical systems. Sometimes this type of damping is introduced deliberately so that relative motion will provide a resisting friction force and also generate heat to dissipate energy. Heat is dissipated in friction, and the friction force opposes the relative motion at the interface. This is a characteristic of slip and stick motion. Great effort has been made in the last two decades to develop analytical procedures to solve active damping near the blade and disk interface.

The physical mechanism of the phenomena that occur between contacting surfaces is not well understood. During dynamic motion the contacting surfaces go through microslip, gross slip, stick, and slip motion. Some of the parameters known to influence this behavior include the normal force trying to hold surfaces together, characteristics of surfaces, material properties, surface roughness, surface treatment, frequency of vibration, and coefficient of dynamic friction. This makes the phenomenon completely nonlinear, although attempts have been made to analyze the phenomenon by approximating it as a linear process. Surfaces resulting from manufacturing are rough and contain asperities. Normal loads crush these asperities, and the contact is taken as the average of the crushed area on which the load is assumed to be acting. Contact will increase with increasing load, thereby further deforming the asperities, and the resistance to slip motion is proportional to the number of these asperities. Energy is dissipated by rupture of the frictional bond, and heat is generated at the cost of energy being input. Utilization of the simple law of friction with tangential force $F = \mu N$, even when applied to a single-DOF system, provides complications. Some advanced methods (Sanliturk and Ewins, 1996; Menq et al. 1991; Griffin and Menq, 1991) have attempted to accurately represent the behavior in a vibrating system and have shown promise.

Many designs have shroud constructions called Z-lock. Figure 6.2 shows such an arrangement. The geometry of a shroud is such that it mates with the adjacent shroud through an interface. This contacting surface is assembled with a preload providing a normal acting force.

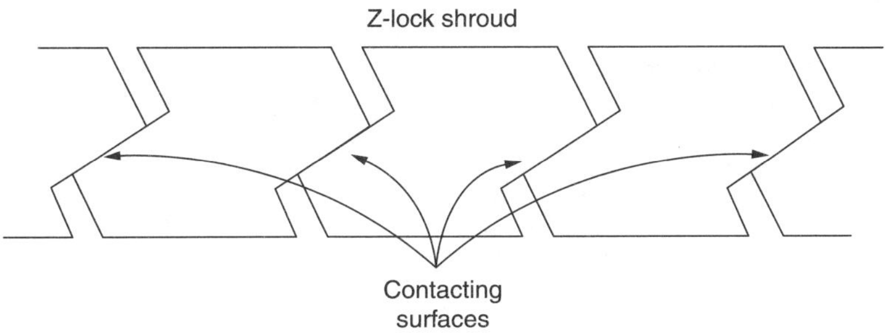

Figure 6.2 Z-lock with contacting surfaces.

A picture of a steam turbine rotor depicting a stage containing blades with Z-lock shroud in successful operation for many years is shown in Fig. 6.3.

The precise conditions and the consequent effects of the contacting surfaces continue to be an unsolved problem. At operating speeds blades untwist, and the interface condition changes from the assembly condition. The dynamic motion of one shroud relative to its neighbor provides tangential resistance and energy dissipation. Many of the analytical approaches mentioned earlier may be used to theoretically evaluate its effectiveness. However, test results should provide a better estimate of the damping magnitude by measurement of the response at any speed or mode of vibration.

It is clear that the physical phenomenon of active damping is not well understood even though many researchers are actively working to understand it. However, for practical applications, the following is a brief description.

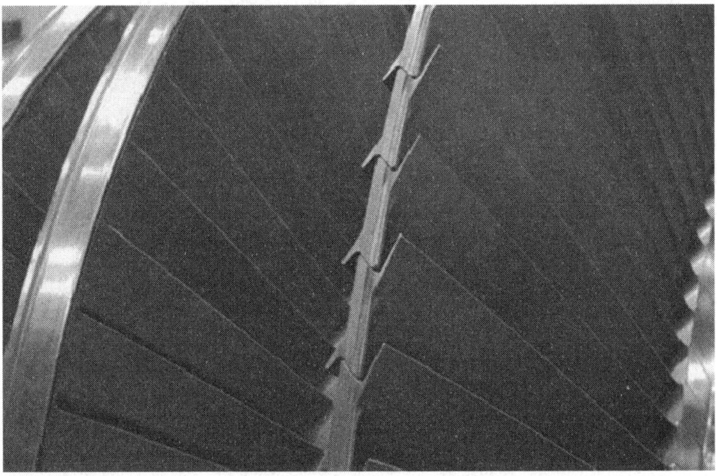

Figure 6.3 Z-lock with contacting surfaces. (*Courtesy Elliott Co.*)

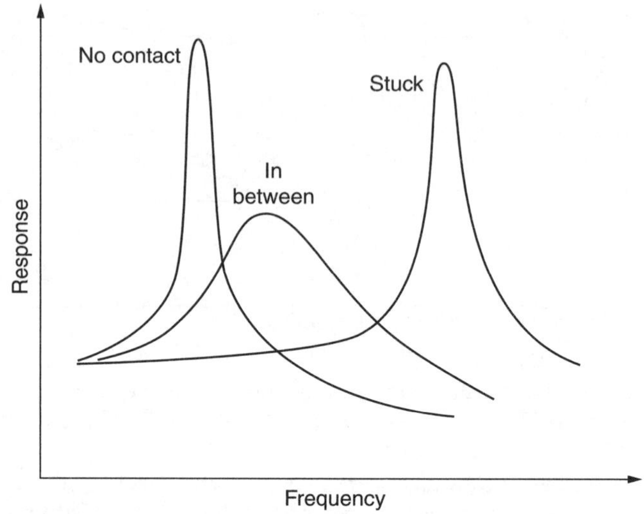

FIGURE 6.4 Response curves for different contact conditions.

Figure 6.4 shows plots of response in a situation of interface friction. The first curve shows the condition for which there is no contact. When the friction force is large enough, then the interface contact might get stuck; i.e., it behaves as a solid interface. In this situation, the frequency is larger than that of the no-contact case. Depending on the contact situation, the response curve is assumed to fall in between these two cases with decreased amplitude depending on the effective damping.

Consider a continuous shroud case where the displacement of the shroud during vibration should be a continuous pattern, as shown in Fig. 6.5. Construction with individual shrouds might displace as shown in Fig. 6.6. During each cycle of vibration, the shroud of one blade will rub against the adjacent shrouds, and the friction force will depend on the contact condition, e.g., the normal force.

The force system acting on the interface is shown in Fig. 6.7. The tangential force provides the resistance to motion.

One can visualize three distinct conditions at the contacting interface—microslip, macroslip, and completely stuck. The tendency might be to design a large interference in an effort to introduce a

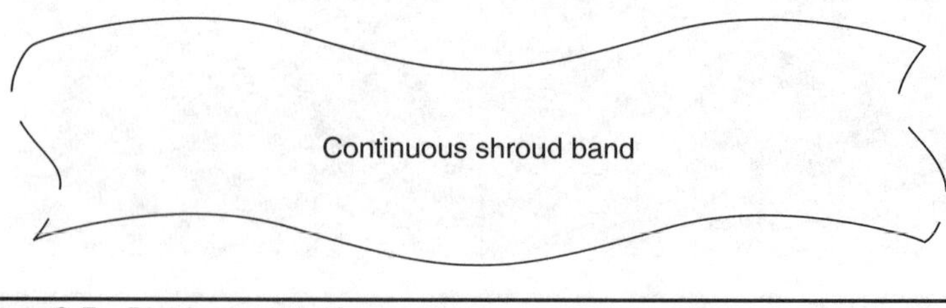

FIGURE 6.5 Displacement of a continuous shroud band.

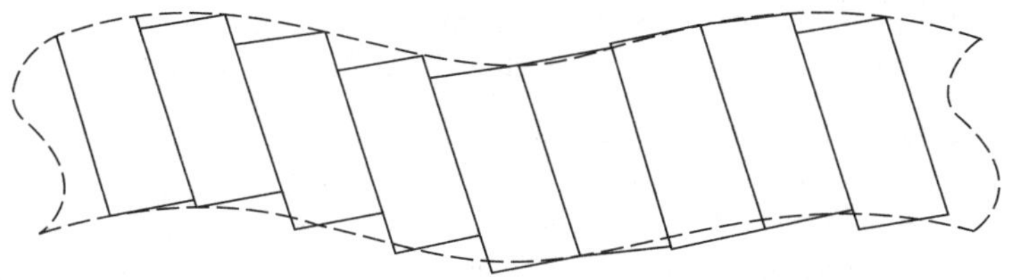

FIGURE 6.6 Displacement of individual mating shrouds.

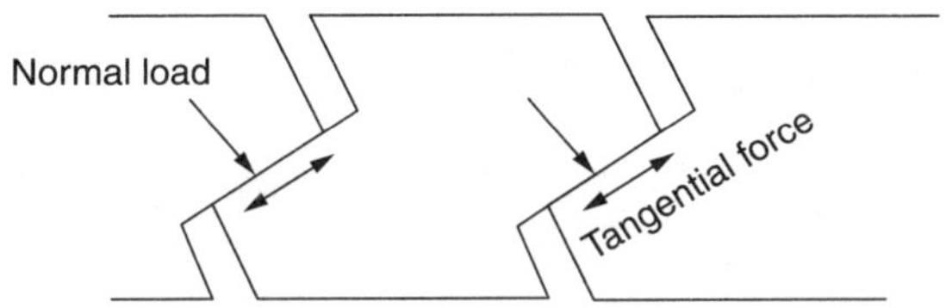

FIGURE 6.7 Loads on mating surfaces.

large friction force. If not done wisely, however, this might not work if the contact surface gets stuck. The only benefit in such a situation might be to increase the natural frequency. Alternatively, there might be a chance of fretting fatigue damage due to relative motion of the contacting surfaces, and in this case, design is a balancing act.

Kielb et al. (1988) described an approach to evaluate a blade damper. The friction type of damper was modeled as attached to blade to the ground. In the Z-lock design, blades make contact with one another at the tips through shroud. Due to this contact a friction type of damper comes into play at the tip of each blade. However, the approach showed the basic insight and correlation of response between the analytical model and bench test data. Kielb et al. mentioned that the tendency is to use a simple dry friction model that may adequately represent the blade-damper system. The important conclusion that can be drawn is that this model represents a macroslip model; i.e., either the interface is stuck or it is completely slipping. The biggest difference between macroslip and microslip models is that even when the interface is stuck, there will be microslippage within the contact zone, and it dissipates energy and thus damping. Microslip is beyond the scope of this book.

6.6.1 Simple Estimation Method—Macromodel

To gain some insight into the behavior of such a construction, some simple relationships from the theory of elasticity and the theory of vibration of mechanical structure may be used. Some simple relationships are detailed next. The first step in the process is to define the magnitude of contact pressure at assembly. This pressure gets reduced due to the effect of speed and temperature of steam in the stage.

Stress Due to Rotation

As the bladed disk system is a complex one, it is even more so with Z-lock construction. To simplify the structure, let us consider a ring to represent the shroud, neglecting its attachment to blades. Expressions of the tangential stress and axial stress are easily obtained from any text on the theory of elasticity. A blade with Z-lock arrangement is shown in Fig. 6.8.

Hoop (tangential stress) σ_θ is given by

$$\sigma_\theta = (3 + \nu)\,\rho\omega^2\{b^2 + a^2 + (a^2b^2)/r^2 - [(1 + 3\nu)/(3 + \nu)]\, r^2\} \qquad (6.10)$$

where a = inner radius of ring
b = outer radius of ring
$c = a/b$
ρ = mass density

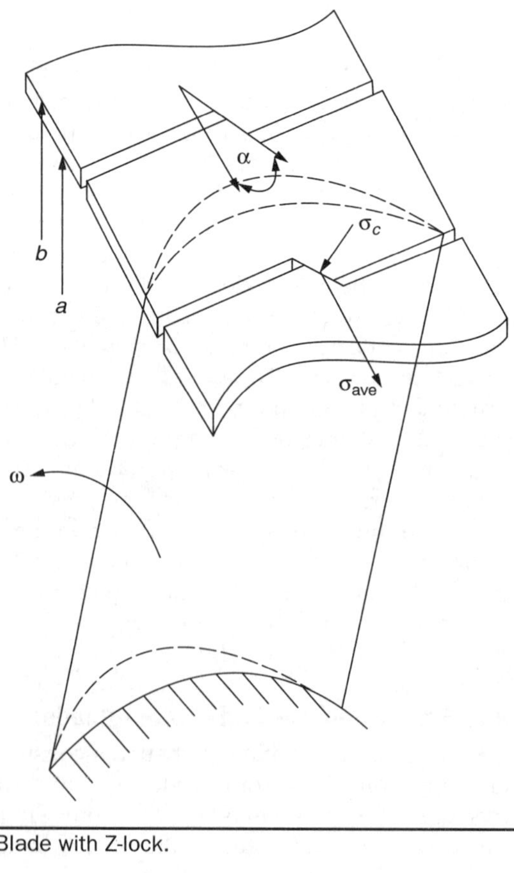

Figure 6.8 Blade with Z-lock.

ω = angular velocity
ν = Poisson's ratio
r = radial distance

For a thin ring where $b - a \approx 0$, the radial stress is zero.

Tangential stress is given by

$$\sigma_\theta \text{ (at } r = a) = [(3 + \nu)/8]\, \rho\omega^2\{b^2 + a^2 + b^2 - [(1 + 3\nu)/(3 + \nu)]\, a^2\} \quad (6.11)$$

$$\sigma_\theta \text{ (at } r = b) = [(3 + \nu)/8]\, \rho\omega^2\{b^2 + a^2 + a^2 - [(1 + 3\nu)/(3 + \nu)]\, b^2\} \quad (6.12)$$

Again for a ring where $b - a$ is small, σ_θ may be assumed to be linear along the radial distance.

Average σ_{ave} is given by

$$\sigma_{ave} = [\sigma_\theta \text{ (at } r = a) + \sigma_\theta \text{ (at } r = b)]/2 = \rho\omega^2 b^2(1 + c^2)/2 \quad (6.13)$$

Contact Stress at Assembly

Let us assume that σ_c is the desired contact stress at assembly and α is the angle that the contact surface plane makes with the tangential direction (Fig. 6.8).

At steady-state condition, the temperature of the ring will be uniform, and the stress due to temperature will be almost zero.

Normal stress on the contact surface due to rotation is given by

$$\sin\alpha\, (\sigma_{ave}) \quad (6.14)$$

The resulting contact stress at the running speed is given by

$$\sigma_{cspd} = \sigma_c - \sin\alpha\, (\sigma_{ave}) \quad (6.15)$$

For positive contact pressure, σ_{cspd} must be greater than zero. The following relationship satisfies this condition.

$$\sigma_c/\sigma_{ave} > \sin\alpha \quad (6.16)$$

Figure 6.9 shows graphically the relationship of Eq. (6.16). It is clear that if a positive contact pressure is desired, then it can be estimated by using a relationship such as given in Eq. (6.15).

Condition of Sliding at Speed without Any Dynamic Load

There is a component of the tangential stress parallel to the contact surface (Fig. 6.8), and this is given by

$$\cos\alpha\, (\sigma_{ave}) \qquad \text{for } \alpha < 90° \quad (6.17)$$

$$-\cos\alpha\, (\sigma_{ave}) \qquad \text{for } \alpha > 90° \quad (6.18)$$

and μ is the static coefficient of friction.

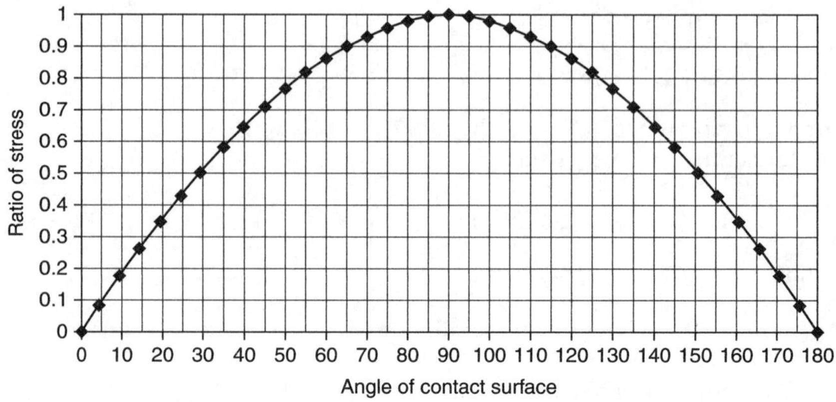

FIGURE 6.9 Stress ratio vs. α.

Resulting resistant force to motion per unit area parallel to the contact surface is given by

$$R_{\text{tang}} = \mu\sigma_{\text{cspd}} - \cos\alpha\,(\sigma_{\text{ave}}) \tag{6.19}$$

For no sliding at speed, $R_{\text{tang}} > 0$; that is, the first term of the equation should be greater than the second term.

$$R_{\text{tang}} = \mu\sigma_c - \mu\sin\alpha\,(\sigma_{\text{ave}}) - \cos\alpha\,(\sigma_{\text{ave}}) = 0$$

or

$$\sigma_c/\sigma_{\text{ave}} > (\mu\sin\alpha + \cos\alpha)/\mu \qquad \text{for } \alpha < 90° \tag{6.20}$$

Similarly for the case when α is greater than 90°, the relationship is given by

$$\sigma_c/\sigma_{\text{ave}} > (\mu\sin\alpha - \cos\alpha)/\mu \qquad \text{for } \alpha > 90° \tag{6.21}$$

The relationships represented by Eqs. (6.20) and (6.21) are shown graphically in Fig. 6.10. This estimation is important in the sense that this provides a means to make sure that contact surface does not slide to change the position of the shroud from equilibrium position during a steady-state condition.

Estimation of Initial Contact Stress Required for No Sliding

For $\alpha < 90°$

$$\sigma_c > \sigma_{\text{ave}}\,[(\mu\sin\alpha + \cos\alpha)/\mu]$$

$$\sigma_c > [\rho\omega^2 b^2(1 + c^2)/2][(\mu\sin\alpha + \cos\alpha)/\mu] \tag{6.22}$$

Figure 6.11 is the graphical depiction of Eq. (6.22).

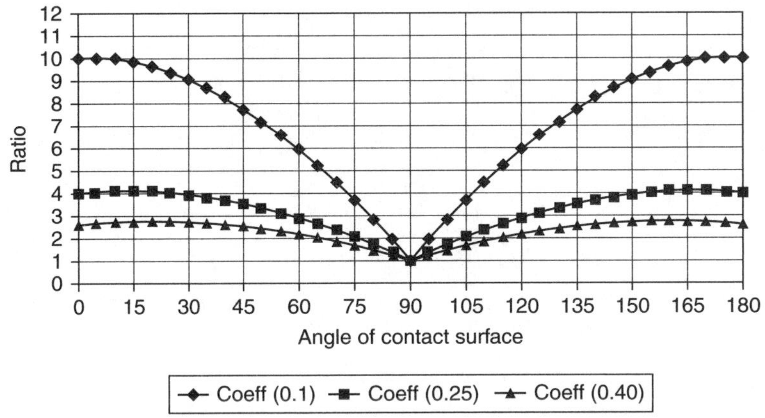

FIGURE 6.10 Ratio of the contact stress at assembly and rotational stress vs. α.

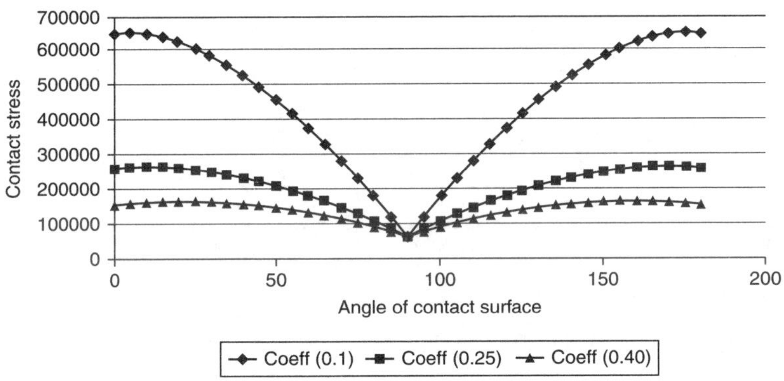

FIGURE 6.11 Contact stress at assembly vs. α.

6.6.2 Dynamic Consideration

A simple one-mode friction model may be used for the Z-lock shroud design shown in Fig. 6.12. A single-mode representation may be made by utilizing modal mass, modal stiffness, and linear viscous damping. Coulomb friction is incorporated at the Z-lock contact surface. This is termed a *macro level* model, and the interface either completely slips or sticks.

Free Vibration

For acceleration a and velocity v,

$$F_{\text{coulomb}} = F_c(v) = \begin{cases} 2\mu_1 N & \text{when } v > 0 \\ 0 & \text{when } v = 0 \\ -2\mu_1 N & \text{when } v < 0 \end{cases} \tag{6.23}$$

and μ_1 is the dynamic coefficient of friction.

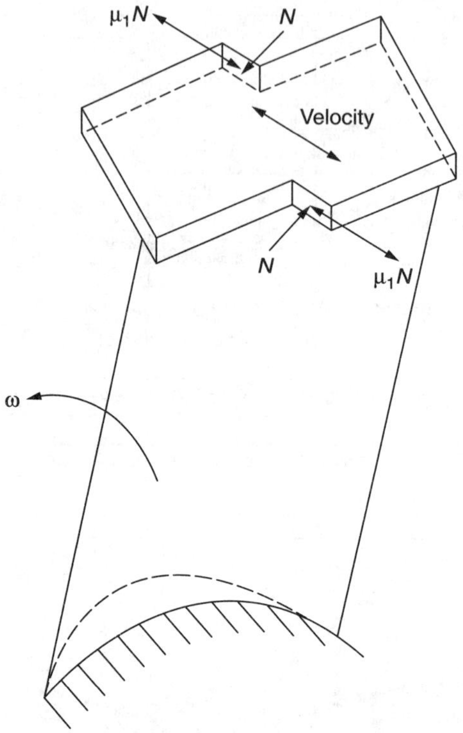

FIGURE 6.12 Single spring-mass system and friction damping force.

Equations of motion are given by

$$ma + kx = 2\mu_1 N \qquad \text{when} \qquad v < 0 \tag{6.24}$$

and

$$ma + kx = -2\mu_1 N \qquad \text{when} \qquad v > 0 \tag{6.25}$$

By using a function sgn(v) that represents a sign based on the direction of v, Eqs. (6.24) and (6.25) can be combined and give

$$ma + 2\mu_1 N \,\text{sgn}(v) + kx = 0 \tag{6.26}$$

This equation is a nonlinear differential equation, and it cannot be solved directly by methods such as variation of parameters. It can be solved, however, by solving for $v \le 0$ and then using the solution to determine the initial condition for $v > 0$.

Solution of the above equation may be written as

$$x(t) = A \sin \omega_n t + B \cos \omega_n t + 2\mu_1 N/k \tag{6.27}$$

where $\omega_n = (k/m)^{1/2}$. Initial conditions to solve the expressions for constants are given by

$$x(0) = x(0)$$
$$v(0) = 0$$

For the above initial conditions, constants A and B may be evaluated, and these are given by

$$A = 0 \qquad B = x(0) - 2\mu_1 N/k$$

Therefore, the solution for $v < 0$ is given by

$$x(t) = [x(0) - 2\mu_1 N/k] \cos \omega_n t + 2\mu_1 N/k \tag{6.28}$$

Solution for $\pi/\omega < t < t_2$ is

$$x(t) = A \sin \omega_n t + B \cos \omega_n t - 2\mu_1 N/k \tag{6.29}$$

The initial conditions to solve Eq. (6.29) may be obtained from Eq. (6.28) as follows:

$$\begin{aligned} x(\pi/\omega) &= [x(0) - 2\mu_1 N/k] \cos \pi + 2\mu_1 N/k \\ &= -x(0) + 4\mu_1 N/k \end{aligned} \tag{6.30}$$

and

$$v(\pi/\omega) = -w[x(0) - 2\mu_1 N/k] \sin \pi = 0 \tag{6.31}$$

Substituting the relationships of Eqs. (6.30) and (6.31) into Eq. (6.29) and its derivatives yields the following:

$$A = 0$$
$$B = x(0) - 6\mu_1 N/k$$

Finally the solution is given by

$$x(t) = [x(0) - 6\mu_1 N/k] \cos \omega_n t - 2\mu_1 N/k \tag{6.32}$$

For the next time step, the initial conditions will be obtained by using Eq. (6.32) and so on.

Forced Response

The situation gets complicated because different forces need to be considered. There are forces due to rotation that act normal and parallel to the contact surface. The basic equation of motion for the spring-mass system in case of forced response analysis can be written as follows:

$$ma + 2\mu_1 N \operatorname{sgn}(v) + kx = F_{\text{rot}} + F_0 \cos \omega t \tag{6.33}$$

6.6.3 Equivalent Viscous Damping

It is convenient to approximate the solution of Eq. (6.33) by solving a viscously damped system that dissipates an equivalent amount of energy per cycle. If the magnitude of the applied force is much larger than the Coulomb force, then this is a reasonable assumption. This approximation gives the steady-state solution of Eq. (6.33) in the following form.

$$x(t) = X_1 \sin \omega t + F_{\text{rot}}/k$$

$$v(t) = \omega X_1 \cos \omega t$$

Energy dissipated (ED) per cycle for viscous damping (for $F = cv$) is given by

$$\text{ED}_{\text{viscous}} = \oint_0^{2\pi/\omega} F dx = \oint_0^{2\pi/\omega} cv\left(\frac{dx}{dt}\right) dt$$

$$= \pi c \omega X_1^2 \tag{6.34}$$

Energy dissipation per cycle for Coulomb damping is

$$\text{ED}_{\text{coulomb}} = 2\mu_1 N \oint_0^{2\pi/\omega} [\operatorname{sgn}(v) v]\, dt$$

$$= X_1 2\mu_1 N \oint_0^{2\pi/\omega} \cos u \, du$$

$$= X_1 2\mu_1 N \left(\oint_0^{\pi/2} \cos u \, du - \oint_{\pi/2}^{3\pi/2} \cos u \, du + \oint_{3\pi/2}^{2\pi} \cos u \, du\right)$$

$$= 8\mu_1 N X_1 \tag{6.35}$$

where $u = \omega t$.

Equating Eqs. (6.34) and (6.35) gives the equivalence energy dissipation.

$$\pi c_{\text{eq}} \omega X_1^2 = 8\mu_1 N X_1 \tag{6.36}$$

$$c_{\text{eq}} = 8\mu_1 N / \pi \omega X_1 \tag{6.37}$$

$$\zeta_{\text{eq}} = 4\mu_1 N / \pi \omega_n \omega X_1 \tag{6.38}$$

Equation (6.33) can be written as

$$ma + c_{eq}v + kx = F_{rot} + F_0 \cos \omega t \tag{6.39}$$

The solution is

$$X_1 = (F_0/k)/[(1 - r^2)^2 + (2\zeta_{eq}\, r)^2]^{1/2} + F_{rot}/k \tag{6.40}$$

where r is the frequency ratio ω/ω_n.

Substituting the expression for ζ_{eq} into Eq. (6.40) yields

$$X_1 = (F_0/k)/[(1 - r^2)^2 + (8\mu_1 N/\pi\omega_n\omega X_1 r)^2]^{1/2} + F_{rot}/k \tag{6.41}$$

Two cases are considered to discuss the solution given by Eq. (6.41).

1. When the tangential force on the contact surface due to rotation can be ignored compared to the applied dynamic force, then

$$X_1 = (F_0/k)\,[1 - (8\mu_1 N/\pi F_0)^2]^{1/2}/(1 - r^2)^2 \tag{6.42}$$

At resonance, i.e., for $r = 1$, X_1 becomes infinite even in the presence of Coulomb damping. It should be recognized that there is always viscous damping or even a small amount of another type of damping that may be approximated as viscous. Therefore, the response never gets to infinite, but it gets controlled by the viscous damping. Also the phase for this case at resonance becomes discontinuous while in the case of a viscously damped system, phase is 90°.

2. If the tangential rotational force F_{rot} at the contact face may not be neglected, then the response near resonance $(1 - r^2 \approx 0)$ may be given as follows:

$$X_1 = (F_{rot}/k)/(1 - \pi F_0/8\mu_1 N) \tag{6.43}$$

Work Done on a Harmonic Vibration by a Harmonic Force

Another simple way to look at the equivalence of the two types of damping for the steady state near resonance is given next. For a single spring-mass system acted on by a force $F \cos \omega t$, the response from the equilibrium position is given by

$$x = X \cos (\omega t - \alpha)$$

Displacement after a small time dt is given by

$$dx = v\,dt = -\omega X \sin (\omega t - \alpha)\, dt \tag{6.44}$$

Work done during this time is given by

$$F \cos \omega t \, dx = -F\omega X \cos \omega t \sin (\omega t - \alpha)\, dt$$

$$\text{Total work per cycle} = -FX\oint_0^{2\pi} \cos \omega t \sin (\omega t - \alpha)\, d(\omega t)$$

$$= \pi FX \sin \alpha \tag{6.45}$$

Work done is proportional to the amplitude F of the force and the amplitude X of the displacement. It is also proportional to the phase angle by which the displacement lags behind the force. When the displacement is in phase with the force, work done is zero while work has maximum value equal to πFX when the displacement is 90° out of phase.

The phase between force and the response is always 90° at resonance, hence work done per cycle at resonance is given by

$$\text{Work per cycle at resonance} = \pi FX \tag{6.46}$$

Energy dissipated by Coulomb damping is $8\mu_1 NX$, where the normal force is $2N$ and the frictional force is $2\mu_1 N$.

The general expression for work done or energy dissipated per cycle by Coulomb damping may be given as

$$\text{Work per cycle} = 4RX \qquad \text{in this case } R \text{ is } 2\mu_1 N \tag{6.47}$$

Equating two expressions gives

$$R = \pi F/4 \tag{6.48}$$

When $R < \pi F/4$, then the work done by the driving force is greater than that dissipated by Coulomb damping. Therefore, the amplitude will build up without limit at resonance. The exact solution in the case of Coulomb damping was provided by Den Hartog (1931). This conclusion has also been reached by examining Eq. (6.42) and discussed earlier.

6.6.4 Macroslip and Microslip

Kielb et al. (1988) presented results for the damper design that should be similar to the final detailed results for Z-lock design. They presented a figure (Fig. 6.13) that is the plot of response, at an excitation level, vs. normal load. For zero normal load, the system's response is controlled by the viscous damping of the system. As normal load is increased, energy is dissipated at the interface and the response decreases. The minimum point of the curve represents the optimal point. Normal load at this point corresponds to when the damper slips and sticks in approximately equal amounts during the vibratory cycle. As the normal load is further increased, the damper starts to

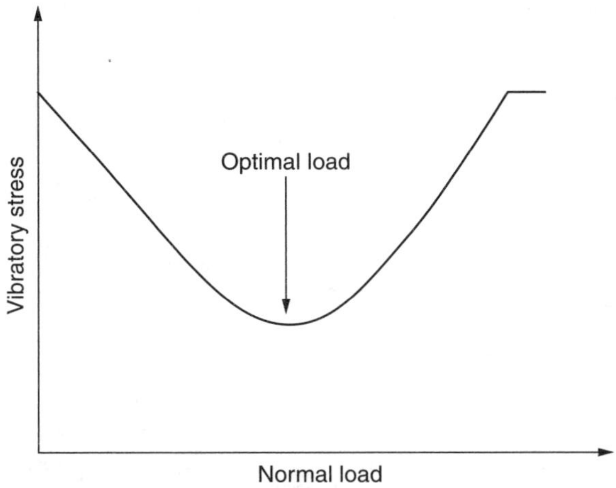

Figure 6.13 Response vs. normal load.

stick more. At a very high normal load, the interface always sticks and there is no damping due to friction.

By using the modified model, a damper performance curve was plotted (Fig. 6.14). Response is estimated as a function of excitation level for a fixed normal load due to centrifugal force. For low excitation level, the friction damper is stuck, the response is linear, and viscous damping controls the response. At large excitation level, the damper slips, but the response is nearly linear since the friction force is small. Response is again controlled by viscous damping. For the

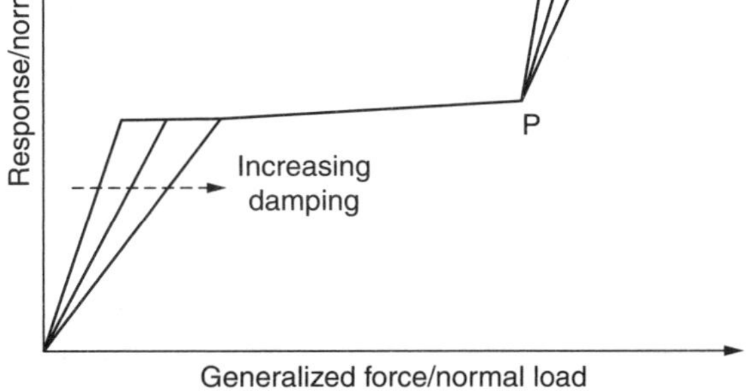

Figure 6.14 Damper performance curve.

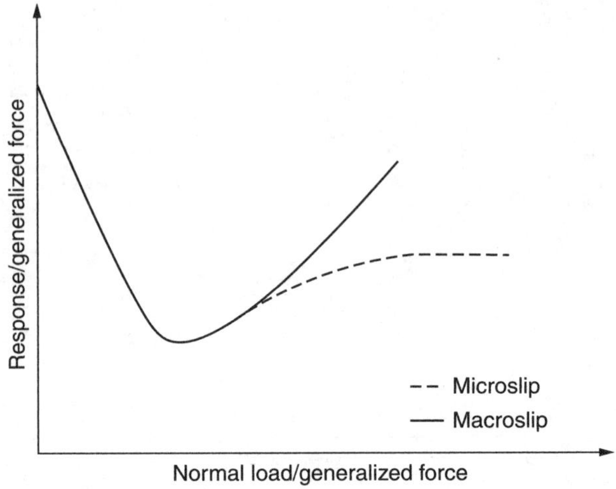

FIGURE 6.15 Damper optimization curve.

intermediate excitation levels, a stuck-slip motion results. In this range the response is nonlinear with friction damping controlling the response. Kielb et al. (1988) pointed out that in this region the response is nearly flat to the point *P*, and location of this point is independent of viscous damping. By a controlled design, the response may be kept below the acceptable level, e.g., endurance strength.

They presented another result to underscore the difference between macro- and microslip in Fig. 6.15.

6.6.5 Summary of Simple Analysis

Design for Z-lock is a complicated process. The exact solution for friction damping during blade vibration is the subject of research. However, these types of designs have been successfully designed, manufactured, and are being used. Simple analysis should be used as guidance for initial design. This is no substitute for detailed analysis using advanced methods, e.g., finite element analysis and test in laboratory or at the test stand on the machine.

No matter what type of analysis is deployed, following steps may help in understanding the design.

1. Estimate rotational stress in the shroud band, assuming it to be continuous, using Eq. (6.13) if desired more accurately, through finite element analysis on the actual design including blades.
2. Estimate contact stress as a ratio of tangential stress on the contact surface using the desired angle of the contact surface, Eq. (6.15).

3. Estimate and adjust the initial contact stress for no sliding under rotational load for an assumed value of the static coefficient of friction. Contact stress angle may be adjusted if required in Eqs. (6.20) through (6.22).

Having estimated stresses, one needs to consider the vibration and energy release due to Coulomb damping.

1. Conduct modal analysis assuming (a) continuous shroud and (b) no contact between blades.
2. Draw SAFE diagrams or Campbell diagrams for the two cases. Determine the mode(s) and frequencies of concern for each case.
3. The continuous case will provide the upper limit of the frequency to avoid, while the no-contact case seems to represent the case of lowest frequency and maximum response.
4. From the SAFE diagram or Campbell diagram the two limits of the frequencies of concern are known, one for no slipping and one for no contact. These two limits have no effect of friction damping on the response of the system.
5. The real situation may be somewhere in between. The mathematical treatment described can be used when generalized mass, generalized stiffness, and generalized force for the mode of concern are known and that mode may be considered as a single spring-mass-damper system.
6. Equivalent viscous damping (Sec. 6.6.3) may be used to understand the energy release due to Coulomb damping.

CHAPTER 7

Vibration Behavior of Bladed Disk System

The assumptions and key points that form the basis for understanding the behavior of a bladed disk system are as follows:

1. The displaced form of the blades in a packet should be similar to the displaced form exhibited by a single blade in a similar mode of vibration.
2. A disk should exhibit similar behavior with or without blades.
3. The displaced form of the cantilevered packet (as a system) is expected to be similar during the vibration of a bladed packeted disk.

7.1 Single Cantilevered Blade

It is instructive to examine the expected natural modes for a single cantilevered blade. The expected shapes of the first three modes are shown in Fig. 7.1. The end of the blade attached to ground has zero displacement, a condition for being joined to ground. For the first mode all the points of the blade at any given time in its one cycle of vibration move in the same direction. For the second mode there is a point of zero displacement (a node) along the blade. This means that during vibration, locations on one side of the node move in the opposite direction to locations on the other side of the node, which is called a *phase change*. The zero displacement point has zero velocity. The third mode changes phase two times. Similarly the fourth mode should change phase three times and so on.

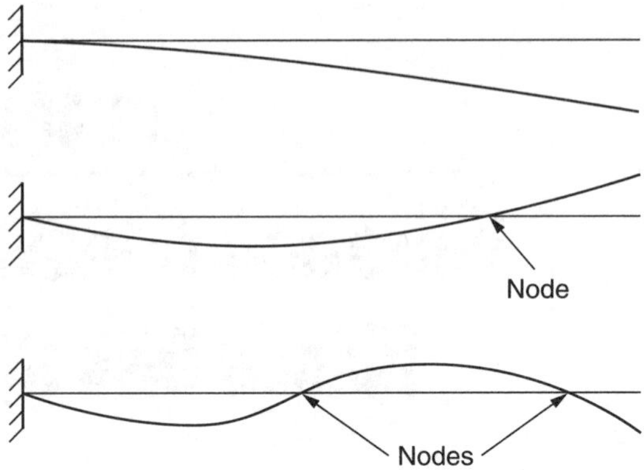

Figure 7.1 Mode shapes of a cantilevered beam.

7.2 Packet of Blades

A packet is a collection of blades that are connected at the tip by a band of metal called a shroud. Due to the coupling through the shroud, a packet exhibits many more modes than a single blade. The number of these modes depends on the number of blades in the packet (see Fig. 7.2 for a description of the modes of a packet with different numbers of blades). The displacement in the plane of the disk is referred to as *tangential*, and that out-of-plane displacement of the disk is called *axial*. It is assumed that any torsional vibration of the blade is coupled with the higher-order axial vibration.

Some key observations are noted from the analysis of a packet of two blades for simplicity, as listed here.

1. The displaced form of the blades in a packet should be analogous to the displaced form exhibited by a single blade in a similar mode of vibration.

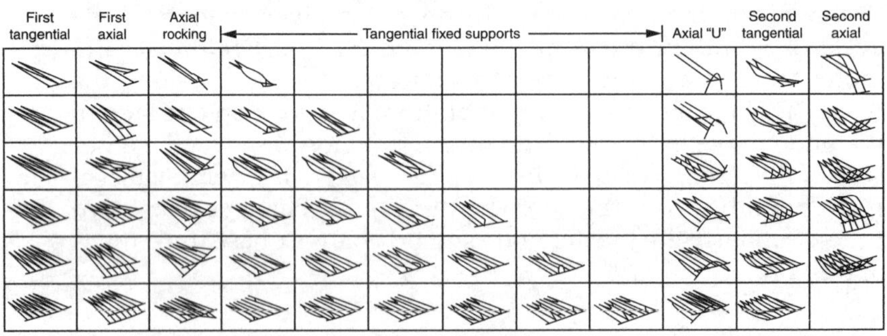

Figure 7.2 Some mode shapes of a packet of blades.

2. There are two modes in which blades move either in phase or out of phase as exhibited in the first tangential and tangential fixed support modes. The displacements of blades in the first tangential and tangential fixed support modes are considered similar.
3. The observations made are also true for axial bending. The blades move in unison in the first axial mode while in axial rocking they move opposite to one another.
4. Statistically, if there are two masses, they will move in two ways in relation to each other. First they move in phase, and then they move out of phase with each other.
5. The implication is that the number of modes that will exhibit the primary displaced shape will be equal to the number of blades. This is also evident in Fig. 7.2 for packets with three blades and packets with four blades.

7.3 Individual Disks

For each natural frequency ω_{mn} there are two corresponding natural modes, except when $m = 0$, where m is the number of nodal diameters and n is the number of nodal circles. When m is not equal to zero, the mode is degenerate. For a vibrating disk, there are two distinct locations where displacements are zero, and these locations form either radial lines or circles. Two radial lines, named *nodal diameters*, are considered to be a diameter for describing a mode. The other is named *nodal circles*. For an axisymmetric structure, the displacement has a sine waveform. These are shown in Fig. 7.3.

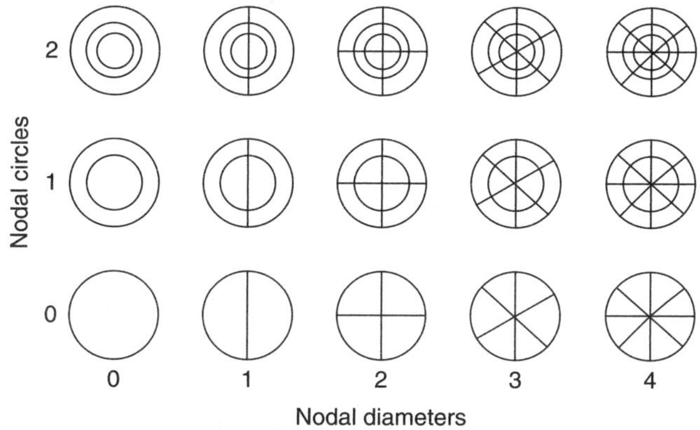

FIGURE 7.3 Modes of vibrating disks.

7.4 Analysis of a Bladed Disk System

The behavior of the bladed disks as one system is described without delving into complex mathematical expressions. Note that there is a coupling between the blades in the packet through the shroud, and there is also a coupling between packets through the disk.

7.4.1 Freestanding (Blades with or without a Shroud But Not Connected to One Another)

Turbines may be designed and manufactured with some stages in which blades are not connected through their shrouds. Two types of construction are prevalent. In one design, the shroud is used as a cover for individual blades, but not for connecting to the next blade. The second type of construction does not have any shroud. If the same disk is used to mount these two types of blades, the natural frequencies are expected to be different, but the mode shapes will be similar. When blades are mounted on a disk, their natural frequencies are influenced by the stiffness of the disk just as the frequencies of a packet were influenced by the shroud's stiffness. Statistically, the total number of modes displaying one type of blade bending mode will be equal to the number of blades. In the vibration of the first mode, all blades move in the same direction at any time in the vibration cycle; i.e., they move in phase. This mode is called the *zero nodal diameter* or *umbrella mode,* and the amplitude of vibration of each blade will be identical at any time.

Figure 7.4 is a plot of the displacements of the tip of the blades and the displacements of the rim of the disk. In this example, there are 120 blades on the disk. Even though this is the mode shape of a packeted bladed disk, it shows the typical behavior of a bladed

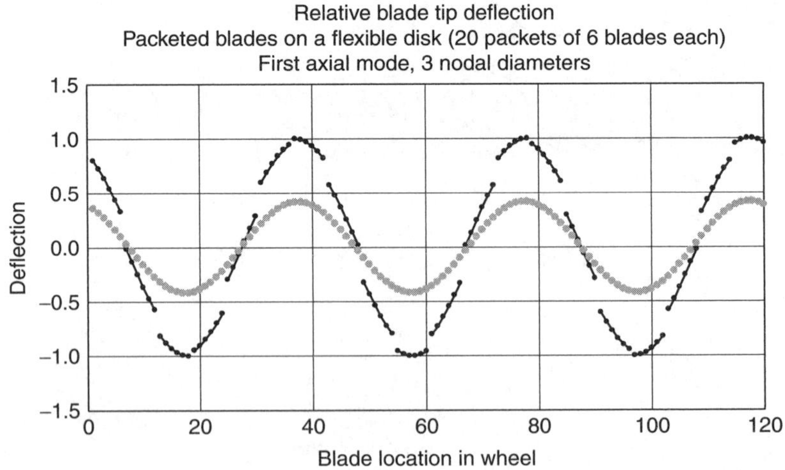

FIGURE 7.4 Three-nodal-diameter mode.

disk with or without a shroud. This particular mode can be described by the function sin(3θ) and contains three full sine waves. It was explained in Chap. 5 that the modes of a disk can be described by a sine function. There is an accompanying mode at the same frequency that can be represented by a cosine function.

It is customary to use the nomenclature of nodal diameter (ND) to describe the modes of a bladed disk (Fig. 7.3). For example, a mode with one sine wave is mathematically described by the function sin θ and is named 1ND (one nodal diameter). One-half the number of blades move in one direction, and the other half move in the opposite direction—there is one phase change in the displacement. The pattern of displacement is similar to a single cosine or a single sine wave.

The natural frequency is plotted on the y axis corresponding to the mode shape that is plotted on the x axis (Fig. 7.5). The disk behavior is similar to the curve as shown. The frequency of the disk will increase as the nodal diameter mode order increases. Natural frequencies have the same magnitude for cantilevered blades. The reason for this is that the base of the blade is anchored to the ground; i.e., the base has infinite stiffness. For blades mounted on a very stiff disk, their characteristics are expected to be similar. This assumption is nearly realized when the vibration of blades on a disk is considered in the tangential direction. When a blade is mounted on the disk, the resultant behavior is as shown. In low-order nodal diameter modes, the disk seems to control the behavior, while at higher order modes the blade seems to control the behavior. There is a transition zone

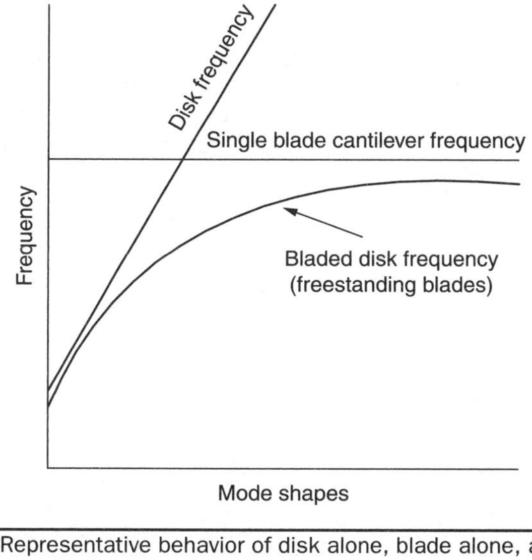

FIGURE 7.5 Representative behavior of disk alone, blade alone, and bladed disk.

where the relative stiffness between the disk and the blade seems to play the controlling role.

7.4.2 Packeted Bladed Disk

Singh and Schiffer (1982) reported a study of the vibrational characteristics of a packeted bladed disk (during the mode of vibration in the tangential direction). Some salient features of the study are worth mentioning, and some key points of the results will be discussed here. Results of a finite element analysis of a turbine stage containing 90 blades are shown. This stage contained 15 packets of 6 blades each mounted on the disk, and the results of a cantilevered packet of blades in the tangential direction are tabulated. The natural frequencies of the first six tangential modes are listed in Table 7.1.

A list of frequencies with associated mode shapes for the packeted blade disk analysis is reproduced in Table 7.2.

The following important points for the packeted bladed disk can be made based on the results presented:

1. The disk is considered infinitely stiff in the tangential direction. Therefore, the natural frequencies in the tangential mode of vibration of the bladed disk are expected to be very close to that of the cantilevered packet.
2. The expected number of modes is 90 for the packeted bladed disk for this tangential vibration because there are 90 blades.
3. There are 15 modes of the bladed disk that exhibit the same mode for each packet, which equals the number of packets on the disk.
4. There will be six groups of modes in which the packet will exhibit the same displacement pattern because there are six blades in a packet. There are six basic modes for a packet containing six blades, which makes a total of 90 ($15 \times 6 = 90$) modes.

Number of Modes	Mode Shapes	Natural Frequency, Hz
1	First tangential	2,612.0
2	Tangential fixed support-1	10,150.0
3	Tangential fixed support-1	11,678.0
4	Tangential fixed support-1	11,948.0
5	Tangential fixed support-1	12,030.0
6	Tangential fixed support-1	12,049.0

Table 7.1 Frequencies and Mode Shapes for a Cantilevered Packet

	Packeted Bladed Disk		Single Packet		Packeted Bladed Disk		Single Packet
Number of Modes	Mode Shapes, ND	Natural Frequencies, Hz	Mode Shapes and Natural Frequencie, Hz	Number of Modes	Mode Shapes, ND	Natural Frequencies, Hz	Mode Shapes and Natural Frequencie, Hz
1	0	1,338.86	First tangential, 2612.0	16	8	10,058.24	Tangential fixed support-1, 10,150.0
2	1	2,534.32		17	8	10,058.24	
3	1	2,534.32		18	9	10,060.23	
4	2	2,584.73		19	9	10,060.23	
5	2	2,584.73		20	10	10,063.98	
6	3	2,596.42		21	10	10,063.98	
7	3	2,596.42		22	11	10,068.76	
8	4	2,601.00		23	11	10,068.76	
9	4	2,601.00		24	12	10,071.97	
10	5	2,604.00		25	12	10,071.97	
11	5	2,604.00		26	13	10,046.87	
12	6	2,606.20		27	13	10,046.87	
13	6	2,606.20		28	14	10,130.79	
14	7	2,606.96		29	14	10,130.79	
15	7	2,606.96		30	15	10,121.95	

TABLE 7.2 Packeted Bladed Disk Frequencies and Mode Shapes (*Continued*)

Number of Modes	Packeted Bladed Disk		Single Packet	Number of Modes	Packeted Bladed Disk		Single Packet
	Mode Shapes, ND	Natural Frequencies, Hz	Mode Shapes and Natural Frequencie, Hz		Mode Shapes, ND	Natural Frequencies, Hz	Mode Shapes and Natural Frequencie, Hz
31	15	11,631.97	Tangential fixed support-2, 11,678.0	46	23	11,913.05	Tangential fixed support-3, 11,948.0
32	16	11,632.38		47	23	11,913.05	
33	16	11,632.38		48	24	11,917.61	
34	17	11,633.01		49	24	11,917.61	
35	17	11,633.01		50	25	11,917.83	
36	18	11,634.16		51	25	11,917.83	
37	18	11,634.16		52	26	11,917.93	
38	19	11,635.02		53	26	11,917.93	
39	19	11,635.02		54	27	11,918.22	
40	20	11,635.66		55	27	11,918.22	
41	20	11,635.66		56	28	11,918.45	
42	21	11,636.07		57	28	11,918.45	
43	21	11,636.07		58	29	11,918.58	
44	22	11,636.27		59	29	11,918.58	
45	22	11,636.27		60	30	11,917.74	

Number of Modes	Packeted Bladed Disk		Single Packet	Number of Modes	Packeted Bladed Disk		Single Packet
	Mode Shapes, ND	Natural Frequencies, Hz	Mode Shapes and Natural Frequencie, Hz		Mode Shapes, ND	Natural Frequencies, Hz	Mode Shapes and Natural Frequencie, Hz
61	30	12,009.45	Tangential fixed support-4, 12,030.0	76	38	12,031.96	Tangential fixed support-5, 12,049.0
62	31	12,004.48		77	38	12,031.96	
63	31	12,004.48		78	39	12,032.06	
64	32	12,008.13		79	39	12,032.06	
65	32	12,008.13		80	40	12,032.07	
66	33	12,008.21		81	40	12,032.07	
67	33	12,008.21		82	41	12,032.09	
68	34	12,008.35		83	41	12,032.09	
69	34	12,008.35		84	42	12,032.10	
70	35	12,008.54		85	42	12,032.10	
71	35	12,008.54		86	43	12,032.11	
72	36	12,008.70		87	43	12,032.11	
73	36	12,008.70		88	44	12,032.12	
74	37	12,009.42		89	44	12,032.12	
75	37	12,009.42		90	45	12,032.06	

TABLE 7.2 Packeted Bladed Disk Frequencies and Mode Shapes (*Continued*)

5. The largest magnitude of the packeted bladed disk frequencies of a group must be less than the frequency of the cantilevered packet blades.
6. Certain mode shapes normally having two equal frequencies for a symmetrical structure will split into two frequencies. This phenomenon is called *splitting of modes* and is due to mistuning arising from breaking shrouds in 15 packets. For example, in this case the 15-ND mode and the 30-ND mode have split into two frequencies.

7.4.3 Completely Shrouded Design

Many turbine stages have blade tips joined by a single shroud band with the two ends of the shroud joined together called a *completely shrouded* (360° shroud) design. The vibration characteristic of this design is similar to that of a single-blade design. The difference is that there is a coupling between blades through the shroud band and the shroud provides stiffness to the system, raising the frequency. The frequencies of a completely shrouded design are higher than those exhibited by a freestanding design. The highest frequency is expected to be less than that of a single cantilevered blade packet containing all blades, as graphically demonstrated in Fig. 7.6.

The results of finite element analysis of a bladed disk with completely shrouded, freestanding packet arrangements are given in the

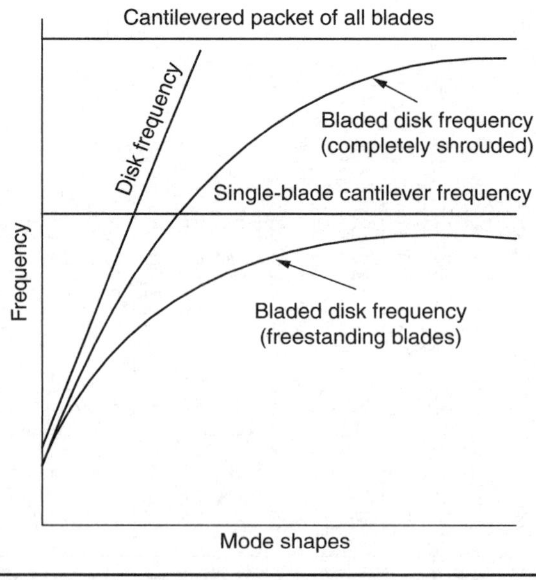

Figure 7.6 Behavior of completely shrouded and freestanding bladed disks.

later part of this section. These results will demonstrate and support the validity pertaining to the argument contained in Fig. 7.6.

Summary of All Bladed Disk Designs

The vibration of bladed disks can be summarized based on the following:

1. A freestanding blade design yields the lower limit of frequencies.
2. A completely shrouded design yields the upper limit of frequencies.
3. A packeted bladed disk design falls between the shrouded and the freestanding designs.
4. A cantilevered single-blade frequency is the upper limit for the freestanding blade frequency.
5. A cantilevered packet frequency is the upper limit for bladed disk frequency.
6. Low-order modes seem to be disk-dominated.

7.5 Evaluation Concepts for Blade Resonant Vibration

7.5.1 Campbell Diagram

The strategy for blade design is to avoid the coincidence of the harmonics of running speed and nozzle passing frequencies with the natural frequency of the blade (Fig. 7.7). The concept of resonance in

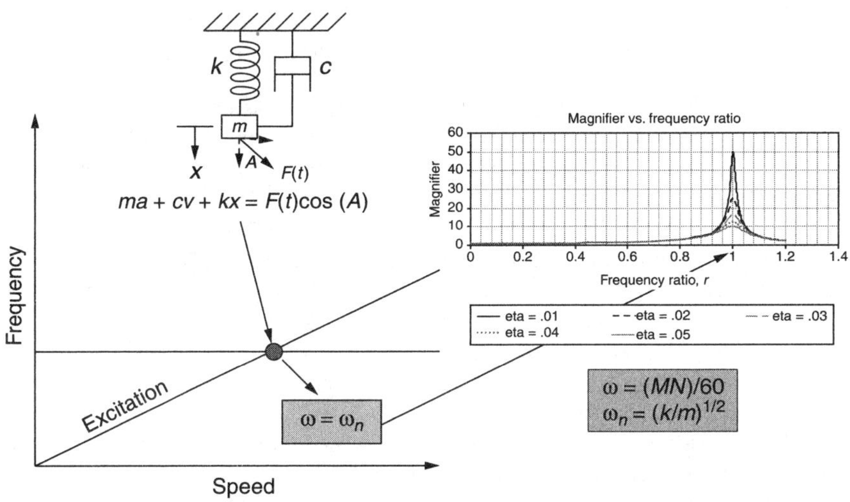

FIGURE 7.7 An explanation of a Campbell diagram.

which the mode shape and the shape of the force also play a defining role is explained later in a section below. The following example of the Campbell diagram is instructive to demonstrate that if (1) the shape of the force and mode shape are identical and (2) the natural frequency is equal to the frequency of the exciting force, the magnitude of the response will be large. However, the blade will also respond to forces even if the frequencies are not equal, but clearly the response will be very small. Resonant response is inversely proportional to damping.

The basis of the validity of a Campbell diagram can be explained by examining the equation of the magnifier presented earlier. The key point is that when the frequency ratio is equal to unity, the response of the system is the largest. The natural frequencies of the system are calculated by suitable analysis or are measured by testing.

7.5.2 Interference Diagram (SAFE Diagram)

The frequency of the exciting force is a function of the speed of rotation and the number of interruptions in the flow field. It is implied in the figure that the shape of the force is identical to the natural mode shape. This assumption might not be realized in every instance of coincidence of frequencies, as might be inferred from the Campbell diagram.

The following two conditions are essential for a true resonance to occur:

1. The frequency of the exciting force equals the natural frequency of vibration.
2. The profile (shape) of the applied force has the same shape as the mode shape associated with that natural frequency.

Mathematical expressions also provided support that each of the above conditions is necessary for resonance to occur, but neither is sufficient by itself. For resonance to occur, both conditions must be satisfied at the same time. The time-varying periodic forces experienced by rotating blades can be resolved in harmonics. The resolution of harmonics is accomplished by performing the Fourier decomposition of the periodic force shape. The frequencies of the harmonics are an integer multiple of the speed of rotation. In general, the force experienced by the blades of a turbine disk during a complete revolution is the consequence of any circumferential distortion in the flow field.

The frequency of the excitation due to nozzle vanes, struts, or any obstruction in the flow field is given by

$$\omega_K = KN/60 \tag{7.1}$$

where ω_K = frequency of exciting force, Hz
K = number of distortions in flow per 360°, e.g., number of nozzles or number of struts in the flow field
N = turbine speed, rpm

Here K represents the shape of the excitation at the k nodal diameter.

The Kth harmonic of the force felt by the blades can be written as

$$P_k(\theta, t) = P_k \sin(\omega_K t + \theta) \tag{7.2}$$

where the frequency of the force is $k\omega$ and θ is the angle on the disk from a reference point.

The mode shape with m nodal diameters and the natural frequency (ω_m) can be expressed as

$$X_m(\theta, t) = -A_m \cos(\omega_m t + m\theta) \tag{7.3}$$

The condition of resonance can occur when the alternating force does positive work on the blade.

7.5.3 Work Done by an Applied Force

The response of a system is the result of work performed by the applied force (Fig. 7.8).

Work done by a constant force W = (force)(distance) = Px

Incremental work done by varying force dW = (force)(distance)
$= P(x)\,dx$

Work done by a linearly varying force $W = 1/2Px$

In the case of a resultant force being zero, there will not be any work done. In essence there is no displacement, hence there is no structural response.

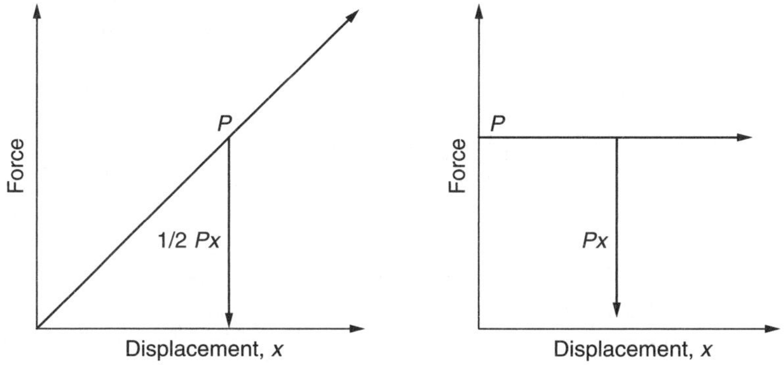

FIGURE 7.8 Definition of work done by a force.

The work done by the kth harmonic of the force acting on a mode shape having m nodal diameters in one complete period T, can be expressed as

$$W = \int_0^{2\pi}\int_0^{T} P_k(\theta,t)\frac{d}{dt}X_m(\theta,t)\frac{N}{2\pi}\,dt\,d\theta$$

$$= \int_0^{2\pi}\int_0^{T} P_k \sin(\omega_K t + K\theta)\omega_m A_m \sin(\omega_m t + m\theta)\frac{N}{2\pi}\,dt\,d\theta \qquad (7.4)$$

$$= N\pi P_k A_m \qquad \text{for } m = K \text{ and } \omega_m = \omega_K$$

or

$$= 0 \text{ for either } m \neq K \text{ and/or } \omega_m \neq \omega_K \qquad (7.5)$$

The first expression of Eq. (7.5) is the positive work ($N\pi P_k A_m$) done by the force. This is true only if the nodal diameter m of the mode shape is the same as the shape of the kth harmonic of the force and the natural frequency ω_m of vibration is equal to the frequency ω_K of the force.

Zero work results when either the nodal diameter m of the mode shape is not the same as the shape of the kth harmonic of the force or the natural frequency ω_m of vibration is not equal to the frequency ω_K of the force. The above argument is the basis of the interference diagram, and it further explains the need to examine the frequency of vibration, the mode shape, and the shape of the force when considering resonance.

A simple way of understanding the implications of Eq. (7.5) is by examining the behavior of a multi-DOF system containing spring and mass.

Let us examine the system of three springs and masses depicted in Fig. 7.9. Each mass represents a degree of freedom; therefore, this is a 3-DOF system. The degree of freedom is displacement (motion) in the vertical direction. There will be three natural modes and three associated natural frequencies (ω_{N1}, ω_{N2}, ω_{N3}). The first mode depicts each mass moving in the same direction at any instant during the vibration cycle. The second mode depicts the displacement of the first two masses as downward, but the third mass moves in an upward direction. The third mode shows the first mass moving downward, the second mass moving upward, while the third mass moves downward. In the first mode there is no phase difference among the motion of masses. During the vibration in the second mode there is one phase change while in the third mode there are two phase changes as one should expect.

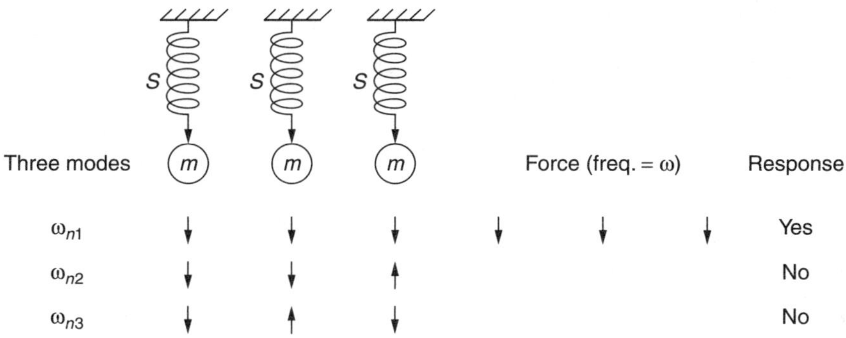

FIGURE 7.9 A 3-DOF system, shapes, and frequencies ($\omega = \omega_{n1}$) match.

Next let us assume a force is being applied to the three masses as shown. Force at each mass is applied in the direction of motion with a frequency of ω. There is expected to be a resonant response of the system when $\omega = \omega_{n1}$. This example meets both conditions as required by the first part of Eq. (7.5), i.e., the shape of the force matches with the mode shape and the excitation frequency, ω, equals the natural frequency, ω_{n1}.

The same condition is shown in Fig. 7.10, but the shape of the force does not match the mode shape. There is no response from the system even though the excitation frequency equals the natural frequency. However, as expected, the second mode responds under the action of this force. If the excitation frequency is not equal to the natural frequency, then there is no response (Fig. 7.11). When $\omega = \omega_{n2}$, there will be a response as depicted in Fig. 7.12. Examples of Figs. 7.9 and 7.12 meet both conditions as expressed in the first part of Eq. (7.5)

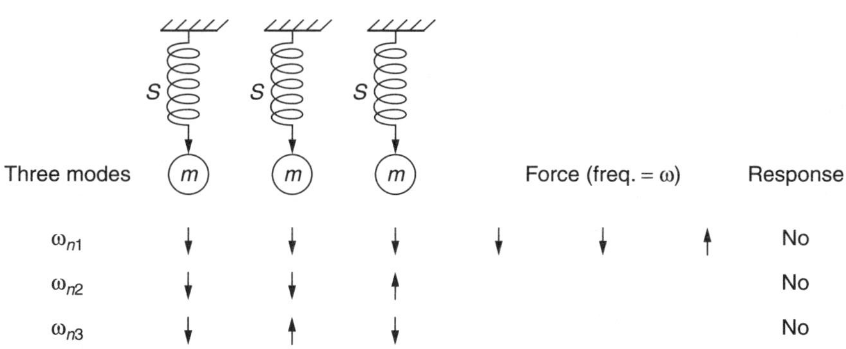

FIGURE 7.10 A 3-DOF system in which shapes do not match and frequencies ($\omega = \omega_n 2$) are equal.

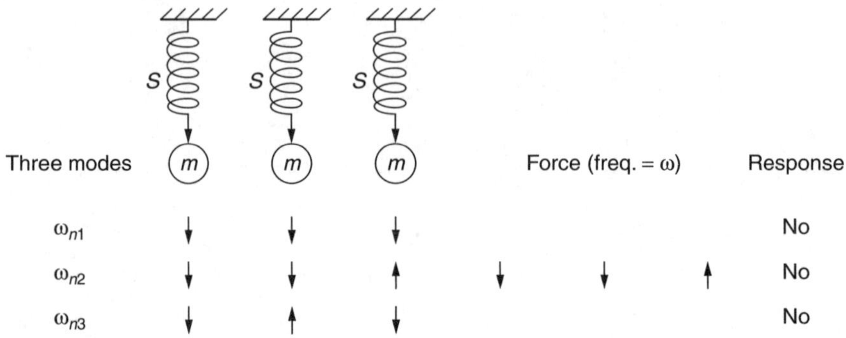

Figure 7.11 A 3-DOF system in which shapes match and frequencies are not equal.

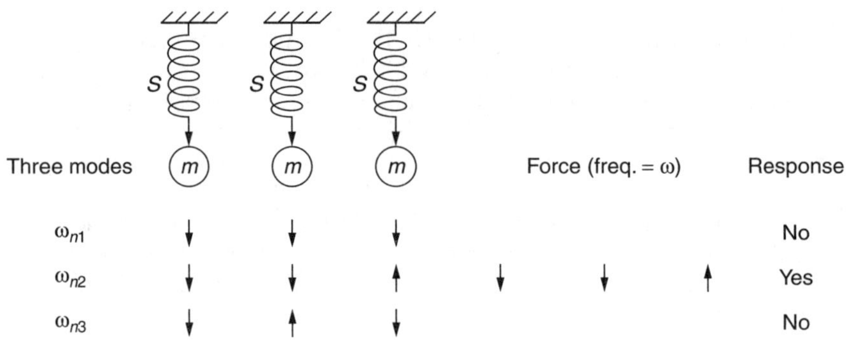

Figure 7.12 A 3-DOF system in which shapes and frequencies ($\omega = \omega_{n2}$) match.

Examples of Figs. 7.10 and 7.11 depict the second part of Eq. (7.5); thereby no response is expected.

To describe the dynamic behavior, two pieces of information are essential: mode shape and natural frequency. Singh and Vargo (1989) described a method of using this information to create an interference diagram, SAFE (Singh's advanced frequency evaluation) diagram, for a packeted bladed disk. The interference diagram combines natural frequencies, mode shapes, excitation frequencies, and operating speeds on one graph. It is an excellent guide to determine the potential of exciting a particular mode of vibration. In any vibration cycle, certain parts of a mechanical structure remain stationary. For a circular symmetric system, these points fall on radii (always an even number) or circles. A pair of radii (opposite each other) is considered as a diameter. Mode shapes can be characterized by specifying the number of nodal diameters (NDs) and nodal circles (NCs). The *x* axis of the diagram represents nodal diameters, and the frequency is plotted on the y axis of the interference diagram. A good design would yield an interference diagram that indicates no coincidence of possible excitation force with a natural mode of the bladed disk. An understanding of the details of the interference diagram for a particular mode shape is essential for determining the likelihood of exciting that particular mode to a degree where it will adversely affect reliability.

The SAFE diagram is based on the description of vibrating characteristics of bladed disk by its natural frequency, mode shape, and the dependence of natural frequency on the speed of the turbine. This can be represented by a plot as shown in Fig. 7.13, which is a 3D representation.

Equation (7.1) describes the forcing function that may act on the moving blades. This equation also contains the same three variables—frequency, shape, and speed. Graphically, Eq. (7.1) is shown in Fig. 7.14.

The probable points of resonance may be found by superimposing Figs. 7.13 and 7.14. As these two intersecting surfaces are three-dimensional, the points of intersection and their locus will be a 3D curve. Figures 7.15 through 7.17 are the three planar views (two

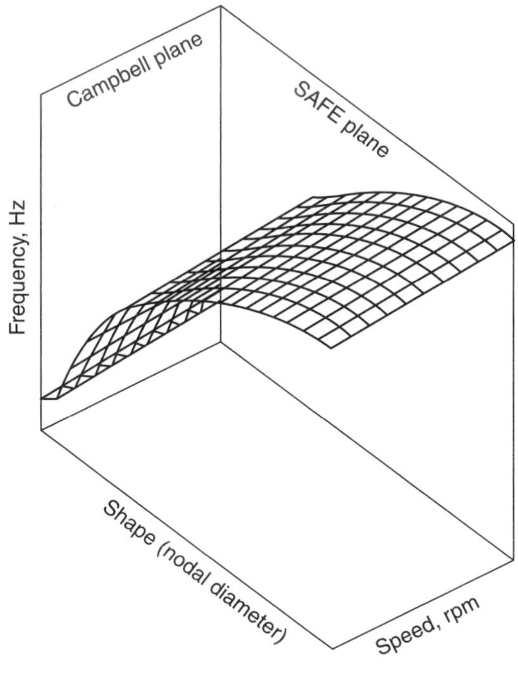

FIGURE 7.13 A 3D representation of vibration characteristic of a bladed disk (completely shrouded or freestanding).

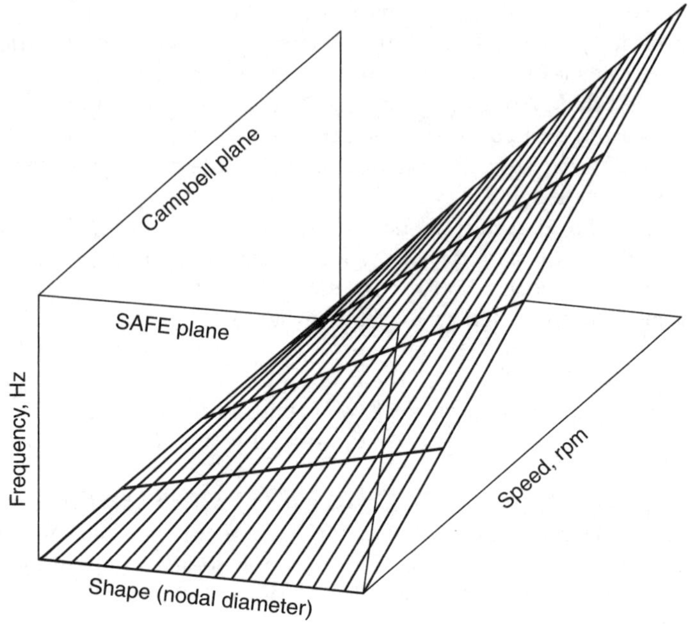

Figure 7.14 A 3D representation of force.

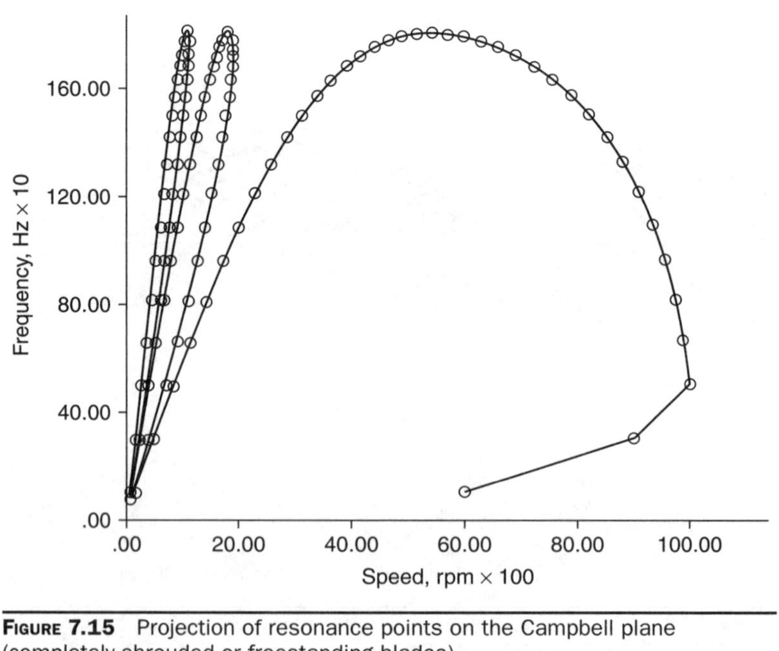

Figure 7.15 Projection of resonance points on the Campbell plane (completely shrouded or freestanding blades).

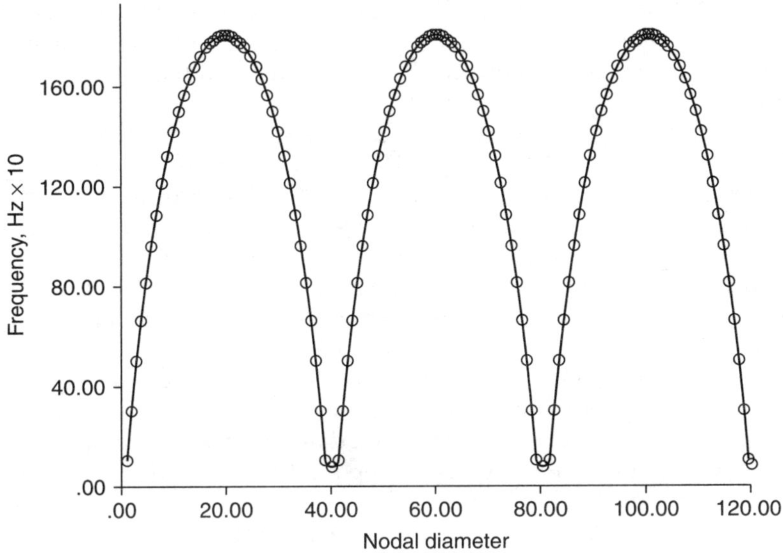

FIGURE 7.16 Projection of resonance points on the SAFE plane (completely shrouded or freestanding blades, 40 blades).

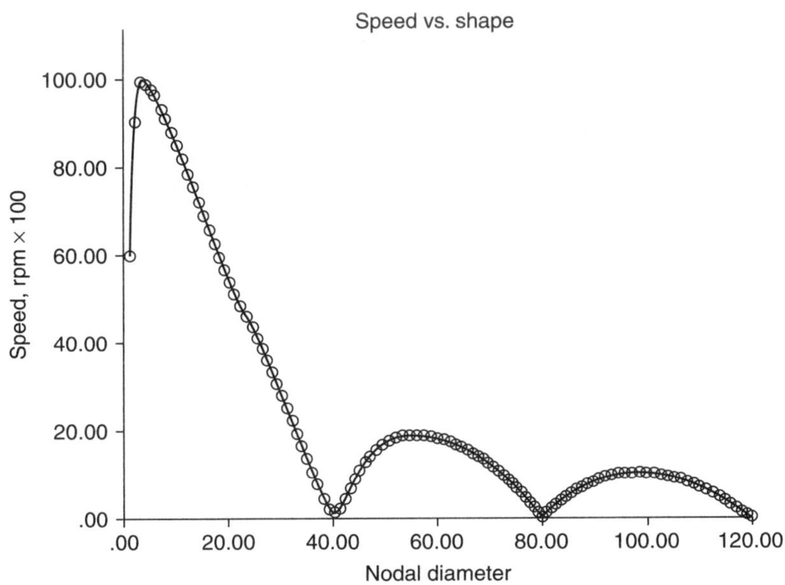

FIGURE 7.17 Projection of resonance points on the horizontal plane (completely shrouded or freestanding blades).

vertical and one horizontal) of these points for a 40-bladed disk. These are the resonance points, which satisfy Eq. (7.5). Frequencies are equal at these points, and the mode shape is identical to the shape of the force.

Plots for resonance points for a packeted bladed disk assembly are shown in Figs. 7.18 through 7.20. The first vertical plot represents the plane of the Campbell diagram. The second vertical plot is the SAFE plane where the horizontal axis represents shape and the vertical axis represents frequency. Inspection of Figs. 7.15 through 7.20 reveals that the projection of intersection points on the SAFE plane shows a remarkable symmetry and repeatability. Information from 0 through 20 ND contains all the essential information. It is easier to construct and represent these points on the SAFE diagram as it will become clear later for any order of excitation. Singh and Vargo (1989) discussed the usefulness of the SAFE diagram through comparison with the results of two published reports (Weaver and Prohl, 1956; Provanzale and Skok, 1973).

Singh and Vargo (1989) presented the method of SAFE diagram to assess the reliability of steam turbine blades utilizing the results of two publications. They demonstrated that the SAFE diagram can estimate the possibility of unreliability in a manner that is simpler than

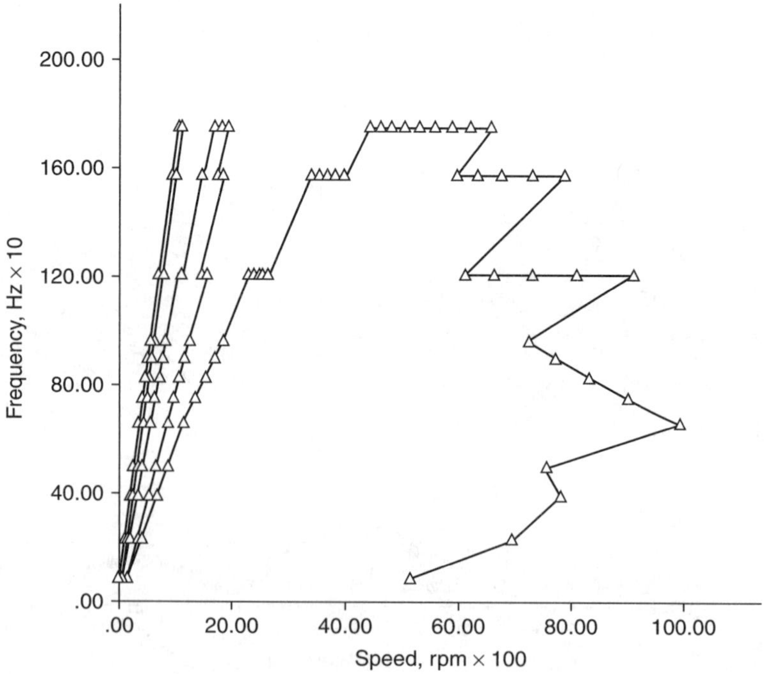

FIGURE 7.18 Projection of resonance points on the Campbell plane (packeted bladed disk).

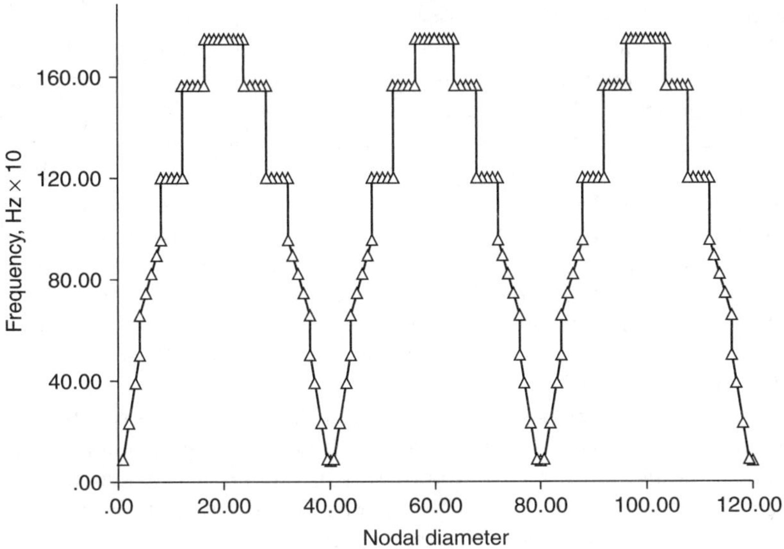

FIGURE 7.19 Projection of resonance points on the SAFE plane (packeted bladed disk, 40 blades).

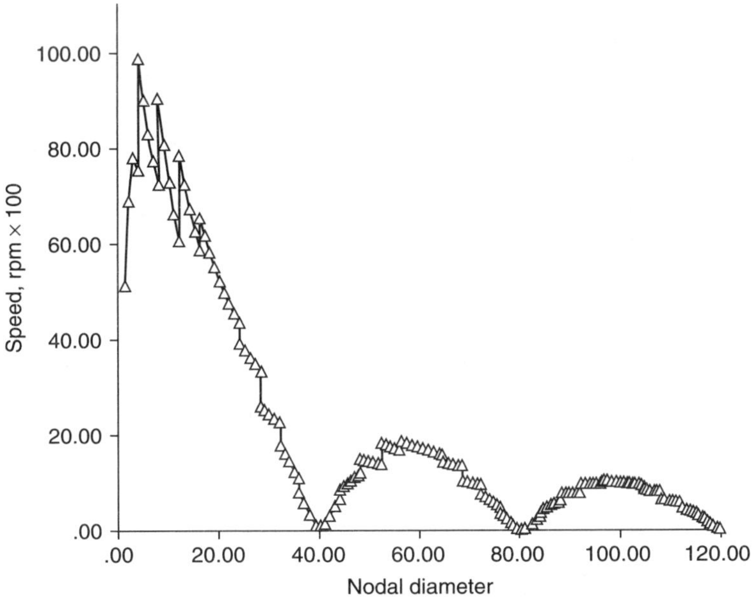

FIGURE 7.20 Projection of resonance points on the horizontal plane (packeted bladed disk).

that obtained after much analysis, as presented by Weaver and Prohl. Also Singh and Vargo were able to show that the results presented by Provanzale and Skok can be explained based on the SAFE diagram, which provides a better explanation of the response points.

Weaver and Prohl used the energy method to calculate the response of a 192-bladed disk. The disk was composed of 32 packets with 6 blades in each packet. The number of nozzles ahead of the stage was 92. They presented their resonant vibration results as shown in Fig. 7.21. Tangential modes are designated by numbers 1 through 7, where number 1 is the first tangential mode, numbers 2 through 6 represent tangential fixed supported modes, and number 7 is the second tangential mode. Roman numbers I and II represent the 1 × NPF and 2 × NPF, respectively.

The SAFE diagram shown in Fig. 7.22 is drawn using the frequency numbers provided by Weaver and Prohl (1956) and assuming that in tangential direction the disk is infinitely stiff to match their cantilever assumption. Probable speeds and the responding modes can now be estimated by examining the SAFE diagram shown in Fig. 7.22. Prediction based on Fig. 7.22 is listed in the Table 7.3. It is easily seen that predictions by the SAFE diagram correlate very well

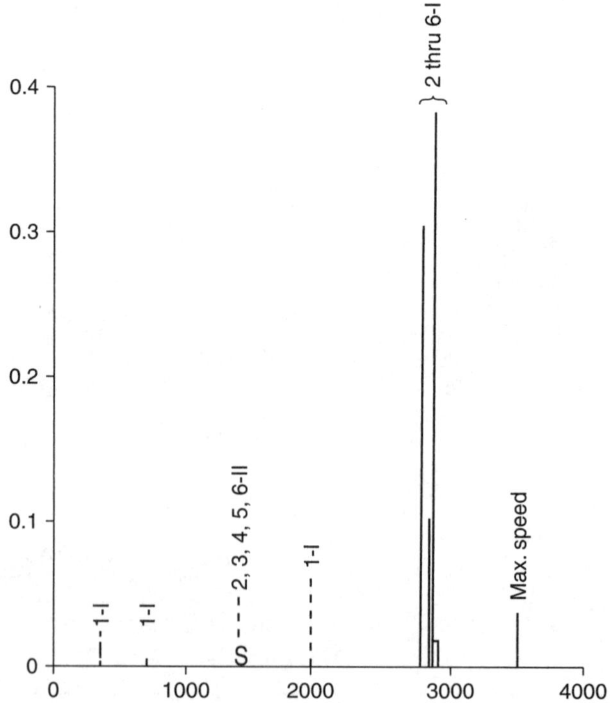

FIGURE 7.21 Resonant vibration stress.

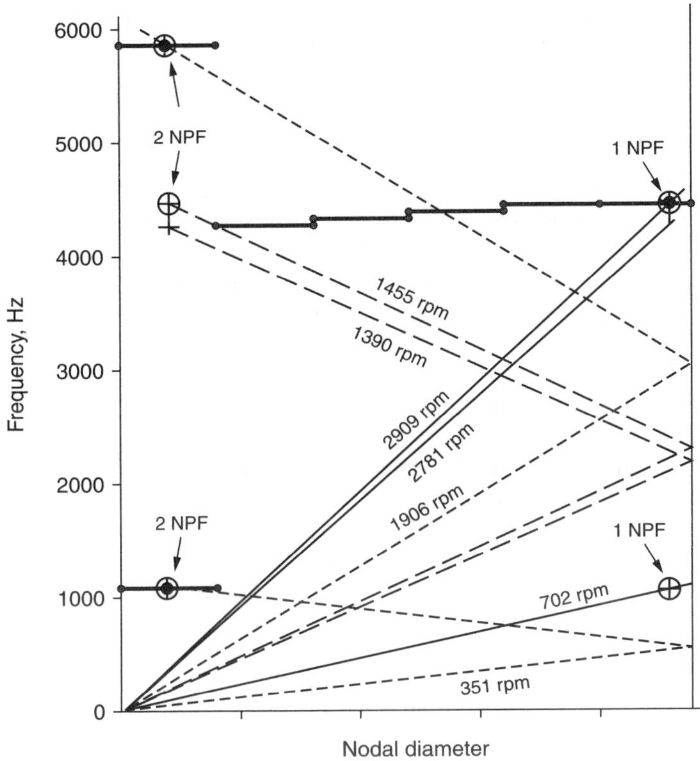

FIGURE 7.22 SAFE diagram using frequency numbers from Weaver and Prohl.

Mode	RPM/Freq (Hz)	Excitation	Response
First tangential	351/1077	2 × NPF	High
	702/1077	1 × NPF	Low
Tangential fixed supports	2909/4461	1 × NPF	High
	1455/4461	2 × NPF	Low
Second tangential	1906/5845	2 × NPF	High

TABLE 7.3 Prediction of High-Response Points by SAFE Diagram

with their results, which were arrived at after numerous calculations. SAFE diagram quickly provides the speed and the mode shapes of concern for a given excitation.

Provanzale and Skok (1973) estimated the response of blades when acted upon by aerodynamic forces. Shape of the force was defined by a parameter β, which is expressed as an angle in degrees. The relationship between β and nodal diameter is given below.

$$\beta = 360^\circ = N \text{ nodal diameters}$$

where N is the total number of blades.

The total number of blades was estimated to be 264. There will be 264 frequencies, and each may be excited when the mode shape matches the shape of the aerodynamic force. They plotted their results as shown in Figs. 7.23 and 7.24. Figure 7.23 is the plot of resonant points on a Campbell diagram, which is similar to Fig. 7.15. Figure 7.24 is similar to Fig. 7.16 representing the SAFE diagram. They arrived at the same result as expected for the SAFE diagram.

The results of the modal analysis are used to draw an interference diagram. The analytical results generate natural frequencies and associated mode shapes. The shapes of deflections around the disk and cover are calculated by using a Fourier decomposition method to identify the mode shapes in terms of nodal diameter.

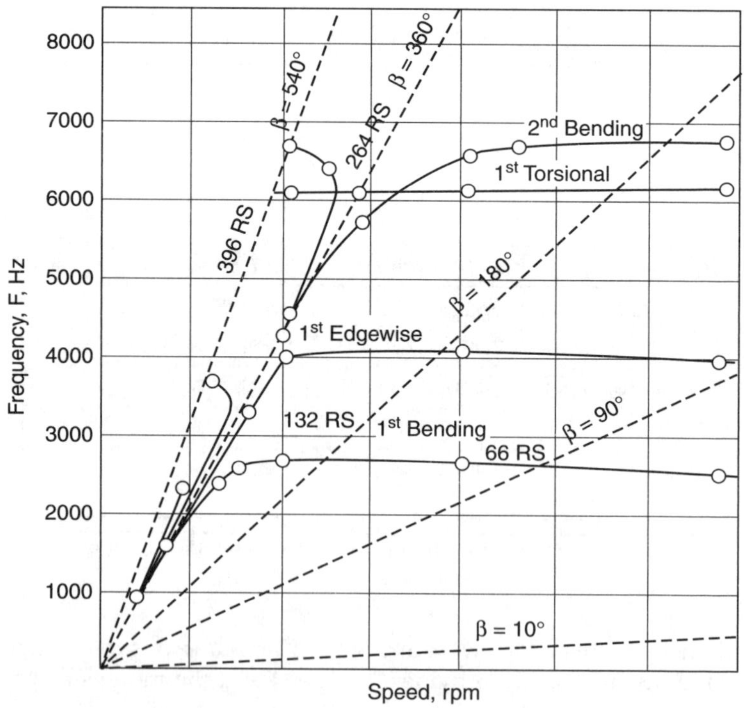

FIGURE 7.23 Resonant points on a Campbell diagram using frequency numbers (Provanzale and Skok, 1973).

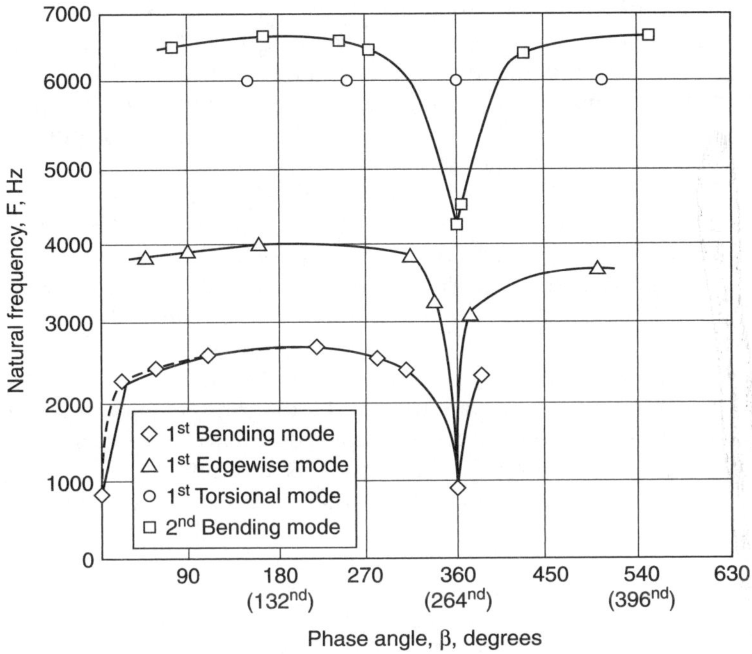

Figure 7.24 SAFE diagram using frequency numbers (Provanzale and Skok, 1973).

In an axisymmetric system, modes occur in pairs except for the first mode, called 0 ND, which is always single, and the total number of modes must be equal to the number of blades on the disk. For constructing any interference diagram (e.g., SAFE diagram) it is important that

1. The maximum number of nodal diameters is equal to one-half of the number of blades when the number of blades is even. For the case of a stage having 90 blades, the largest nodal diameter mode will be 45 (90/2).
2. The maximum number of nodal diameters is one-half of 1 less than the number of blades when the number of blades is odd. If the stage has 91 blades, the largest nodal diameter mode will be 45 [(91 – 1)/2)].
3. The last nodal diameter mode will be single in the case of an even number of blades.
4. The last nodal diameter mode will be double in the case of an odd number of blades.

The results of the Fourier analysis on displacement can be divided into axial and tangential directions, depending on the dominant deflection of each mode. The description of mode shapes by nodal diameters is a key element in the construction of the interference diagram. It is essential for understanding how a structure can be excited when introducing a periodic stimulus to the blades. The mode shapes in the first family will consist of each blade displaying the first bending mode of a single blade.

For true resonance, the coincidence of frequencies is required and the shape of the exciting force must match the mode shape at the same time. The examination of a Campbell diagram, as usually drawn, can only infer coincidence of frequency. Thus, the Campbell diagram is overly conservative and can predict the intersection of excitations that is not physically possible. Avoidance of such intersections can restrict the design without any real value.

The interference diagram (e.g., SAFE diagram) includes information about frequencies as well as the shapes. Hence, it depicts a much more accurate condition for true resonance interference.

To draw an interference or SAFE diagram, the following information is required:

1. Natural frequencies of bladed disk. (Note that these frequencies might depend on the speed of rotation and temperature of the blade and that the effect of speed is not easily displayed on SAFE diagram.)
2. Mode shape of the bladed disk, defined as nodal diameter.
3. Sources of excitation must be known, e.g., number of nozzles, number of struts, etc.

SAFE diagram can be constructed as follows:

1. The mode shape as nodal diameter is plotted on the horizontal axis, and the natural frequencies are plotted on the vertical axis.
2. The equation of the force (mentioned earlier $\omega = KN/60$) is used to draw the frequency of the excitation line for a given speed as a function of the nodal diameter. This is a radial straight line starting from the origin.
3. The frequency of the exciting force is drawn as a vertical line between the operating speed range at the ND denoting the shape of the force, e.g., for 36 nozzles the $1 \times$ NPF is drawn at 36 ND.

The data in Table 7.2 were used to construct the SAFE diagram in Fig. 7.25. Frequencies are plotted on the vertical axis, and the corresponding mode designations (e.g., 1 ND, 2 ND, etc.) are plotted on the horizontal axis. The excitation lines are the radial lines representing the number of nozzles and turbine speed using the equation

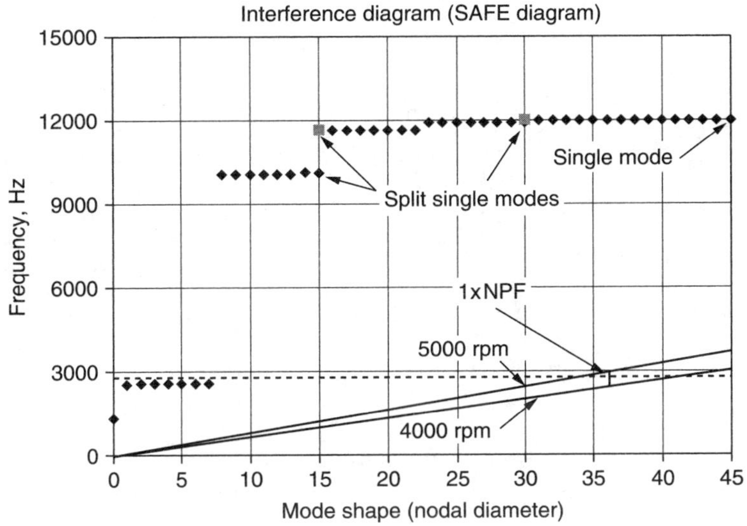

FIGURE 7.25 Interference diagram (SAFE diagram).

$$\omega = MN/60$$

where ω = natural frequency
M = designation of mode
N = turbine speed, rpm

There are two basic pieces of information displayed on the interference (SAFE) diagram. Points on the diagram represent the mode and frequency of the bladed disk system. The exciting force is displayed as a vertical line between the operating speed range at the nodal diameter representing the shape of the force. For "resonance" to occur, the frequency and shape of the exciting force and the bladed disk must be identical. Simply stated, there should not be any concern if the vertical line representing force does not pass through a point representing the bladed disk vibration, as shown in Fig. 7.25. The same data when plotted on a Campbell diagram discussed earlier (Fig. 7.7) raise concern. The shape of the force and the mode shape of the vibration are not taken into account in the Campbell diagram.

7.5.4 Interference Diagram When Harmonics of Excitation Are Larger Than One-Half of the Number of Blades

The SAFE diagram plots modes up to one-half of the number of blades. There are situations in which the harmonic of excitation may be larger than the maximum mode shape plotted on the diagram. For example, 2 × NPF excitation relates to a harmonic of excitation of 72 ND. The maximum mode shape displayed on the

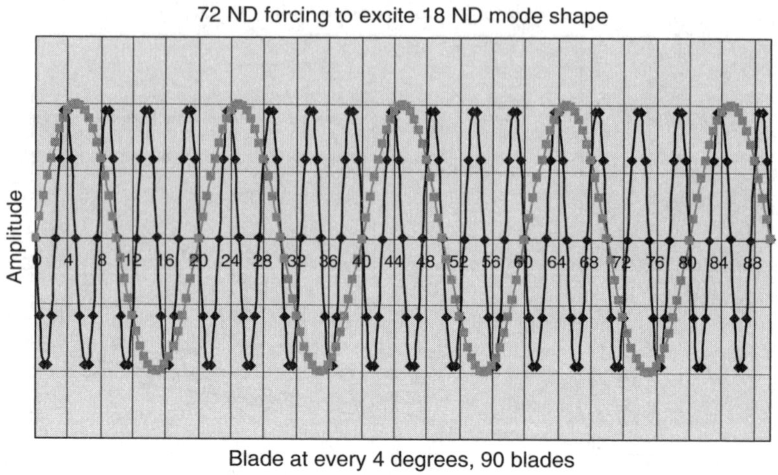

FIGURE 7.26 Sin 18θ and sin 72θ.

diagram is 45 ND. The question is whether this excitation may excite any mode displayed on the diagram. The answer is provided next.

Let us examine the two sine plots of Fig. 7.26. The subtended angle between blades is 4° for a stage that has 90 blades. It is clear from the plot that these two waves intersect at precisely 4°. The implication is as follows: If each blade is instrumented to measure and thereby to determine the shape of the fluctuating force in each revolution, then both waves are equally probable. This means that an 18-nodal diameter mode of the bladed disk may be excited by a force that has a 72-ND shape, provided that the frequencies are the same. There are many combinations for which this assertion is true, but the combination is unique for each individual bladed disk. The combination, however, depends on the number of blades in the stage. Another such combination for this disk is a 10-ND mode that may be excited by a force of an 80-ND shape if the frequencies are the same (Fig. 7.27).

A mode shape and the shape of the excitation should be the same for resonance to occur. Consider a case in which the nodal diameter (the shape of the excitation) is larger than the maximum nodal diameter of the mode shape as discussed above. For example, let us assume that the number of nozzles used for the design is 36 and the shape of the 2 × NPF excitation is 72. The question is, Will the force resulting from 2 × NPF excite any bladed disk mode that is less than 45 nodal diameters? This situation is graphically shown in Fig. 7.28. Even though 45 ND is the maximum shape representation required for a bladed disk vibration of 90 blades, the shape of the exciting force may be larger than 45 ND, as is the case in Fig. 7.28. The 2 × NPF

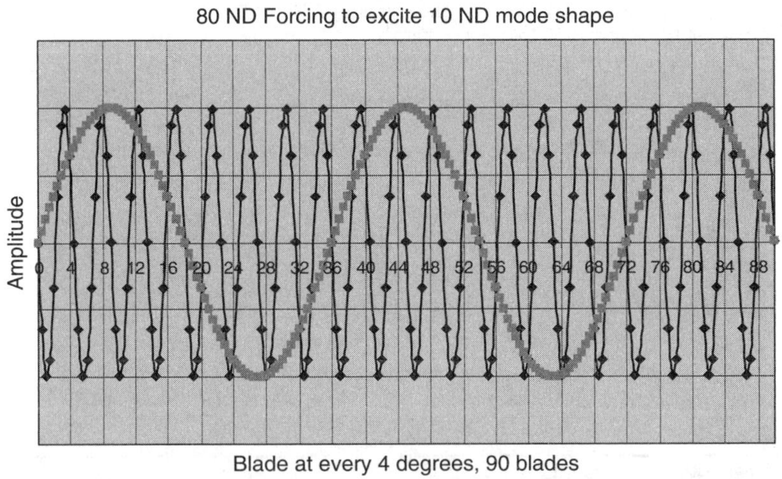

FIGURE 7.27 Sin 10θ and sin 80θ.

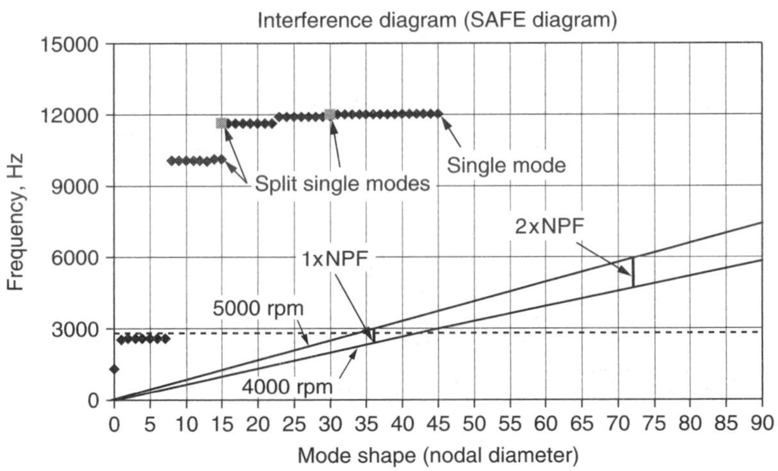

FIGURE 7.28 Excitation shape larger than blade mode shape.

excitation shape is 72 ND. This phenomenon (aliasing) is taken care of on the SAFE diagram by folding the radial speed lines as shown in Fig. 7.29.

Singh (2006) provided the following relationships for possible combinations between mode shape and the harmonic of force. The possible combinations for excitation are given in Figs. 7.30 and 7.31

$$K = \text{abs}(nL + M) \tag{7.6}$$

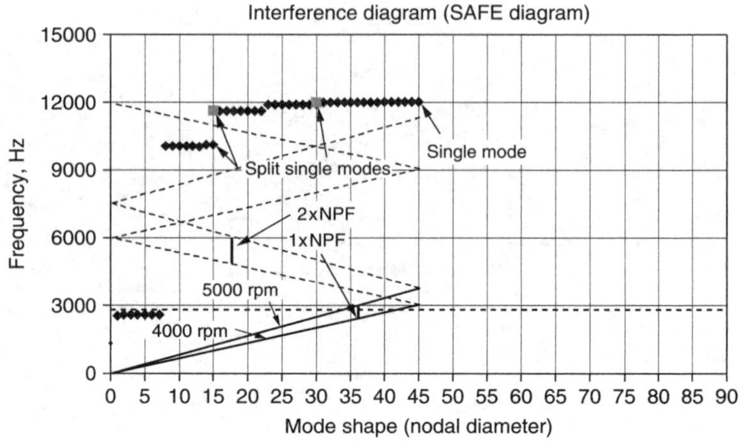

FIGURE 7.29 Reflected speed lines.

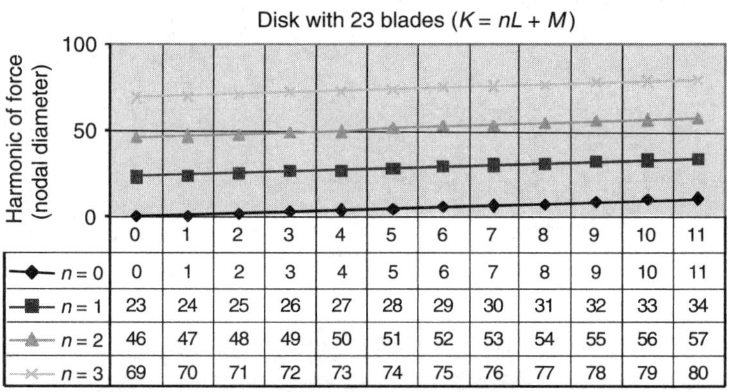

	0	1	2	3	4	5	6	7	8	9	10	11
$n = 0$	0	1	2	3	4	5	6	7	8	9	10	11
$n = 1$	23	24	25	26	27	28	29	30	31	32	33	34
$n = 2$	46	47	48	49	50	51	52	53	54	55	56	57
$n = 3$	69	70	71	72	73	74	75	76	77	78	79	80

Mode shape (nodal diameter)

FIGURE 7.30 Possible combinations for disk with 23 blades.

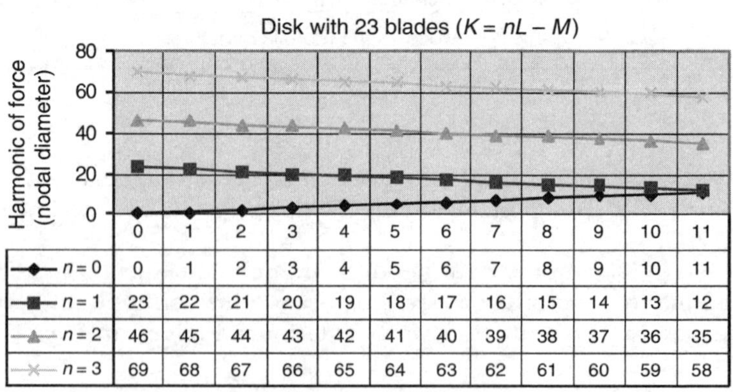

	0	1	2	3	4	5	6	7	8	9	10	11
$n = 0$	0	1	2	3	4	5	6	7	8	9	10	11
$n = 1$	23	22	21	20	19	18	17	16	15	14	13	12
$n = 2$	46	45	44	43	42	41	40	39	38	37	36	35
$n = 3$	69	68	67	66	65	64	63	62	61	60	59	58

Mode shape (nodal diameter)

FIGURE 7.31 Possible combinations for disk with 23 blades.

and

$$K = \text{abs}(nL - M) \tag{7.7}$$

where K = shape of force (nodal diameter)
$n = 0, 1, 2, 3, \ldots, L$
L = total number of blades
M = nodal diameter of mode shapes

7.6 Effect of Temperature and Speed on Natural Frequencies

$$\omega_{T1} = \omega_{T2}(E_{T1}/E_{T2})^{1/2} \tag{7.8}$$

where ω_{T1} = frequency at temperature T_1
ω_{T2} = calculated frequency at T_2
E_{T1} = modulus of elasticity at T_1
E_{T2} = modulus of elasticity at T_2

7.7 Effect on Natural Frequency due to Centrifugal Stiffening

Stress stiffening effects can be easily included for specific speeds in a Campbell diagram. Correction for speed can be made by estimating the natural frequencies at zero speed ω_0 and by calculating the natural frequencies with centrifugal stiffening at the rated speed ω_{spd}. For each mode a stiffness correction A is calculated as shown:

$$A = 3600(\omega_{\text{spd}}^2 - \omega_0^2)/N^2 \tag{7.9}$$

where N = rated speed, rpm.

The speed correction is applied at any speed N_1 during plotting using the following equation:

$$\omega_{\text{spd1}} = [\omega_0^2 + (AN_1^2)/3600]^{1/2} \tag{7.10}$$

For example, a blade has analyzed frequencies, for a mode at zero speed and at the rated speed, of 1000 and 1500 Hz, respectively. The rated speed is 5000 rpm, and the number of nozzles is 30. Then

$$A = 3600(1500^2 - 1000^2)/5000^2 = 180$$

A Campbell diagram of this example can be drawn as in Fig. 7.32.

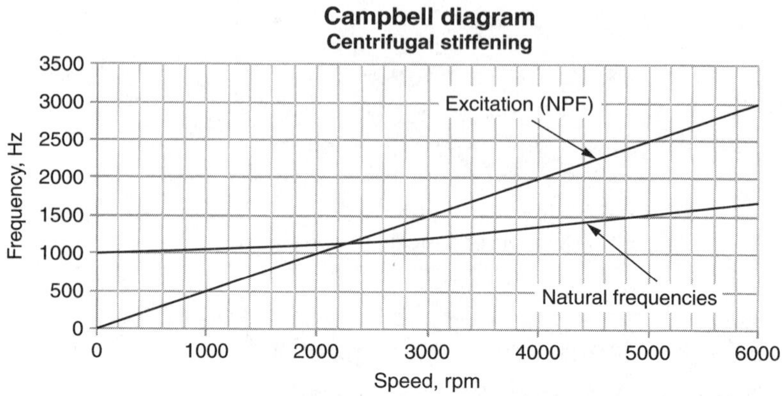

FIGURE 7.32 Campbell diagram with centrifugal stiffening.

7.8 Lacing Wire Construction

The design of a long reaction blade without shroud with large enough stiffness to increase the natural frequency becomes nearly impractical. Blade construction utilizing standard design guidelines may not yield an acceptable design in such a situation. The target is to keep stress at a low level in the blade and to increase the natural frequency of the blade to a higher level. However, many times this is not feasible, and in such a situation a lacing wire design becomes attractive. This type of design is very effective when the requirement is to provide more damping in a particular mode. Sometimes this type of design has been shown to reduce vibratory stress by 10-fold.

There are some detriments to this type of construction, such as obstruction in flow path that may reduce performance, higher noise level, and stress concentration due to hole in the blade that must be dealt with. This area may be subjected to erosion, fretting wear, fretting fatigue, and loss of blade weight, and it may change some natural frequencies of the blade.

Many types of construction have been employed in blade designs. Some of these designs use single continuous wire, multiple wires at different heights of blade, and wires covering certain numbers of blades as packets. Some typical ones are demonstrated in Figs. 7.33 through 7.38.

Due to many competing factors, the design of lacing wire is a complicated effort. Some of the guidelines based on lessons learned by many researchers are given here.

The location of wire on the blade is largely dependent on the mode shape that should be targeted to be damped. It is realized that lower modes have large amplitude of vibration, and these modes are the target for reducing the vibration amplitude of long blades.

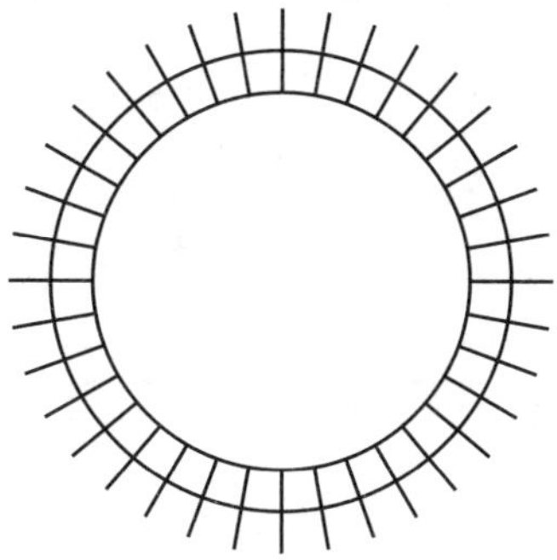

Figure 7.33 Single continuous lacing wire.

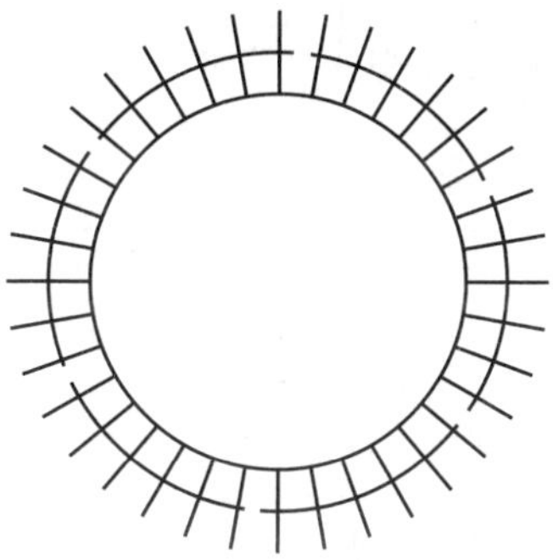

Figure 7.34 Single lacing wire with blades in packets.

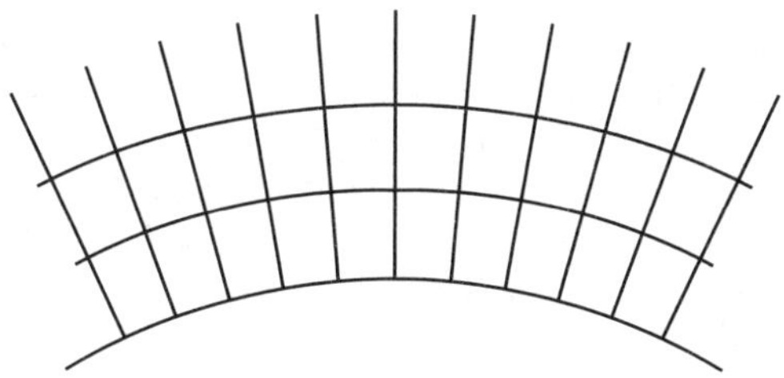

Figure 7.35 Multiple lacing wires at two radial locations.

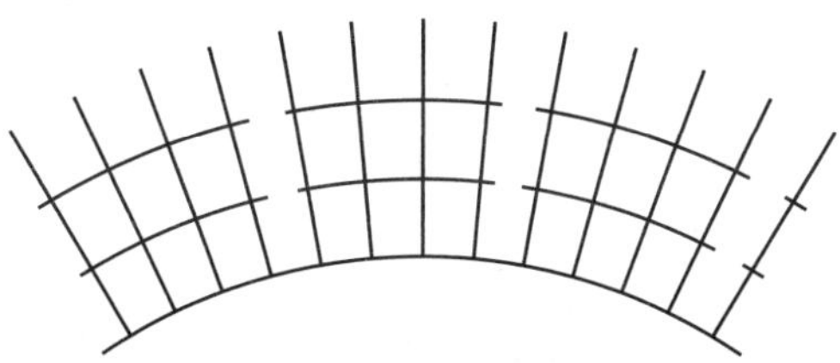

Figure 7.36 Packets with multiple wires.

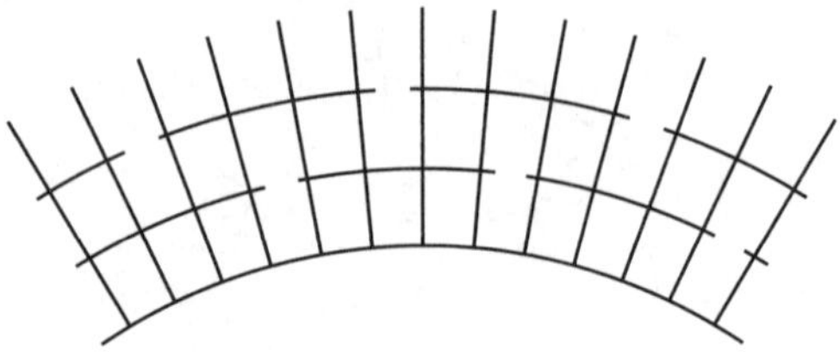

Figure 7.37 Staggered packets with multiple wires.

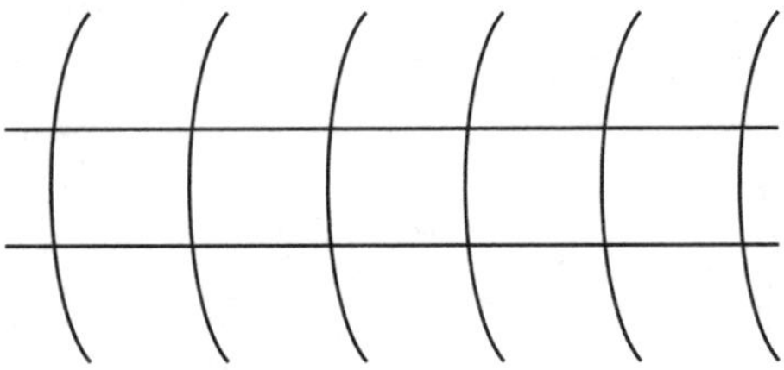

Figure 7.38 Multiple wires in a parallel arrangement.

Higher-order modes have low amplitude of vibration at the higher frequencies. Lower-order modes, such as first tangential, first axial, and first torsional modes, should have highest vibration at the tip of the blade. Placing wire near the tip will impose high centrifugal load in the hole due to the lacing wire. The usual locations seem to be from 67 percent to 87 percent of the height of the blade.

If stiffness is not required to raise frequency, the loose wire type of lacing wire is reported to reduce vibratory stress about 10 times (Arkadyev, 1968; Yampalskaya and Arkadyev, 1965; Hager et al., 1965). Loose wire construction seems not to have considerable effect on the natural frequency, and a brazed type of construction may change the frequency. Designs with loose wires reduce the area of the blade near the hole, and the hole also introduces stress concentration.

The wire should be designed (1) to have low weight and (2) to reduce aerodynamic losses. Larger-size wire is used when change in frequency is desired, but care must be taken about stresses near the hole due to the centrifugal load. The wire is subjected to centrifugal stress and bending stress during vibratory motion. Some constructions have employed increasing thickness of the blade near the hole to reduce stress near the hole, which has a negative impact on performance. The first choice of material for the wire is that of the blade. Less-dense material for the wire has also been used. Dissimilar materials should be handled carefully to avoid corrosion in the wet stage.

7.8.1 Determination of Effects of Number of Blades in a Packet

When blades are grouped in packets through lacing wire in either a loose or a rigid connection, a large reduction in vibratory stress may be achieved compared to a single blade without wire. Arkadyev (1968) provided the following equation to estimate a reduction factor μ_R, and the target is to reduce this factor to zero if possible.

$$\mu_R = \sigma \text{ without wire}/\sigma \text{ with wire}$$

The final expression for μ_R is

$$\mu_R = \text{Abs}\{\sin(\pi Km/Z)/[m.\ \sin(\pi K/Z)]\} \tag{7.11}$$

where K = harmonic of speed
m = number of blades in a packet
Z = total number of blades in the stage

Reduction factor μ_R is listed for various harmonics of speed in Table 7.4.

Figure 7.39 shows the effect of a packet of blades with lacing wire due to running harmonics on the response of the blade.

7.8.2 Quick Check for Requirement of a Lacing Wire Construction

The first step in the construction process should be to check if a lacing wire is required. If lacing wire is required, then the number of blades in a packet should be determined for an acceptable design. The stress

	Reduction Factor, μ_R			
m (Blades in a Packet)	$K=1$	$K=2$	$K=3$	$K=4$
36	3.90472E-17	3.91964E-17	3.94468E-17	3.9801E-17
18	0.637428514	3.91964E-17	0.214650184	3.9801E-17
12	0.828043929	0.415603461	3.94468E-17	0.211007407
9	0.901460049	0.639863387	0.303561201	3.9801E-17
4	0.981060262	0.925416578	0.836516304	0.71984631
3	0.989871835	0.959795081	0.910683603	0.844029629
2	0.996194698	0.984807753	0.965925826	0.939692621
1	1	1	1	1

Table 7.4 Reduction Factor vs. Number of Blades in Packet for First through Fourth Harmonics of Running Speed

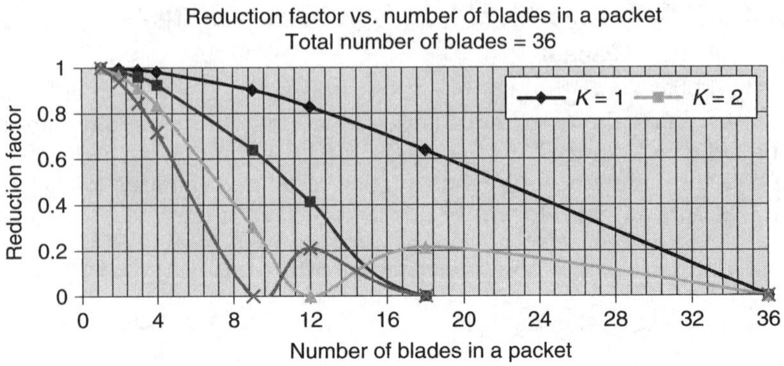

FIGURE 7.39 Reduction factor vs. number of blades in a packet.

due to rotation, bending stress imposed by steam forces, and vibratory stress due to running speed excitation should be evaluated. A Goodman-type analysis may indicate if the damping mechanism (lacing wire) is desirable, and to demonstrate the process, an estimation procedure is outlined next. However, a more accurate analysis should be performed for a final decision with modern tools, such as a finite element analysis (FEA).

Example

Height of blade	11.00 in
Number of blades in stage	36
Turbine speed	5000 rpm
Blade natural frequency ω_n	465 Hz
Material's ultimate strength σ_{ult}	110,000 psi
Material's fatigue strength $\sigma_{fatigue}$	55,000 psi

Estimated stress at the base of blade:

Tensile stress due to centrifugal load σ_T	35,000 psi
Bending stress due to steam forces σ_b	2500 psi

The blade's natural frequency (465 Hz) lies between the third (399 Hz) and the fourth (532 Hz) harmonics of running speed. The magnification factor for running speed excitation may be estimated by the approximate formula discussed earlier.

The magnification factor due to running speed excitation, $\mu' = \beta\,(\text{rpm})/\omega_n$.

$$\beta = 1.0 \quad \text{worst case}$$

$$\mu' = 1.0 \times 5000/465 = 10.75$$

$$\sigma_v = \mu'\sigma_b = 10.75 \times 2500 \text{ psi} = 26{,}882 \text{ psi}$$

The factor of safety using Goodman's equation is

$$1/\text{FS} = \sigma_{\text{static}}/\sigma_{\text{ult}} + \sigma_v/\sigma_{\text{fatigue}}$$

$$\sigma_{\text{static}} = \sigma_T + \sigma_b = (35{,}000 + 2500) \text{ psi} = 37{,}500 \text{ psi}$$

$$1/\text{FS} = 37{,}500/110{,}000 + 26{,}882/55{,}000 = 0.341 + 0.489 = 0.830$$

$$\text{FS} = 1.20$$

If the magnitude of FS is less than the acceptable limit, then a damping arrangement, such as lacing wire, may be required. If we assume that a lacing wire type of damping device is needed, then the next step is to decide what packet arrangement may be more beneficial with the help of Fig. 7.39 or Table 7.4. Figure 7.39 shows that a continuous single wire is most effective, but it poses difficulty in manufacturing. Therefore, more often a packet-type arrangement is employed. As mentioned earlier, the natural frequency lies between the third and the fourth harmonics of running speed excitation, and it appears that 12 blades in a packet should be a good choice. The reduction factor will range from about 0 to 0.21. For a design assume a reduction factor of 0.21; then an acceptable magnitude for vibratory stress for the design can be estimated.

$$\sigma_v = 0.21 \times 26{,}882 \text{ psi} = 5645 \text{ psi}$$

$$\text{FS} = 2.25$$

7.8.3 Sizing and Positioning of a Lacing Wire

Figure 7.40 shows portion of a lacing wire between two blades. Rotational force acts in radial direction on the wire.

For simplicity, it is assumed that the wire is rigidly fixed at two ends to the blades. An idealized view is shown in Fig. 7.41. This is done to show the various components of stress that may exist in the wire.

It was mentioned earlier that the primary use of lacing wire is to introduce active damping into the blade system. However, it also introduces stiffness that may change the natural frequency. This aspect should be taken into account during design. Another important concern is the reliability of the wire itself. The design of wire must be strong enough to resist stresses imposed by rotational forces. Stress in the wire depends on its distance from axis of the rotor and the size of its cross section. For a circular cross section, stress will be function of its diameter.

Equations of stresses are developed next. Various parameters required for the development of equations are listed below.

Number of blades	M
Diameter of lacing wire	d
Cross-sectional area	$A = \pi d^2/4$

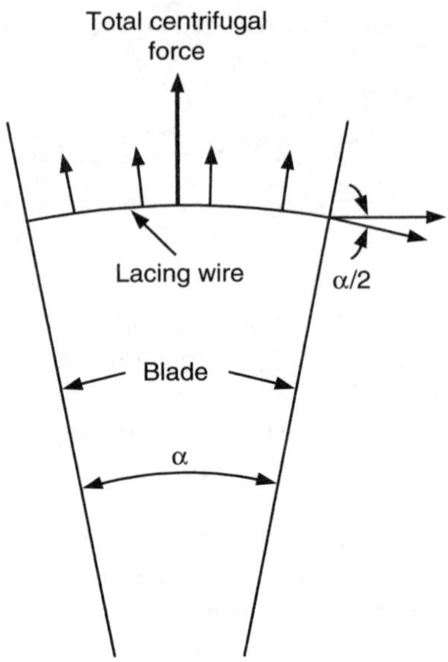

FIGURE 7.40 Single lacing wire between two blades.

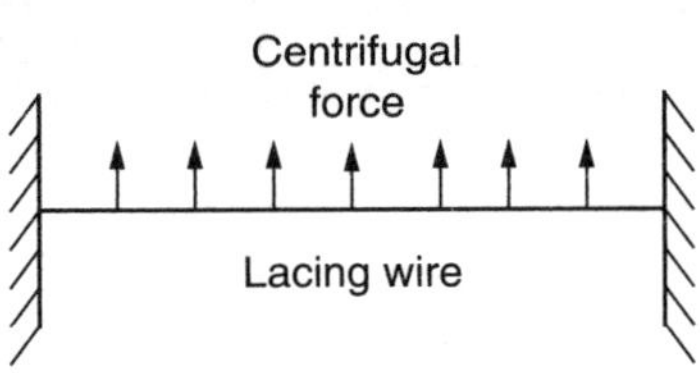

FIGURE 7.41 Fixed-fixed boundary conditions.

Radial location of wire	D
Unsupported length	$L = \pi D/M$
Angle of one blade pitch	$\alpha = 2\pi/M$
Section modulus	$Z = \pi d^3/32$
Turbine speed	N rpm
Centrifugal force	$\text{CF} = (AL\rho/g)(2\pi N/60)^2(D/2)$

The maximum bending stress in the wire may be estimated as follows:

If it is assumed that the ends of the wire are simply supported, then the maximum bending stress occurs at the middle of the span, and it may be evaluated by

$$\sigma_b = (\text{CF})L/(8Z) \tag{7.12}$$

For the fixed-end case, the maximum stress will occur at the support, and the expression for stress is given by

$$\sigma_b = (\mathrm{CF})L/(12Z) \qquad (7.13)$$

There will also be tension in the wire, and the tensile stress or hoop stress is given by

$$\sigma_T = (\mathrm{CF}/2)/[\sin(\alpha/2) \times \mathrm{A}] \qquad (7.14)$$

$$D = 45$$

$$N = 36$$

$$\rho = 0.285 \text{ lb/in}^3$$

$$\alpha = 2\pi/M = 0.175 \text{ rad}$$

$$L = 3.927 \text{ in}$$

$$\text{Turbine speed} = 3000 \text{ rpm}$$

Table 7.5 lists the results for bending stresses and hoop stress. The values of parameters are listed above. Stresses are evaluated by using Eqs. (7.12) through (7.14).

Bending and hoop stresses as a function of pin diameter are plotted in Fig. 7.42.

7.8.4 Check of Stress in the Hole in the Blade

$$\sigma_{\text{tension}} = (\mathrm{CF}_{\text{near hole}} + \mathrm{CF}_{\text{wire}})/\text{area of blade} = (22{,}700 + 710)/1.10 = 25{,}645 \text{ psi}$$

$$\sigma_{\text{bending}} = M/Z = 21 \times 2.0/.070 = 600 \text{ psi} \qquad \text{taking hole into account}$$

Consider the stress concentration for the hole:

$$K_T = 2.5 \qquad K_B = 2.0$$

$$\text{Peak tensile stress} = K_T\,\sigma_{\text{tension}} = 2.5 \times 25{,}645 \text{ psi} = 64{,}114 \text{ psi}$$

$$\text{Peak bending stress} = K_B\,\sigma_{\text{bending}} = 2.0 \times 600 \text{ psi} = 1200 \text{ psi}$$

$$\text{Notch sensitivity } q = (K_f - 1)/(K_B - 1)$$

Magnifier due to resonance = μ

Pin Dia.	Area	*Z*	CF	Sb at Center	Sb at End	St (Hoop)
0.1875	0.027612	0.000647	177.5997	134.712689	89.8084593	36.89979
0.25	0.049087	0.001534	315.7329	101.0345167	67.3563445	36.89979
0.3125	0.076699	0.002996	493.3326	80.82761339	53.8850756	36.89979
0.375	0.110447	0.005177	710.3989	67.35634449	44.9042297	36.89979
0.4375	0.15033	0.008221	966.9319	57.73400956	38.4893397	36.89979
0.5	0.19635	0.012272	1262.931	50.51725837	33.6781722	36.89979

TABLE 7.5 Stress in the Lacing Wire vs. Wire Diameter

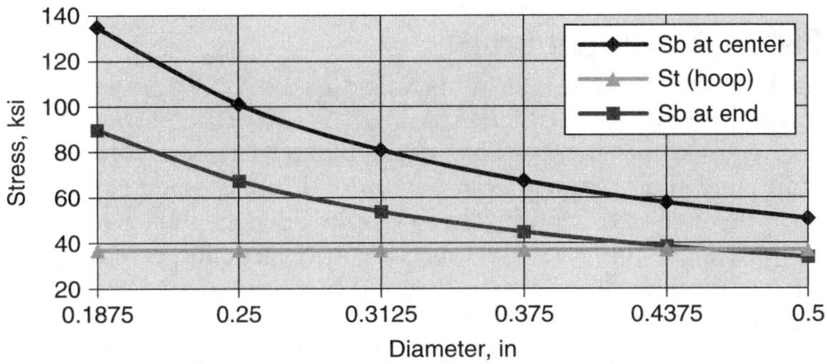

FIGURE 7.42 Stress in wire vs. wire diameter.

For $q = 0.5$ and $K_B = 2.0$, $K_f = 1.5$.

$$1/\text{FS} = \sigma_{\text{static}}/\sigma_{\text{ult}} + \sigma_v/\sigma_{\text{fatigue}}$$

$$1/\text{FS} = K_T\,\sigma_{\text{tension}}/110{,}000 + \mu K_B\,\sigma_{\text{bending}}/55{,}000$$

$$= 64{,}114/110{,}000 + (10.75 \times 1200)/55{,}000$$

$$= 0.583 + 0.235$$

$$\text{FS} = 1.23$$

Due to lacing there will be a reduction in the dynamic response by a factor of at least 0.21, which is an improvement.

$$1/\text{FS} = K_T\,\sigma_{\text{tension}}/110{,}000 + (\text{reduction factor})(\mu K_B\,\sigma_{\text{bending}})/55{,}000$$

$$= 64{,}114/110{,}000 + (0.21 \times 10.75 \times 1200)/55{,}000$$

$$= 0.583 + 0.049$$

$$\text{FS} = 1.58$$

In summary following steps need to be considered:

1. Estimate the stress in the single blade if a lacing wire type design is needed.
2. Determine the size and location, materials, clearance, and packeting, if required.
3. Hertzian contact stress in the hole imposed by the centrifugal load of the lacing wire should also be investigated. The contact stress should be within the acceptable limit, and this stress also implies fretting fatigue damage near the hole.
4. Fretting fatigue near the hole in the blade and any loss of performance must be carefully determined before a final decision is made.

7.9 Partial Admission Stage

A turbine stage in which steam is introduced in only a portion of the complete rotation of the blade is known as a *partial admission stage*. It is understood that the response of the blade to the steam forces will be different than where steam is admitted to the stage completely. This type of construction has been employed since the early days of turbine design. For low loads this type of construction produces high efficiency and for high temperature and high pressure has resulted in blade and shroud failures.

Many years have elapsed since the method described by Kroon (1940) and Allen (1940). Designs generally were conducted by increasing the magnifier to estimate vibratory stress. Pigott (1980) presented a comprehensive treatment in which the combined effect of shock load and nozzle wake was considered (Fig. 7.43). The vibratory response of the blade for these two effects was analyzed. Pigott considered a packeted bladed disk assembly, and the effect of the disk was not included in the analysis. By separating the effects of two types of load, Pigott demonstrated the dependence of blade response on the natural frequencies, running speed, number of nozzles, number of blades, number of blades in a packet, and damping. The response can be simulated on a computer with the help of the detailed equations provided.

Figure 7.43 depicts a partial admission force profile experienced by a blade in one revolution, which includes the shock load effect superimposed with the force profile due to each nozzle opening. Figure 7.44 is a simplified depiction of the force profile due to the

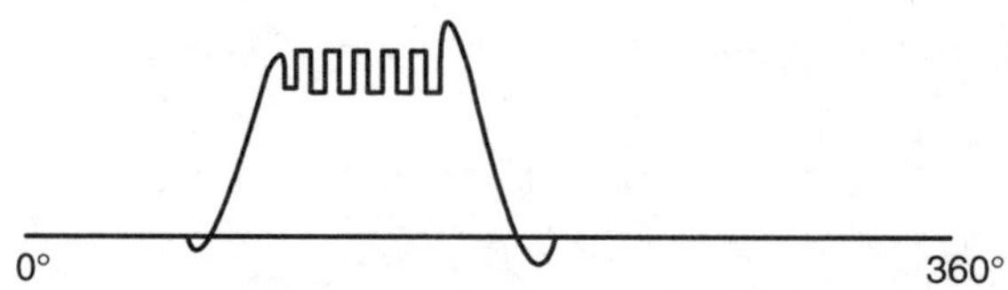

FIGURE 7.43 General forcing for partial admission includes nozzle wake force.

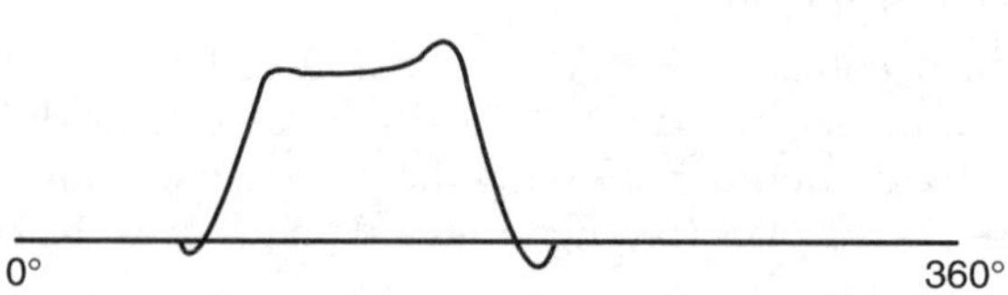

FIGURE 7.44 General forcing for partial admission only.

FIGURE 7.45 General forcing for nozzle wakes only.

partial admission while Fig. 7.45 shows the forces felt by a blade due to nozzle openings. The response or stress of a blade can be thought of as comprising of two components, one due to partial admission and the other due to nozzle wake.

$$\sigma_{\text{total}} = \sigma_{\text{nozzle wake}} + \sigma_{\text{partial admission}} \tag{7.15}$$

The response terms on the right side of Eq. (7.15) may be estimated by multiplying the static stress by magnifiers suitable to these two situations

$$\sigma_{\text{total}} = \mu_1(\sigma_{\text{steady}}) + \mu_2(\sigma_{\text{steady}}) \tag{7.16}$$

where μ_1 and μ_2 are magnification factors for nozzle wakes and partial admission, respectively.

The method described by Kroon is very instructive and is discussed here. His method to examine the response behavior of the blade assumes that any blade mode can be analyzed as a single spring-mass-damper system. He used simplified assumptions about the nature of loading to assess the partial admission effects on the blade stress as a function of the blade's natural frequency, turbine speed, and damping. In design, it is important to estimate the maximum amplitude or magnifier during the vibration due to the steam force. A simplified blade load pattern is shown in Fig. 7.46. In each revolution, the rotor blade will go through successive

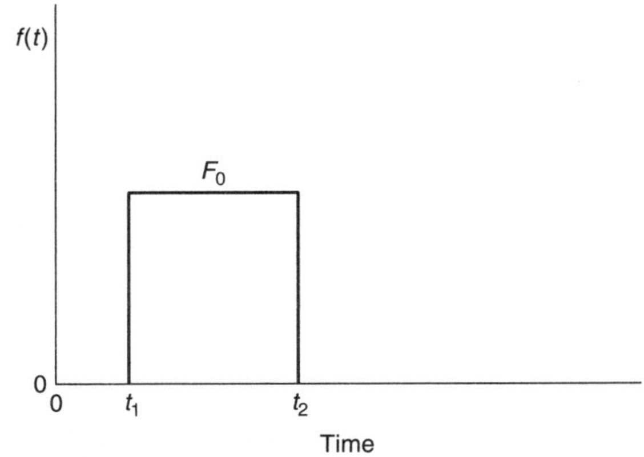

FIGURE 7.46 Simplified load pattern.

instantaneous loading and unloading. Maximum motion will occur when the load is applied when the blade is in its extreme backward position, and the load is removed when the blade is in its extreme forward position. Due to the complex nature of the equation for response, the expression for the magnifier was approximated in successive steps ending in an expression for low damping. The three main equations are Eqs. (4.17) through (4.19). Kroon assumed that n is the number of cycles of vibration, n_2 is the loaded portion, n_1 is the cycle of vibration in the unloaded portion, y_{osc} is the amplitude of vibration, y_{st} is the amplitude due to a steady load, and δ is the logarithmic damping.

$$n = 60\omega/N$$

where ω is the frequency and N is the rotor speed, rpm.

$$y_{osc} = y_{st}\{0.5 + (2 + e^{-n_1\delta} + e^{-n_2\delta})/[2(1 - e^{-n\delta})\, e^{\delta/2}]\}$$

or

$$\mu_1 = y_{osc}/y_{st} = 0.5 + (2 + e^{-n_1\delta} + e^{-n_2\delta})/[2(1 - e^{-n\delta})\, e^{\delta/2}] \qquad (7.17)$$

After expanding the exponential and using the relationship $n_1 + n_2 = n$, Eq. (7.17) reduces to

$$y_{osc} = y_{st}\{0.5 + (2 - n\delta/2)/[n\delta(1 + \delta/2)]\}$$

$$\mu_2 = y_{osc}/y_{st} = 0.5 + (2 - n\delta/2)/[n\delta(1 + \delta/2)] \qquad (7.18)$$

For low damping, terms containing $\delta/2$ and $n\delta/2$ can be neglected, resulting in

$$y_{osc} = y_{st}[2/(n\delta)]$$

$$\mu_3 = y_{osc}/y_{st} = 2/(n\delta) \qquad (7.19)$$

Three expressions of the magnifier in Eqs. (7.17) through (7.19) are plotted in Figs. 7.47 and 7.48. Figure 7.47 is the plot of the magnifiers as a function of turbine speed while Fig. 7.48 shows the magnifier as a function of frequency. It is worthwhile to note that magnifiers μ_2 and μ_3 are almost equal.

The accurate magnitude of the magnifier is represented by Eq. (7.17). The magnitude of magnifier μ_1 is plotted as a function of turbine speed and frequency in Figs. 7.49 and 7.50, respectively.

Kroon developed an expression by using a simplified load profile of Fig. 7.51 for a *severity factor* given below.

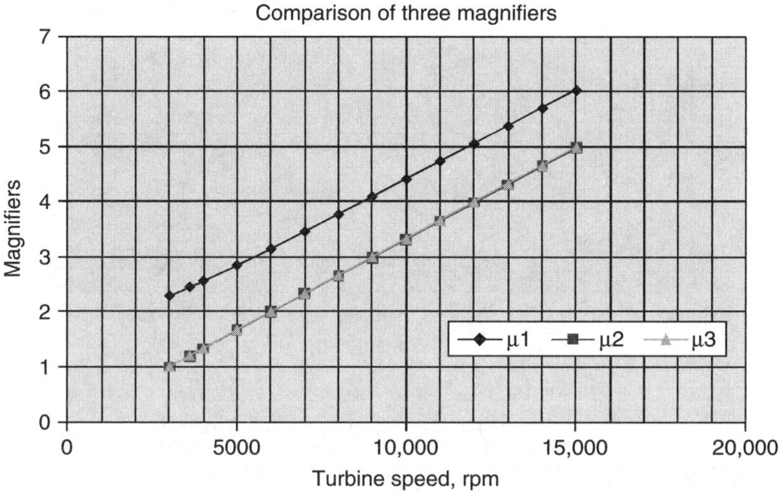

Figure 7.47 Comparison of magnifiers vs. turbine speed.

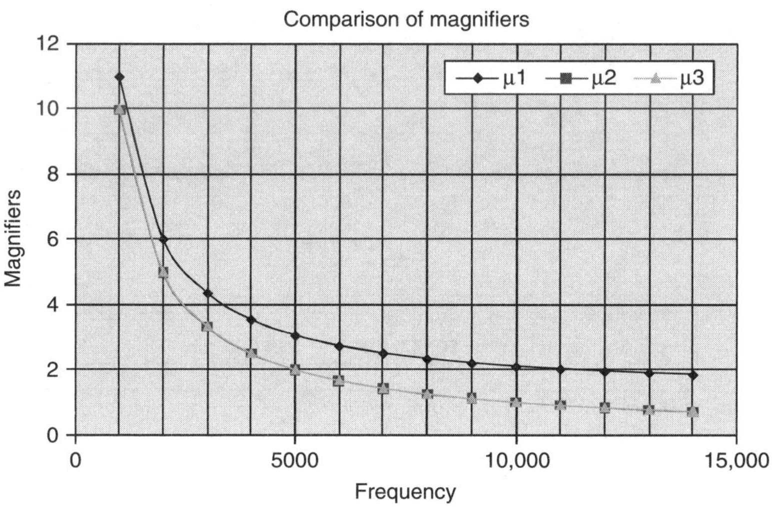

Figure 7.48 Comparison of magnifiers vs. frequency.

The load as shown in Fig. 7.51 is applied to a spring-mass system.

The equation of motion is written when P_0 is increased gradually from $t = 0$ to $t = \tau$. The magnitude of τ will determine how slow or fast the load is applied.

$$m(d^2y/dt^2) + ky = P_0(t/\tau) \tag{7.20}$$

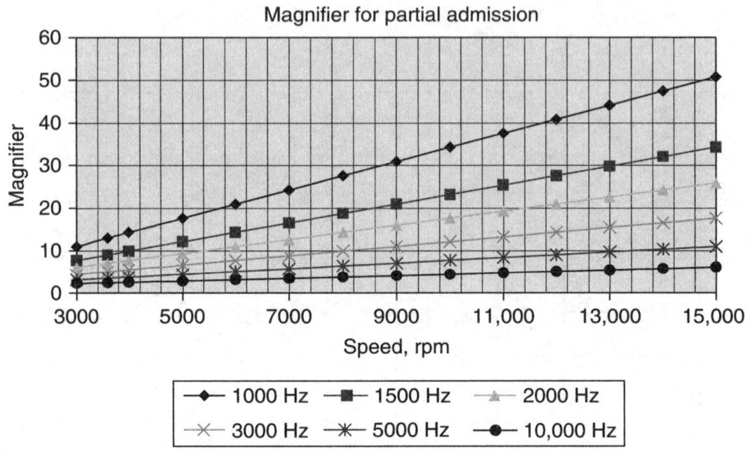

FIGURE 7.49 Magnifier μ_1 vs. turbine speed.

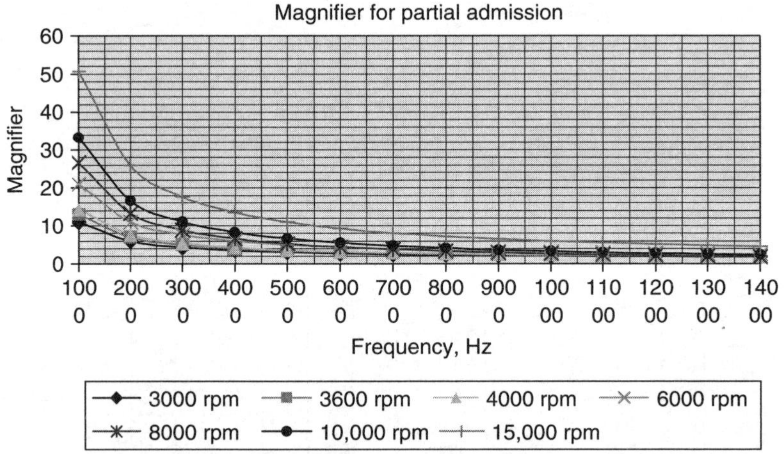

FIGURE 7.50 Magnifier μ_1 vs. natural frequency.

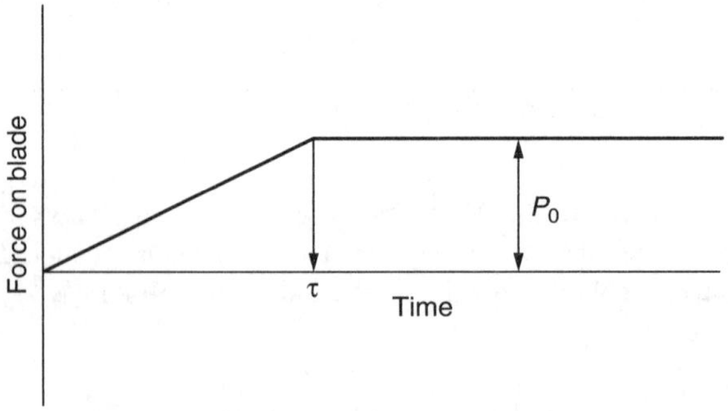

FIGURE 7.51 Simplified loading profile.

where m and k are the mass and stiffness, respectively.

$$\text{Natural frequency } \omega = (k/m)^{1/2}$$

After solving the equation, Kroon defined the severity factor η as follows:

$$\eta = (T/\pi\tau)\sin(\pi\tau/T) \tag{7.21}$$

where T is the period of free undamped vibration $= 2\pi/\omega$.

The plot of η vs. steam force loading time τ is given in Fig. 7.52.

The large motions are predicted when the number of vibrations per revolution is not an integer (Fig. 7.52). Another important observation is that the lowest motion can be expected when the loading time τ is an integer multiple of the vibration period T of the blade.

Another similar simplified method of analysis may use the concept developed for analyzing the vibration characteristics of a spring-mass-damper (SMD) system. The assumption is that the response of any natural mode of a blade can be expressed and analyzed as an SMD system. Generalized mass, generalized stiffness, and modal damping of any mode of the blade may be used as mass, spring, and damping for an equivalent SMD system.

After the turbine is started, the blade outside the partial admission zone will respond as a free damped system. At the time it enters the partial admission zone, the initial displacement and velocity are the displacement and velocity experienced at the end of the free vibration case. At the time the blade exits the partial admission zone,

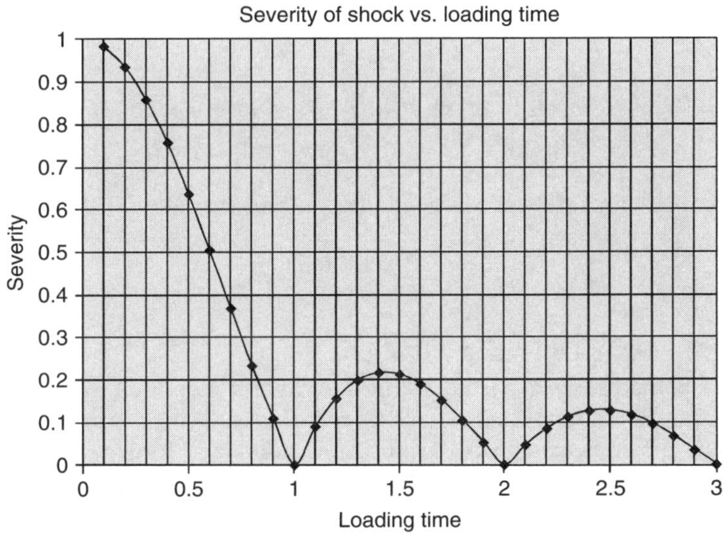

FIGURE 7.52 Severity factor vs. loading time.

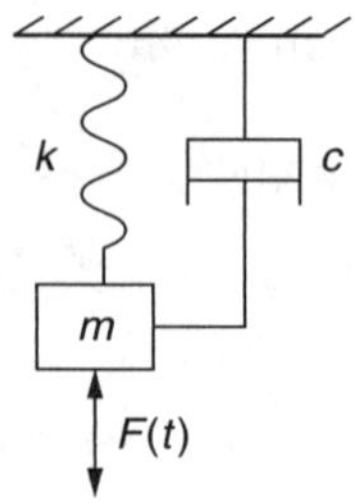

Figure 7.53 Spring-mass-damper system.

the displacement and velocity will become the initial conditions for free vibration analysis.

Consider a simple SMD system as in Fig. 7.53 representing a blade mode with generalized mass, generalized stiffness, and modal damping.

The equation of motion is expressed as

$$ma + cv + kx = F(t) \tag{7.22}$$

7.9.1 Damped Free Vibration

$$F(t) = 0 \tag{7.23}$$

The solution of the equation is

$$x(t) = e^{-\zeta\omega_n t}\{[v(0) + x(0)]\sin \omega_d t/\omega_d + x(0) \cos \omega_d t\} \tag{7.24}$$

and

$$\begin{aligned} v(t) &= \zeta\omega_n e^{-\zeta\omega_n t}\{[v(0) + \zeta\omega_n x(0)] \sin \omega_d t/\omega_d + x(0) \cos \omega_d t\} \\ &\quad + e^{-\zeta\omega_n t}\{[v(0) + \zeta\omega_n x(0)] \cos \omega_d t - x(0) \sin \omega_d t\} \end{aligned} \tag{7.25}$$

where $v(0)$ and $x(0)$ are initial conditions and $\omega_d = (1 - \zeta^2)^{1/2}\omega_n$.

$$\zeta = C/C_c = C/2m\omega_n \tag{7.26}$$

7.9.2 Damped Forced Vibration

As a blade enters a partial admission area, it responds to the force that is suddenly applied. The force remains constant during the partial arc; then it exists with no force. An idealized force is shown in Fig. 7.54.

$$F(t) = F_0$$

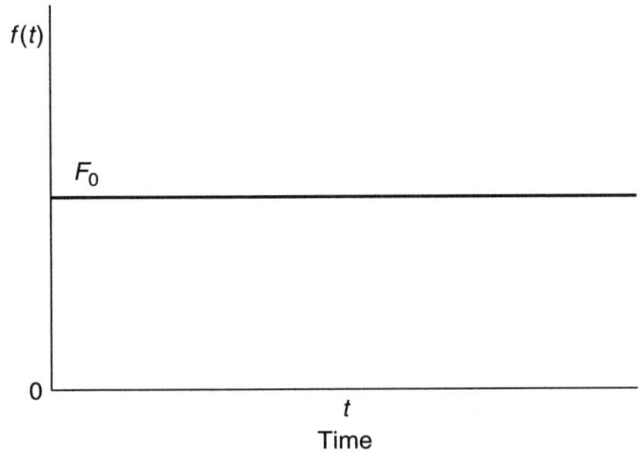

FIGURE 7.54 Idealized step force.

The solution of the differential equation for this case is given below.

$$x(t) = Xe^{-\zeta\omega_n t}(\sin \omega_d t - \phi) + F_0/m\omega_n^2 \quad (7.27)$$

$$v(t) = X(-\zeta\omega_n)e^{-\zeta\omega_n t}(\sin \omega_d t - \phi) - Xe^{-\zeta\omega_n t}(\cos \omega_d t - \phi)\omega_d \quad (7.28)$$

Using the given initial conditions [$x(0) = X_0$ and $v(0) = v_0$], $x(t)$ and $v(t)$ can be solved, and the solution is given below.

$$\phi = \tan^{-1}[\omega_d/(\zeta\omega_n - v_0/F_1)] \quad (7.29)$$

where

$$F_1 = F_0/m\omega_n^2 - X_0 \quad (7.30)$$

$$x(t) = e^{-\zeta\omega_n t}F_1(\sin \omega_d t/\tan\phi - \cos \omega_d t) + F_0/m\omega_n^2 \quad (7.31)$$

and

$$v(t) = -e^{-\zeta\omega_n t}F_1[\zeta\omega_n(\sin \omega_d t/\tan\phi - \cos \omega_d t) + \omega_d(\cos \omega_d t/\tan\phi + \sin \omega_d t)] \quad (7.32)$$

Now that the equations for the basic scenario have been developed, the response of a blade under a partial admission load can be analyzed. Figure 7.55 shows an idealized partial admission situation. Assume that a blade starts rotating at time zero when the turbine is started. A blade will enter the arc of the admission at time t_1 and then exit at time t_2. After this the applied force is zero, and the blade goes through a free vibration mode until it enters the partial admission again during the second revolution. This process will repeat upon every rotation of the disk.

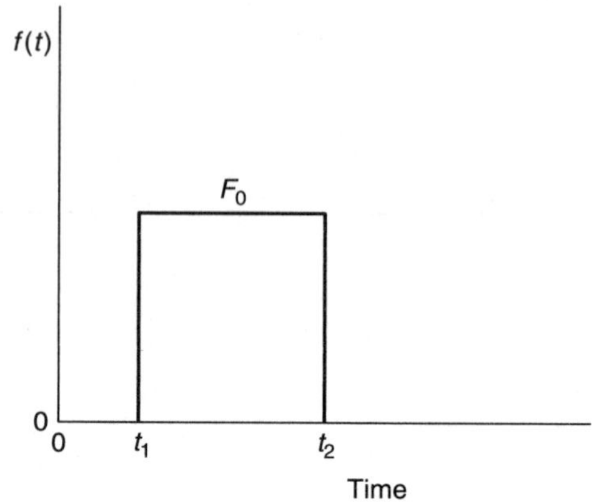

FIGURE 7.55 Idealized one partial admission zone.

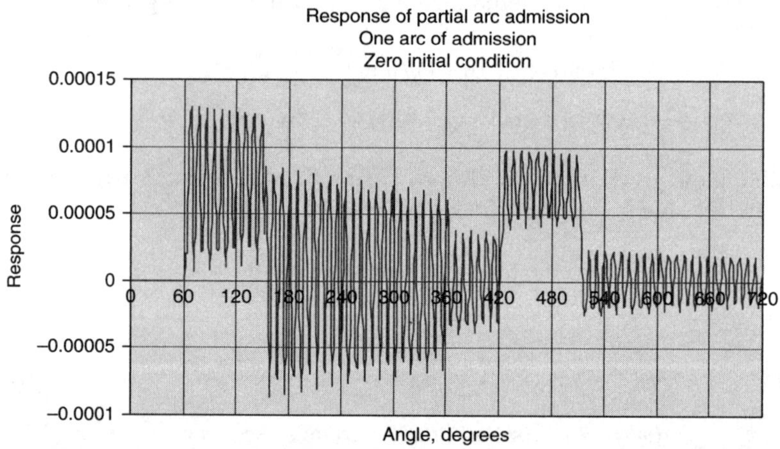

FIGURE 7.56 Variation in displacement for one port vs. angular position of rotation.

Displacements as a function of angular rotational position for two complete rotations are plotted in Fig. 7.56. Displacement before entering the partial arc is assumed to be zero together with the initial velocity. Figure 7.57 shows the variation in velocity as a function of angular position.

Figure 7.58 shows a case of two arcs of admission, and the free vibrating blade enters the first zone at time t_1 and exits at time t_2. It goes through free vibration when at time t_3 it enters the second zone and finally exits at time t_4. The same cycles will be repeated in the next revolution.

Figure 7.58 shows an idealized two parts of admission.

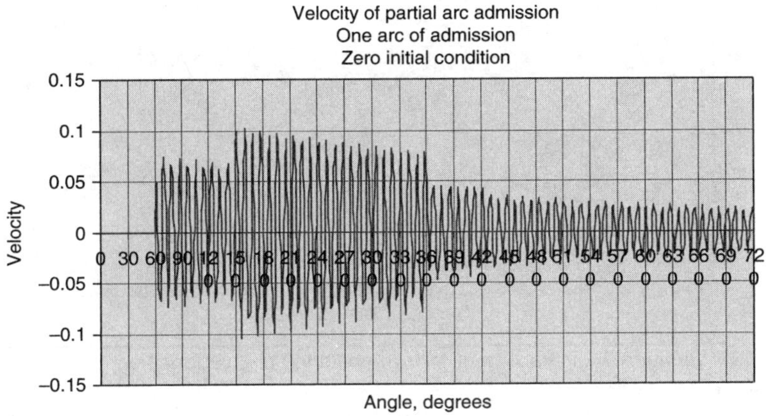

Figure 7.57 Variation in velocity for one port vs. angular position of rotation.

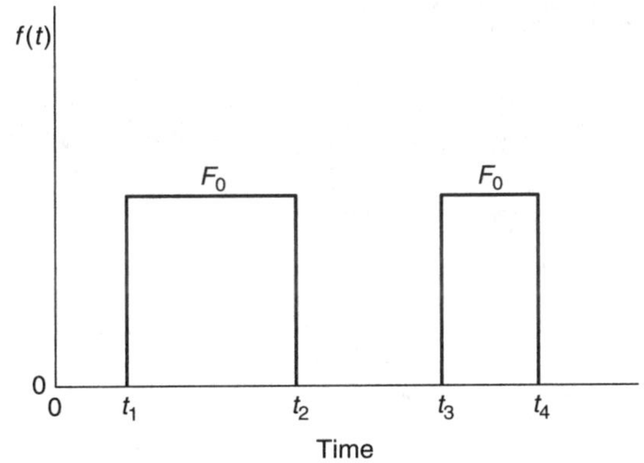

Figure 7.58 Idealized two partial admission zones.

Figure 7.59 shows the displacement as a function of angular position of blade during rotation. It shows the results of the analysis for three complete rotations.

The variation in the velocity as a function of angular position during rotation is shown in Fig. 7.60. It shows the variations in three complete rotations.

The expanded view of response for the first rotation is shown in Fig. 7.61.

The expanded views of response for the second and the third rotational cycles are given in Figs. 7.62 and 7.63, respectively.

Damped forced vibration equations are used to estimate the displacement and velocity as a blade enters and leaves the arc of admission. These are shown graphically in Figs. 7.59 through 7.63. Note,

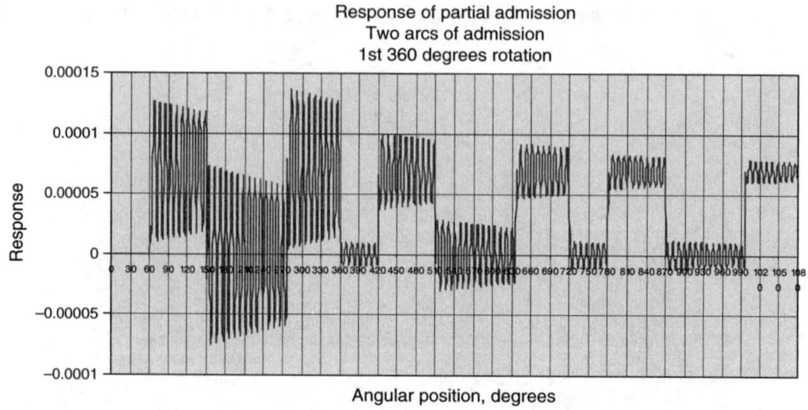

FIGURE 7.59 Variation in displacement for two ports vs. angular position in three rotations.

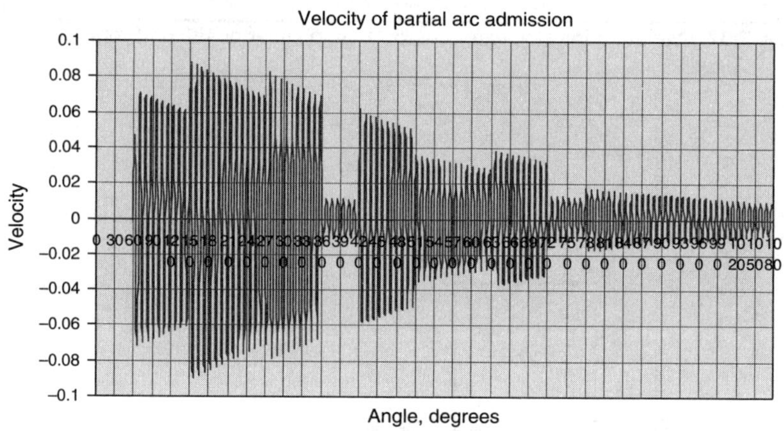

FIGURE 7.60 Variation in velocity for two ports vs. angular position in three rotations.

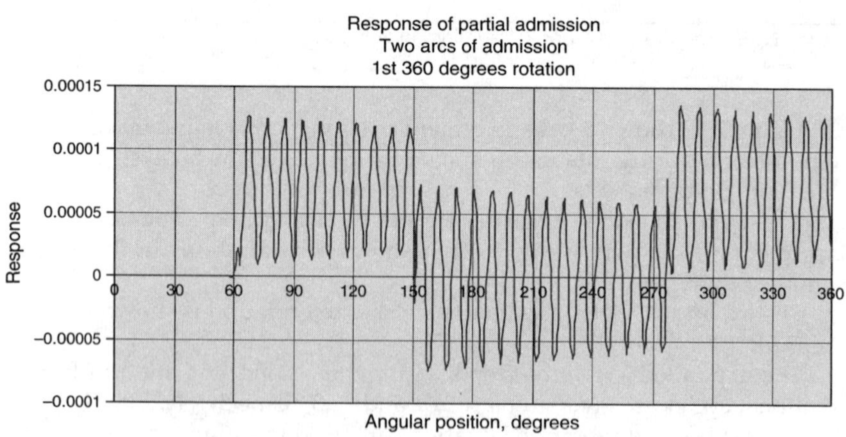

FIGURE 7.61 Variation in displacement for two ports vs. angular position in first rotation.

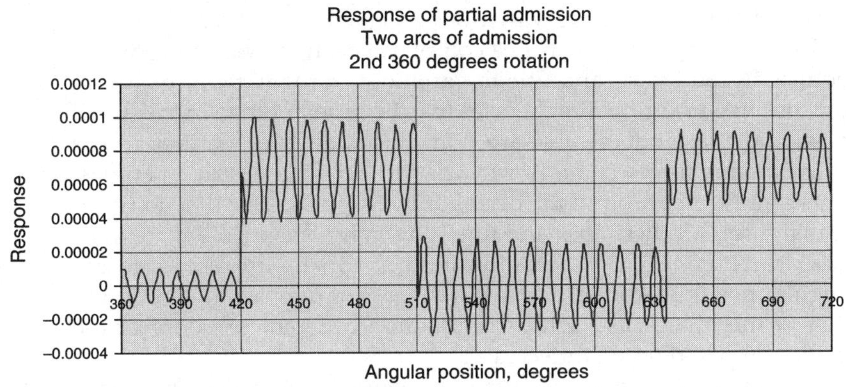

Figure 7.62 Variation in displacement for two ports vs. angular position in second rotation.

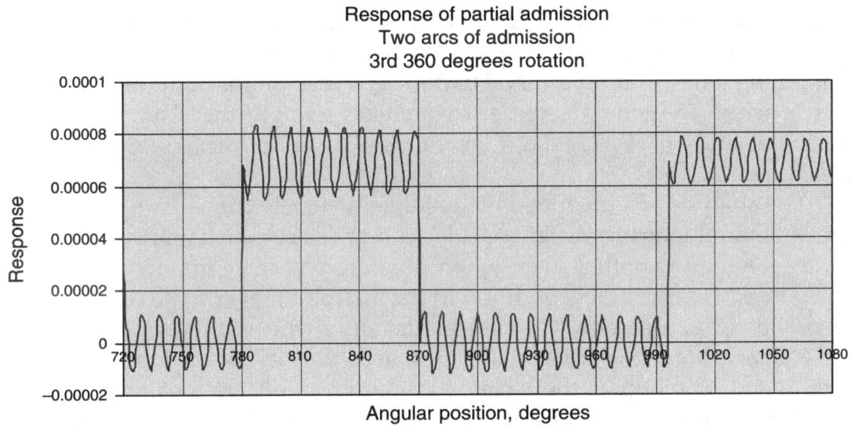

Figure 7.63 Variation in displacement for two ports vs. angular position in third rotation.

however, that sometimes the amplitude of vibration might not go to maximum value. This may be due to the fact that the blade might not have entered the loaded portion of the arc in its extreme backward position, and it might not have left the loaded portion in its extreme forward position during vibration. A careful matching of those conditions may show the maximum amplitude.

7.10 Effect of Mistuning of a Bladed Disk System on Vibration Response

Mistuning is any irregularity in the bladed disk system that makes the response of the system asymmetric. These irregularities can be caused by structural and/or aerodynamic means. Structural irregularities arise due to variations in geometric dimensions from blade to blade; they can also be complicated due to variations in the boundary conditions and due to variations in the fixity of blades to the disk or shrouding. Frequencies may be different than in the tuned case, and the mode shapes are nonsinusoidal. Irregularities in the aerodynamic forces come from variation between nozzle openings. This causes force around the periphery of the disk to be of a nonsinusoidal form. Mode shapes and forces though nonsinusoidal are periodic in nature. Fourier decomposition of these with respect to the angular position around the disk will yield harmonics in addition to that of the tuned case, which results in a different response of the blade compared to the tuned case.

Structural mistuning results in a phenomenon called *splitting of frequency*. For a tuned system, frequencies occur in duplicate; i.e., there are two identical frequencies for a given mode shape. These shapes are graphically the same, but differ from each other by a phase angle. Splitting separates these frequencies for each mode, and the amount of separation of frequencies depends on the amount of mistuning. For example, a 3-ND pattern of a disk might occur at two frequencies. In general, these frequencies crowd the Campbell diagram, and the SAFE diagram yields a better depiction of this phenomenon.

The analysis of a mistuned bladed disk system becomes complex. Due to manufacturing variations, blades on the same disk will be different from one another even when they are within manufacturing tolerances. Their contact condition in the disk slot may also be different. Therefore, assembly of these blades on a disk will be random unless each blade is individually measured and its location on the disk noted. Sometimes the location is chosen in order to balance a disk containing long blades. However, even known locations pose difficulty in analysis. Each bladed disk produced in such a manner from the same lot of blades will be different, and therefore, their vibration characteristics will also be different. The complexity of the analysis is compounded by the presence of a large number of blades on the disk.

Ewins (1969) reported that for given excitation and damping conditions, the response in one mode (or pair of modes) depends on the "split" of the two natural frequencies. The presence of slight mistuning, small blade imperfections, can cause resonant stress levels up to 20 percent above the optimum. This kind of increase depends on the specific arrangement of the blades around the disk. Ewins (1976) reported the results of a study of a slightly mistuned bladed disk. The study considered a group of nominally identical blades. Various dimensional variations in blades nominally result in about ±2 to 3 percent variation in a 1-F frequency. Even though there is a slight change in frequency, the result showed that the effect on mode shapes is large. This phenomenon of variation in the mode shape has considerable effect on the response of the blade. The main interest in the design of a blade is its response under load; hence mode shape should be critically examined. Singh and Ewins (1988) and Singh (1992) presented probabilistic analysis results by considering a random arrangement of blades on the disk, utilizing Monte Carlo simulation to arrive at the results.

The analysis reported earlier for the packeted bladed disk may be considered as deliberate mistuning. This was achieved by breaking a continuous shroud band into packets at some predetermined locations, which permits known modes to be split, and it has been used successfully in packeted bladed disk design.

For a mistuned bladed disk system, the vibration energy is unevenly distributed among all blades of the disk. The mode shapes may be far from a regular sinusoidal shape. The examination of modes in such a situation is done using Fourier decomposition of the displacement pattern in harmonic content. The mode is traditionally named in a SAFE diagram in accordance with the largest content of the harmonics of the Fourier decomposition.

Analysis of a mistuned system is a complicated process, but it may have considerable effect on the dynamic response of the blade. The papers listed in the Bibliography should provide greater insight into this phenomenon and analytical methods for interested readers.

7.11 Impure Mode Shapes (Packeted Bladed Disk)

The continuous shrouded bladed disk and bladed disk without shroud are only a tuned system if each blade has the same form and dimension having the same fixity in the disk. Due to the dimensional variations from blade to blade, these identical frequencies for a mode split into two frequencies for the same mode. The shape of the mode may contain other than primary harmonics. This phenomenon will be examined for a packeted disk assembly. It was mentioned in Sec. 7.10 that this is a case of deliberate mistuning; i.e., mistuning is introduced by breaking the shroud at a specified interval. This breaks the geometric symmetry of the bladed disk. An example is presented to demonstrate the implication of irregularities in the mode shape due to packeting and asymmetry of the aerodynamic forces. An assumption is made that each blade is identical so that no mistuning is introduced due to a variation in the blades.

A finite element model of a bladed disk with 60 blades is shown in Fig. 7.64 and includes 10 packets with 6 blades in each packet.

Three different modal analyses were performed to estimate natural frequencies and associated mode shapes. For the first analysis, a disk with a single continuous shroud band, i.e., one packet, was considered. The second analysis was conducted for an individual bladed disk with individual shroud on the tip of each blade. Essentially, this

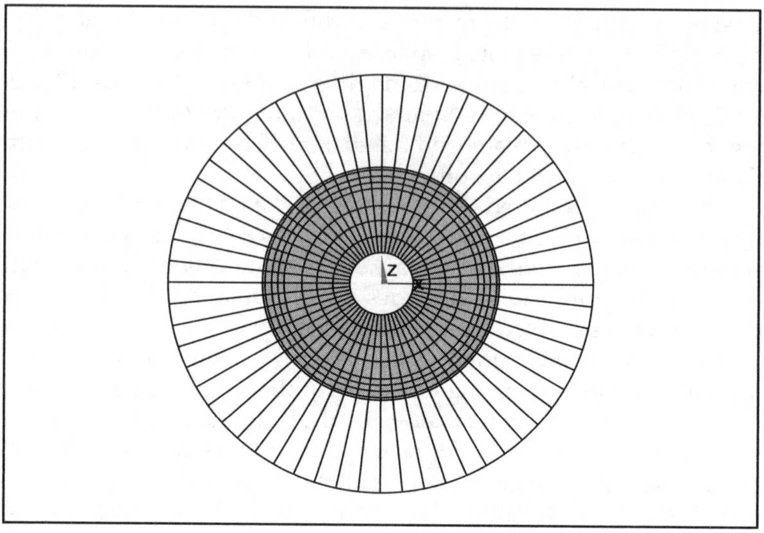

Figure 7.64 Model of a packeted bladed disk, 60 blades, 10 packets.

is the case of 60 packets with only 1 blade in a packet. Finally, the third analysis was performed for 10 packets with 6 blades in a packet.

Figures 7.65 through 7.67 display plots of 3-ND mode shapes for these three cases. The mode shapes for continuous shroud and for freestanding blades show a regular sinusoidal form. The mode shape for the 10-packet case looks like a sinusoidal form, but it has some irregularities. It is expected that a Fourier analysis of this form will yield the third harmonic as a major contributor, but it will also have some other harmonics. It has been observed in such cases that low-order modes do not seem to deviate much from the primary shape; however, for high-order modes the deviation is considerable. High-order modes appear to have many harmonics other than the expected primary harmonic.

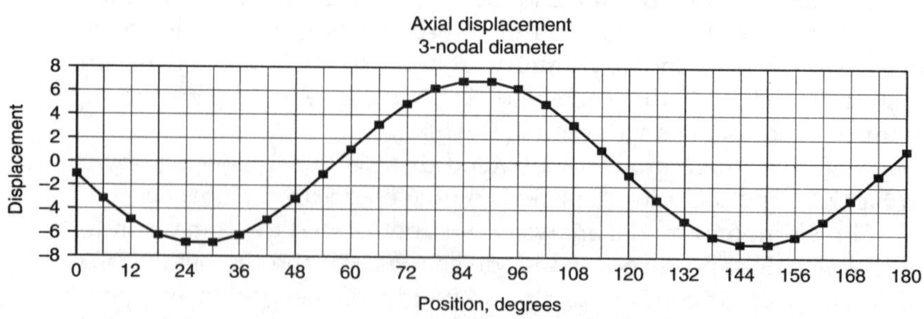

Figure 7.65 A 3-ND mode shape for single continuous shroud.

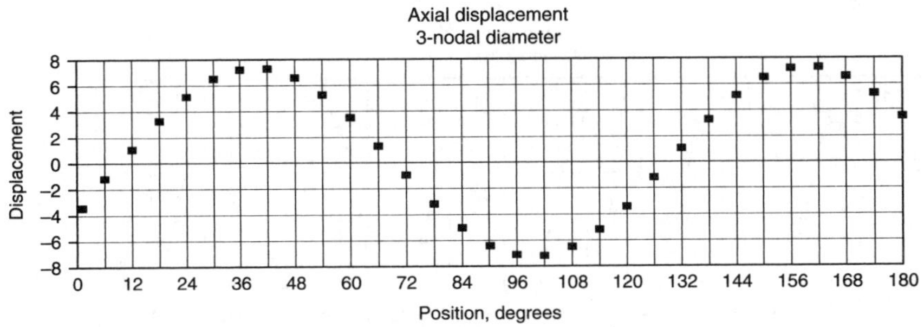

FIGURE 7.66 A 3-ND mode shape for blades with individual shrouds.

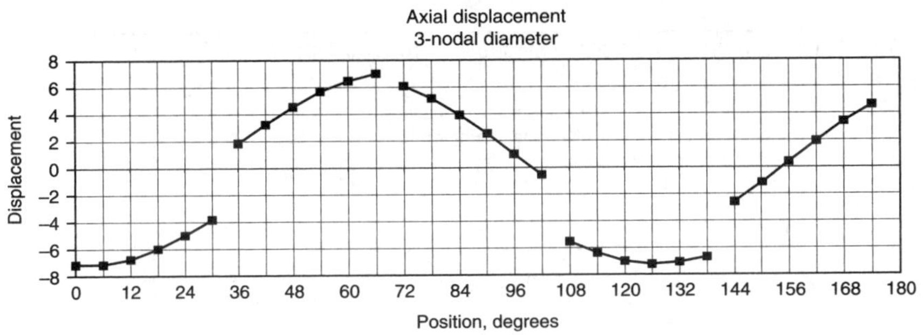

FIGURE 7.67 A 3-ND mode shape for 10 packets, 6 blades in a packet.

Mode shapes for 13 ND for these three cases are shown in Figs. 7.68, through 7.70. As expected, the forms for the continuous and individual blade cases have pure sinusoidal forms, but for the packeted case it is far from pure sinusoidal. It is expected, however, that the 13th harmonic will be predominant but there will be contribution from other harmonics.

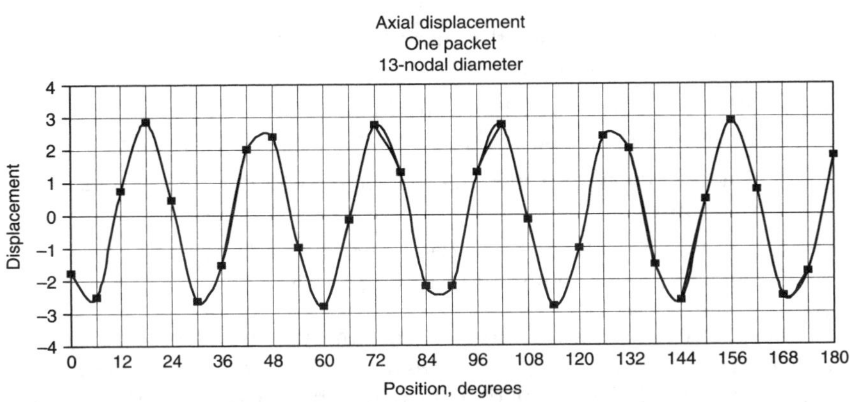

FIGURE 7.68 A 13-ND mode shape for single packets.

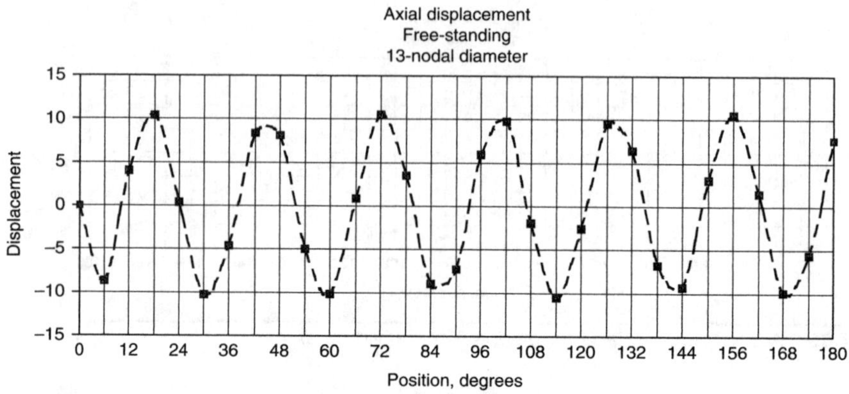

FIGURE 7.69 A 13-ND mode shape for individually shrouded blades.

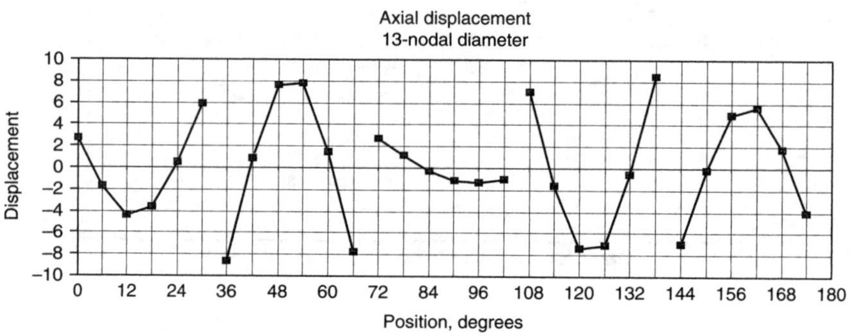

FIGURE 7.70 A 13-ND mode shape for 10 blade packets, 6 blades in a packet.

The results of these three cases have been plotted on the SAFE diagram in Fig. 7.71.

The following observations should be mentioned:

1. The frequencies for the completely shrouded blades are largest for every mode shape.
2. The frequencies for the individually shrouded blades are lowest for every mode shape.
3. The frequencies for the packeted blades fall in between these two limits.
4. The 5-ND, 10-ND, 15-ND, 20-ND and 25-ND modes for the packet case have split into two frequencies.

The 15-ND mode shape for the continuous case is plotted in Fig. 7.72. There are two identical modes at the same frequency and for an individually shrouded bladed disk, which will show the same form.

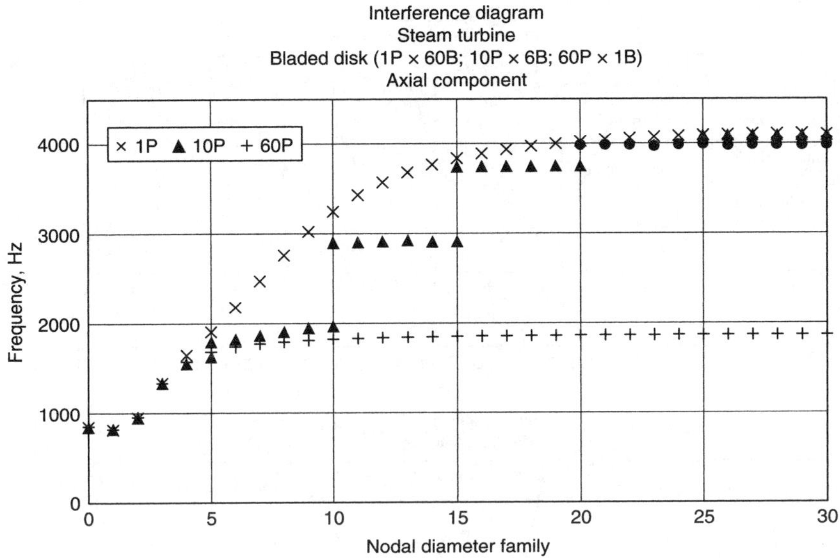

FIGURE 7.71 Interference diagram for single packet, individual blades, and 10 blade packets, 6 blades in a packet.

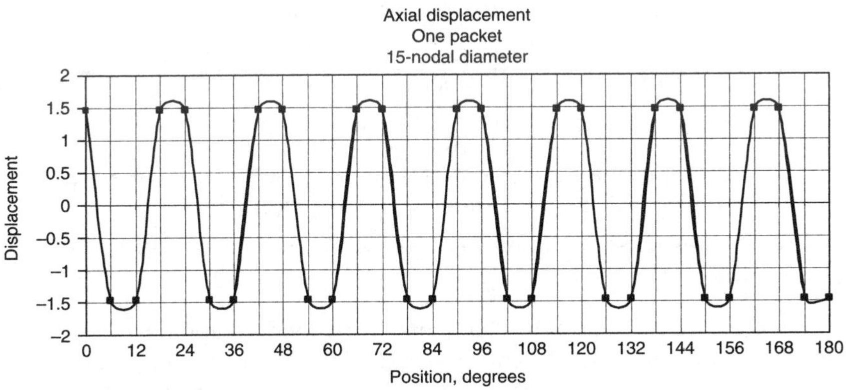

FIGURE 7.72 A 15-ND mode shape for single packets.

Figures 7.73 and 7.74 show two mode shapes of 15 ND for the packeted case.

The following observations are important to note:

1. The frequency has split in two, and the mode shapes are quite different.
2. The displacement patterns of the blades within a packet are quite different between these modes.

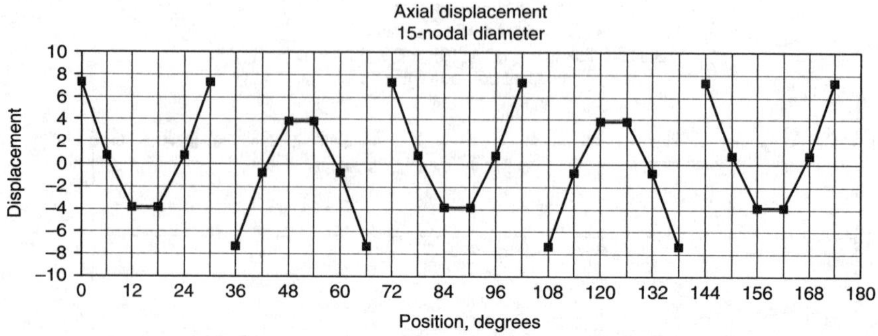

FIGURE 7.73 First split 15-ND mode shape for 10 blade packets, 6 blades in a packet.

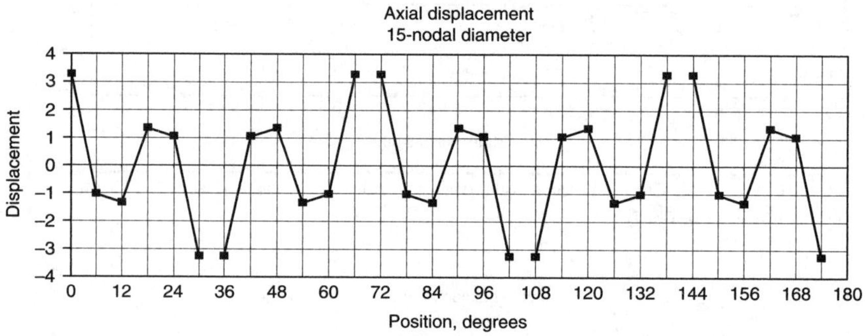

FIGURE 7.74 Second split 15-ND mode shape for 10 blade packets, 6 blades in a packet.

3. The first mode shows axial-U type of displacement while the second one displays axial-S type of displacement.
4. The response of blades in these two cases is expected to be different.

To determine the harmonic contents of any mode shape, a Fourier analysis utilizing the displacement values of the blade obtained from the modal analysis should be performed. It is a tedious task, but an approximate method that has been developed using spring-mass system by observing results of many simple analyses can be used. This method identifies harmonics that will participate in a particular mode for a given construction. However, it does not provide the actual relative contribution of each harmonic.

There are two methods to identify the harmonic contents; the first is graphical and the second gives a mathematical expression.

7.11.1 Graphical Method

A checkerboard type of graph is constructed. The horizontal axis is designated as harmonic content. If the number of blades on the disk is N, then the maximum designation is equal to $N/2$ if there are an even number of blades or $(N - 1)/2$ if there are an odd number of blades. The vertical axis is the mode shape to be considered, and it is designated as nodal diameter. The diagram can be constructed for any given number of packets n in the disk. The radial lines at ±45° are drawn at each integer multiple of n from the horizontal and vertical axes. Such a graph is drawn for the completely shrouded construction and for the individually shrouded bladed disk, which have the same drawing. Figure 7.75 is the drawing for a disk with 60 blades.

For the case of 10 packets the diagram is shown in Fig. 7.76, and the ±45° radial lines are drawn at integer multiples of 10.

Now let us consider a 13-ND mode shape for this construction. For the continuous or individually shrouded case from Fig. 7.75, one moves across horizontally at 13 ND, and there is only one crossing at the 13th harmonic. Hence, this mode will have only the 13th harmonic.

The same exercise is performed for the 13 ND for the 10 packets case using Fig. 7.76. The resulting harmonic contents are estimated to be 3d, 7th, 13th, 17th, 23d, and 27th harmonics. The 13th harmonic is expected to be the largest, and that is the reason this mode is referred to as 13 ND, even though it contains other harmonics and is thus an impure mode.

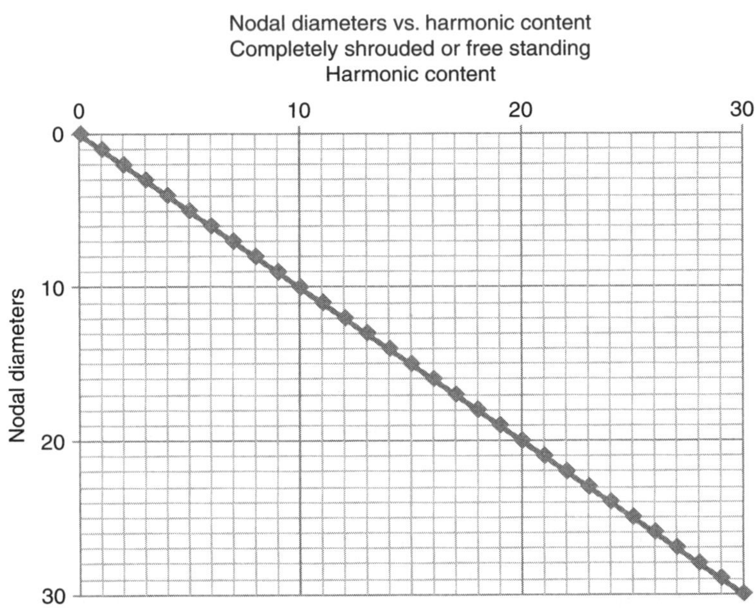

FIGURE 7.75 Harmonic contents diagram for continuously shrouded and individually shrouded disks.

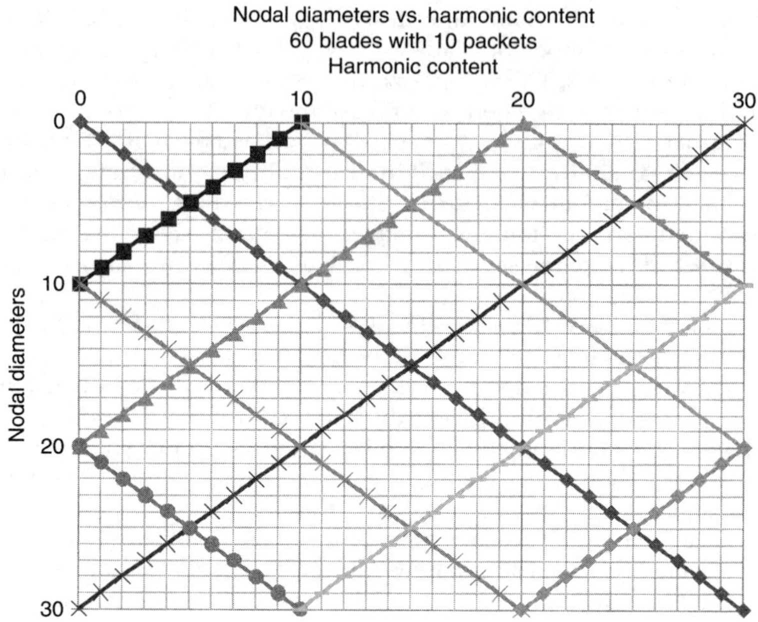

FIGURE 7.76 Harmonic contents diagram for bladed disk with 60 blades and 10 packets.

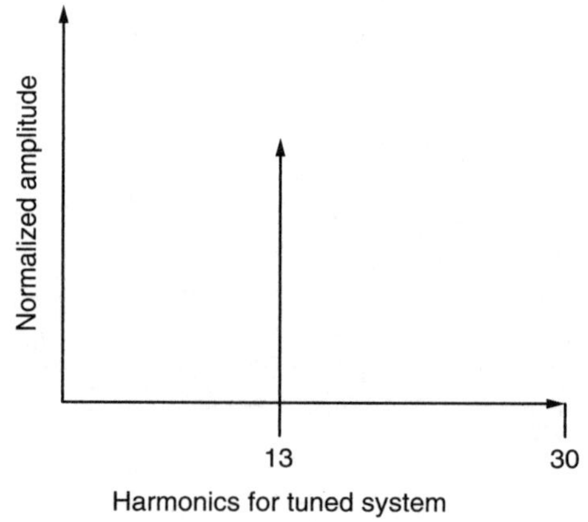

FIGURE 7.77 Estimated harmonic content in 13 ND for continuously or individually shrouded disk.

The harmonic contents of the 13-ND mode for the completely shrouded disk (single packet) are shown in Fig. 7.77. Figure 7.78 shows the harmonic contents for the 10 packets case. It is clear that for the single packet case there is only 13th harmonic while for the 10 packets case there are others in addition to the 13th harmonic.

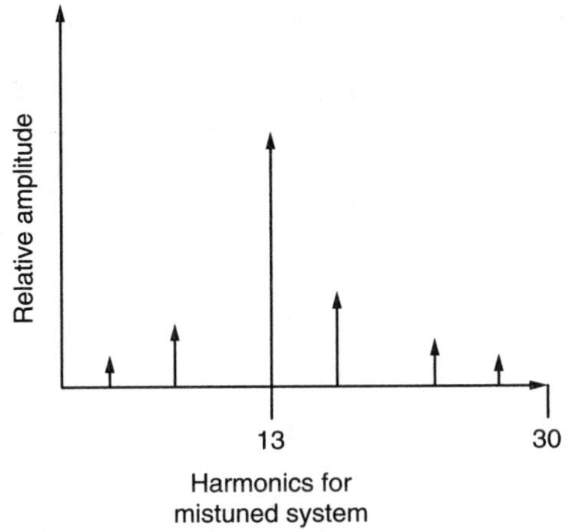

FIGURE 7.78 Estimated harmonic content in 13th for 10 packets.

The amplitude shown in Fig. 7.78 is not real, but it shows the existence of harmonic contents. The relative magnitudes can be estimated by using a Fourier decomposition of the eigenvector.

7.11.2 Mathematical Expression

By examining many results like this, a mathematical expression may also be used. The mode shape for M nodal diameter in a general sense can be expressed as follows, where L is the number of harmonic contents.

$$X(M) = \sum A_1 \sin(L\theta + \phi_L) \tag{7.33}$$

The harmonic content L can be estimated as follows for the packeted bladed disk:

$$L = \text{abs}(ln \pm M) \tag{7.34}$$

where $l = 0, 1, 2, 3, 4, \ldots$
n = number of packets
N = number of blades

and

$$0 \le L \le N \quad \text{or} \quad 0 \le L \le (N-1)/2 \quad \text{if } N \text{ is odd} \tag{7.35}$$

A simple way of examining the effect of irregularities in mode shape and the shape of the aerodynamic forces can be done by estimating work done by the forces during the motion of a particular mode shape. Figure 7.78 is the plot of the 13-ND mode shape. The primary harmonic content of the shape is 13th, but there are other harmonics. A Fourier decomposition of the shape will estimate the contents of all harmonics. The harmonic contents of this mode can be estimated either from Eq. (7.34) or from Fig. 7.75.

For this case,

$$n = 10 \qquad M = 13 \qquad N = 60$$

$$L = \text{abs}[l(10) \pm 13]$$

$$\text{For } l = 0, \qquad L = 13$$

$$\text{For } l = 1, \qquad L = 23 \text{ and } 3$$

$$\text{For } l = 2, \qquad L = 33 \text{ and } 7$$

$$\text{For } l = 3, \qquad L = 43 \text{ and } 17$$

$$\text{For } l = 4, \qquad L = 53 \text{ and } 27$$

$$\text{For } l = 5, \qquad L = 63 \text{ and } 37$$

The harmonic contents should be less than 30 ($N/2 = 60/2 = 30$) according to Eq. (7.35). Therefore, the expected harmonic contents for this mode are 3, 7, 13, 17, 23, and 27.

Now the expression for the mode shape can be written as

$$\begin{aligned} X(M) &= A_3 \sin(3\theta + \varphi_3) + A_7 \sin(7\theta + \varphi_7) + A_{13} \sin(13\theta + \varphi_{13}) \\ &\quad + A_{17} \sin(17\theta + \varphi_{17}) + A_{23} \sin(23\theta + \varphi_{23}) + A_{27} \sin(27\theta + \varphi_{27}) \\ &= A_3(\sin 3\theta \cos\varphi_3 + \cos 3\theta \sin\varphi_3) + A_7(\sin 7\theta \cos\varphi_7 + \cos 7\theta \sin\varphi_7) \\ &\quad + A_{13}(\sin 13\theta \cos\varphi_{13} + \cos 13\theta \sin\varphi_{13}) \\ &\quad + A_{17}(\sin 17\theta \cos\varphi_{17} + \cos 17\theta \sin\varphi_{17}) \\ &\quad + A_{23}(\sin 23\theta \cos\varphi_{23} + \cos 23\theta \sin\varphi_{23}) \\ &\quad + A_{27}(\sin 27\theta \cos\varphi_{27} + \cos 27\theta \sin\varphi_{27}) \end{aligned} \tag{7.36}$$

The actual relative magnitude of the coefficients should be estimated by performing a Fourier analysis.

7.11.3 Effect on Response

Assume that the shape of the force is given by the following expression:

$$\begin{aligned} F(N) &= F_0 + F_4 \cos(4\theta + \phi_4) + F_7 \cos(7\theta + \phi_7) + F_{13} \cos(13\theta + \phi_{13}) \\ &\quad + F_{14} \cos(14\theta + \phi_{14}) + F_{24} \cos(24\theta + \phi_{24}) + F_{28} \cos(28\theta + \phi_{28}) \end{aligned}$$

$$
\begin{aligned}
&= F_0 + F_4(\cos 4\theta \cos\phi_4 - \sin 4\theta \sin\phi_4) \\
&\quad + F_7(\cos 7\theta \cos\phi_7 - \sin 7\theta \sin\phi_7) \\
&\quad + F_{13}(\cos 13\theta \cos\phi_{13} - \sin 13\theta \sin\phi_{13}) \\
&\quad + F_{14}(\cos 14\theta \cos\phi_{14} - \sin 14\theta \sin\phi_{14}) \\
&\quad + F_{24}(\cos 24\theta \cos\phi_{24} - \sin 24\theta \sin\phi_{24}) \\
&\quad + F_{28}(\cos 28\theta \cos\phi_{28} - \sin 28\theta \sin\phi_{28})
\end{aligned} \tag{7.37}
$$

$$
dW = F(N)\, d\,[X(M)]
$$

$$
\begin{aligned}
d[X(M)] = [&3A_3(\cos 3\theta \cos\varphi_3 - \sin 3\theta \sin\varphi_3) \\
&+ 7A_7(\cos 7\theta \cos\varphi_7 - \sin 7\theta \sin\varphi_7) \\
&+ 13A_{13}(\cos 13\theta \cos\varphi_{13} - \sin 13\theta \sin\varphi_{13}) \\
&+ 17A_{17}(\cos 17\theta \cos\varphi_{17} - \sin 17\theta \sin\varphi_{17}) \\
&+ 23A_{23}(\cos 23\theta \cos\varphi_{23} - \sin 23\theta \sin\varphi_{23}) \\
&+ 27A_{27}(\cos 27\theta \cos\varphi_{27} - \sin 27\theta \sin\varphi_{27})]\, d\theta
\end{aligned} \tag{7.38}
$$

$$
\begin{aligned}
dW = [&F_0 + F_4(\cos 4\theta \cos\phi_4 - \sin 4\theta \sin\phi_4) + F_7(\cos 7\theta \cos\phi_7 - \sin 7\theta \sin\phi_7) \\
&+ F_{13}(\cos 13\theta \cos\phi_{13} - \sin 13\theta \sin\phi_{13}) \\
&+ F_{14}(\cos 14\theta \cos\phi_{14} - \sin 14\theta \sin\phi_{14}) \\
&+ F_{24}(\cos 24\theta \cos\phi_{24} - \sin 24\theta \sin\phi_{24}) \\
&+ F_{28}(\cos 28\theta \cos\phi_{28} - \sin 28\theta \sin\phi_{28})] \\
\times [&3A_3(\cos 3\theta \cos\varphi_3 - \sin 3\theta \sin\varphi_3) \\
&+ 7A_7(\cos 7\theta \cos\varphi_7 - \sin 7\theta \sin\varphi_7) \\
&+ 13A_{13}(\cos 13\theta \cos\varphi_{13} - \sin 13\theta \sin\varphi_{13}) \\
&+ 17A_{17}(\cos 17\theta \cos\varphi_{17} - \sin 17\theta \sin\varphi_{17}) \\
&+ 23A_{23}(\cos 23\theta \cos\varphi_{23} - \sin 23\theta \sin\varphi_{23}) \\
&+ 27A_{27}(\cos 27\theta \cos\varphi_{27} - \sin 27\theta \sin\varphi_{27})]\, d\theta
\end{aligned} \tag{7.39}
$$

$$
\begin{aligned}
\int_d^{2\pi} dW = \int_d^{2\pi} [&F_0 + F_4(\cos 4\theta \cos\phi_4 - \sin 4\theta \sin\phi_4) \\
&+ F_7(\cos 7\theta \cos\phi_7 - \sin 7\theta \sin\phi_7) \\
&+ F_{13}(\cos 13\theta \cos\phi_{13} - \sin 13\theta \sin\phi_{13}) \\
&+ F_{14}(\cos 14\theta \cos\phi_{14} - \sin 14\theta \sin\phi_{14})
\end{aligned}
$$

$$
\begin{aligned}
&+F_{24}(\cos 24\theta \cos\phi_{24} - \sin 24\theta \sin\phi_{24})\\
&+F_{28}(\cos 28\theta \cos\phi_{28} - \sin 28\theta \sin\phi_{28})]\\
&\times [3A_3(\cos 3\theta \cos\varphi_3 - \sin 3\theta \sin\varphi_3)\\
&+7A_7(\cos 7\theta \cos\varphi_7 - \sin 7\theta \sin\varphi_7)\\
&+13A_{13}(\cos 13\theta \cos\varphi_{13} - \sin 13\theta \sin\varphi_{13})\\
&+17A_{17}(\cos 17\theta \cos\varphi_{17} - \sin 17\theta \sin\varphi_{17})\\
&+23A_{23}(\cos 23\theta \cos\varphi_{23} - \sin 23\theta \sin\varphi_{23})\\
&+27A_{27}(\cos 27\theta \cos\varphi_{27} - \sin 27\theta \sin\varphi_{27})]\, d\theta \qquad (7.40)
\end{aligned}
$$

Using the complete integral relationship given in the Appendix, Eq. (7.40) reduces to the following:

$$
\begin{aligned}
\int_d^{2\pi} dW &= \int_d^{2\pi} 7A_7F_7 \cos\varphi_7 \cos\phi_7 \cos^2 7\theta\, d\theta\\
&\quad + \int_d^{2\pi} 13A_{13}F_{13} \cos\varphi_{13} \cos\phi_{13} \cos^2 13\theta\, d\theta\\
&= \pi(7A_7F_7 \cos\varphi_7 \cos\phi_7 + 13A_{13}F_{13} \cos\varphi_{13} \cos\phi_{13}) \qquad (7.41)
\end{aligned}
$$

The following observations should be noted:

1. The blade will respond to the harmonics of force that match the harmonic contents of the mode shape.
2. The resulting response distribution will be different from the mode shape.
3. It is quite possible that there will not be a true resonance, but a forced response in such a situation.

CHAPTER 8
Reliability Evaluation for Blade Design

8.1 Loads, Stress, and Evaluation

The rotating blades of a steam turbine convert the thermal energy of steam to mechanical energy. The power produced can be in excess of 200 hp per blade with tip speeds of about 1500 ft/s, and the force due to rotational speed can be about 10,000 times the force of gravity. Despite these large forces imposed on the blades, blades have been designed to operate trouble-free for many years of continuous service.

The damage in the blades of mechanical drive turbines is almost always due to fatigue of the material. The reliability analysis of turbine blades consists of essentially two parts—stress analysis and dynamic analysis of the blades.

The stress analysis calculates the stresses due to steam forces and centrifugal forces acting on the blade at the maximum operating conditions. The blade dynamics analysis compares the blade's natural frequencies with exciting force frequencies to determine the vibratory stresses in the blades. Fatigue damage is due to the alternating stress imposed on the mechanical structure and is also influenced by the magnitude of mean stress imposed on the blades. The centrifugal stress is the mean stress at the operating speed of the turbine, and the alternating stress in the blade is the result of unsteady forces that exist in the turbine. The magnitude of the alternating stress depends on the nature and magnitude of unsteady forces, damping, and the resonance condition that might exist. Traditionally, the reliability of a steam turbine blade design is judged based on the estimated magnitude of the factor of safety that is determined by using a Goodman-type diagram.

Some of, but not all, the factors that contribute to the damage of blades are listed here:

Centrifugal stress
Stress due to steam forces
Steady stress
Alternating stress
Stress due to resonant vibration
Low cycle fatigue
Thermal fatigue
Creep damage
Environmental effect
Stress corrosion
Corrosion fatigue
Impact stress

8.2 Stress due to Centrifugal Load

Centrifugal stress for most part is steady in nature, can be estimated accurately, and may be large, but is of secondary importance. It is almost never a principal cause of blade failure, but may become important for numerous start-up and shutdown cycles (low cycle fatigue damage). The blade failure due to centrifugal stress might occur above approximately 75 percent over speed. The appearance of the surface generated during fracture is quite different from that due to fatigue. Some dimensional changes near the rim of the disk or near the tip of the blade might occur. Centrifugal stress is, however, a contributing factor in the fatigue damage because the material's fatigue property is influenced by mean stress.

A quick estimation of stress due to centrifugal load can be performed (Stodola, 1905). For a constant section blade, the centrifugal force and stress can be estimated very accurately as follows:

$$\text{Centrifugal force } F = m\omega^2 r \tag{8.1}$$

where m = mass of blade
ω = rotational speed
r = distance to center of gravity of blade

Equation (8.1) can be simplified as follows:

$$F = (W/g)\,(2\pi \text{RPM}/60)^2(D/2) \tag{8.2}$$

where W = weight of blade, lb
g = gravity constant, ft/s^2
RPM = speed of turbine
D = diameter of center of gravity, in (pitch diameter of blade)

$$F = (V\rho/g)\,(2\pi \text{RPM}/60)^2\,(D/2) \tag{8.3}$$

After inserting the values for g and ρ (= 0.283 for steel) into the equation, it reduces to

$$\begin{aligned} F &= (VD)(\text{RPM}/500)^2 \text{ lb} \\ &= 4VD \text{ lb for 1000 rpm} \end{aligned} \tag{8.4}$$

Thus, the centrifugal stress σ_{CF} at the base of the blade at 1000 rpm is given by

$$\sigma_{CF} = 4VD/A \qquad A = \text{cross-sectional area}$$

$$V = LA \qquad L = \text{length of blade}$$

Finally,

$$\sigma_{CF} = 4LD \qquad \text{psi (at 1000 rpm)} \tag{8.5}$$

Stress at any speed N can be estimated as

$$\sigma_{CF} = 4LD(N/1000)^2 \text{ psi} \tag{8.6}$$

It is evident from this simple, though exact, formula for a constant section blade that a blade of a different size, but of the same geometry and operating at the same tip speed, experiences the same centrifugal stress at the base as at all other sections.

For example, a blade 15 in long on a 30-in disk operating at 3000 rpm produces the same stress as a 7.5-in blade of similar geometry on a 15-in disk running at 6000 rpm.

$$\begin{aligned} \sigma_{CF} \text{ for 15-in blade at 3000 rpm} &= 4(15)(30+15)(3000/1000)^2 \text{ psi} \\ &= 24{,}300 \text{ psi} \end{aligned}$$

$$\begin{aligned} \sigma_{CF} \text{ for 7.5-in blade at 6000 rpm} &= 4(7.5)(15+7.5)(6000/1000)^2 \text{ psi} \\ &= 24{,}300 \text{ psi} \end{aligned}$$

8.3 Stress due to Steam Forces

Steady Stress

Steady stress is more predictable and is due to the bending of the blade. Bending stress generated by steam is added to the centrifugal stress.

Alternating Stress

The predominant source of force is steam, and most blade fatigue failures are due to the alternating stress induced by steam bending. The steam force may be large enough to cause damage in a nonresonant case, which is a case of forced response (for high pressure ratio, single-stage turbines where a series of single jets are used, the first stage of a multistage turbine may also have a high steam load). Ideally, steam flow in the 100 percent admission stage (full admission) is supposed to be uniform, but many construction and structural features cause variations in flow as listed below.

Stationary vanes preceding the rotating blades (nozzle passing frequency)
Interrupted arc of admission or partial arc admission
Structural members in the inlet and exhaust hood
Many inlet ports
Steam extraction ports
Nonuniform spacing of stationary vanes at the horizontal split
Manufacturing variation in the pitch of stationary vanes
Manufacturing variation in the geometry of rotating blades
Manufacturing variation in the pitch of rotating blades
Struts or braces behind the stage
Turn in the exhaust hood, applicable to last-stage blades

The magnification factor (measure of vibration amplitude) vs. the frequency ratio curve is reproduced in Fig. 8.1. The amplitude of vibration is a function of the ratio of the exciting frequency to the natural frequency of the blade. However, for small damping the maximum magnitude of response occurs when the frequency ratio is equal to 1.

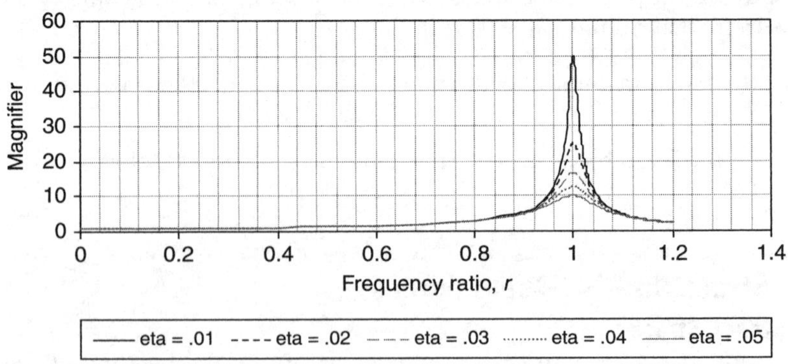

FIGURE 8.1 Magnifier vs. frequency ratio.

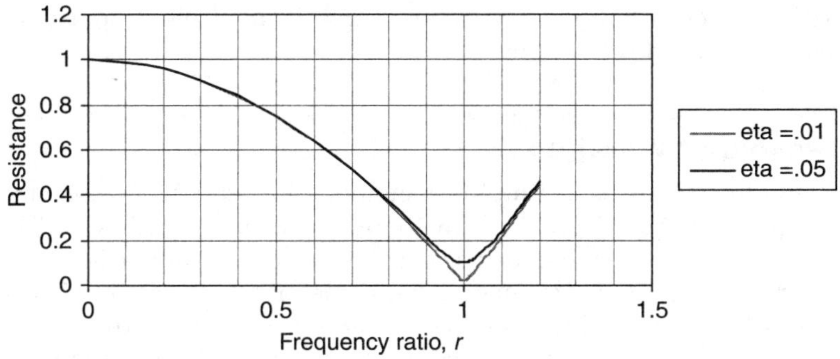

Figure 8.2 Resistance vs. frequency ratio.

Another instructive way to look at the same data is to examine the plot of resistance vs. the frequency ratio, which is reproduced in Fig. 8.2. The minimum resistance of the structure to the forcing occurs when the value of the frequency ratio is unity.

The strategy for blade design is to avoid the coincidence of harmonics of running speed and nozzle passing frequencies with the natural frequency of the blade. The concept of resonance in which mode shape and the shape of the forcing also play a defining role was explained earlier. The following example of a Campbell diagram (Fig. 8.3) is instructive in demonstrating that if the shape of the force and the mode shape are identical and the natural frequency is equal to the frequency of the exciting force, then the magnitude of the response will be large. However, the blade will also respond to forces even if the frequencies are not equal, but the response will be very

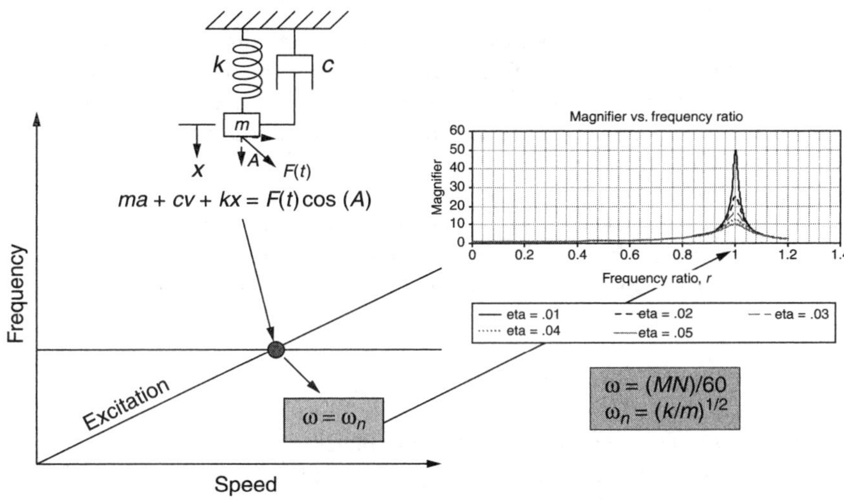

Figure 8.3 An explanation for the validity of a Campbell diagram.

small. It is clear that the resonant response is inversely proportional to the magnitude of damping.

8.4 Resonant Vibration

The industry has used simplified methods to estimate stresses and designs that, based on these methods, have been successful. The impetus for simple methods has been the unavailability of fast computers in the past and the experience gained from a large number of successful designs. The basic logic has been that the bending stress caused by the steam loading on the blade gets modified depending on the dynamic characteristics of the rotating blade as demonstrated from the equation discussed below.

A more common equation used to calculate blade vibratory stress is

$$\sigma_{vib} = S\,(\pi/\delta)\,(\mathrm{RRF})\,\sigma_{hp} \tag{8.7}$$

where σ_{vib} = calculated blade vibratory stress
σ_{hp} = steady steam bending stress in blade
S = stimulus factor; percent of steam force that is unsteady (ranges from 0.1 to 0.3)
π/δ = resonant amplification factor. If a critical damping ratio of 0.003 is used, then the log decrement damping δ is equal to $0.003(2\pi) = 0.01885$ and the resonant amplification factor is equal to $\pi/0.01835 = 166.66$
RRF = resonant response factor depending on mode of vibration and other factors (ranges from 0 to 1).

Weaver and Prohl (1956) defined the resonant response factors (RRFs) that range from 0.1 to 0.9. RRF depends on the mode shape of the blade vibration and phasing of the exciting force, which is consistent with the concept of the SAFE diagram. When the mode shape and shape of the forcing are identical, RRF will equal 1.0. Any value of RRF less than 1.0 signifies a forced response.

A simplified equation can also be used to calculate blade vibratory stress as shown below. This equation includes the effects of steam force, resonant response magnification, and a stress concentration factor and is a simplification of the previously discussed equation.

$$\sigma_{vib} = \mu.K_{\sigma}\sigma_{hp} \tag{8.8}$$

where σ_{vib} = calculated vibratory stress in blade
μ = resonant response magnification factor
K_{σ} = stress concentration factor
σ_{hp} = steady steam bending stress in blade

The term K_σ in Eq. (8.8) can be considered as a fatigue strength reduction factor due to the edges, fillets, and corners that occur in the blade. The typical value for K_σ can be taken to be equal to 1.50, which in reality is a large reduction in fatigue strength of the blade material or alternatively a substantial increase in the vibratory stress and will conservatively account for all notches occurring in the normal blade design. The value to be used for μ is based upon the evaluation of a Campbell diagram and an interference diagram. The largest calculated resonant response magnifier value of μ from the applicable excitation should be used in the vibratory stress calculation. The value of σ_{hp} is the steady steam bending stress being calculated at the maximum stage horsepower and maximum operating speed condition. The calculated vibratory stress includes the material degradation effects on the blade material fatigue strength for the normal steam environment (wet or dry). A further reduction in blade material fatigue strength should only be included if the steam quality is known to be below standard and is achieved by increasing the value of K_σ.

The steam force acting on the blade produces the stage output horsepower. The magnitude of steady steam bending stress in an airfoil depends on the steam forces acting on the blade. A simple method based on beam theory can be utilized to estimate the blade steady stresses. These stresses should be estimated at the maximum speed and stage horsepower operating conditions.

When simple calculations are made for stresses, then the limiting values of stresses (steam bending stress and centrifugal stress) are kept low. This should be consistent with the OEM's experience with similar designs, and a record of validation of such a design should be produced.

8.5 Blade Frequency Evaluation

Most of the mechanical drive steam turbine stages have blades banded in groups by connecting the tips of the blades by a riveted-on shroud band. The number of blades per group can vary, but in general many manufacturers use six blades per packet. The dynamic behavior of the packeted bladed disk assembly was discussed earlier. The dynamic behavior, meaning the frequency and the associated mode shapes of a shrouded packet of blades, can be calculated by using the finite element analysis (FEA). Typically a blade packet is modeled by using a series of beams for the blade and the connecting shroud. Assumptions are made that the blades are rigidly attached to the disk at the upper root lock and the shroud is rigidly attached to the blade tip. Due to the powerful computers and availability of many finite element programs it is convenient to use FEA to analyze vibration of the blading system.

The calculated blade frequencies and mode shapes are then plotted on Campbell and interference diagrams. A description of the packeted blade mode shapes was discussed earlier. The interference and Campbell diagrams are used to determine the possibility of the resonance of blades in each stage due to the exciting force within the operating speed range. The interference diagram (SAFE) complements the Campbell diagram in determining if any true resonances exist, and a true resonance exists only when the blade's natural frequency and exciting force frequency match in both magnitude and shape. Interference diagram (SAFE) is preferred since it has a clearer representation of these two features and is better in pinpointing where true resonances exist.

8.6 Exciting Forces

The resonant response magnifier (discussed earlier) depends upon the type of exciting force and the blade's natural frequency. The three primary types of exciting forces that occur in a steam turbine are running speed harmonic excitation, nozzle passing frequency excitation, and the partial admission excitation force.

A brief description of each type of exciting force is given next with an explanation of how their vibratory stress resonant response magnifier is determined.

8.6.1 Running Speed Harmonic Excitation

Running speed harmonic excitation occurs due to any circumferential variation or interruption in the steam flow in a full admission stage in one complete revolution. Some of the sources of running speed harmonic excitation are exhaust or extraction ports, diaphragm's axial or concentric misalignment, diaphragm vane manufacturing or spacing variations, mismatching of diaphragm vanes at the horizontal split, struts, steam splitters, or supports in the exhaust casing. Per revolution exciting forces are generally the largest in the exhaust stages with taller blades. The reason for this is that they are closer to the exhaust hood. The other reason is that the fundamental frequency of these blades is often below 15 × RS. According to the API Standard 612, fifth edition, one time through 15 times the running speed (15 × RS) of excitation are to be considered in blade design.

The resonant response magnification factor μ caused by running speed harmonic excitation can be estimated by

$$\mu = \beta(\text{RPM})/\omega_n \tag{8.9}$$

where μ = resonant response magnification factor
RPM = turbine speed
ω_n = fundamental blade frequency, Hz
β = a constant, typically 0.9 to 1.0

Equation (8.9) provides the expression for a resonant response magnification factor as a function of the untuned blade frequency and the harmonic of running speed. This magnifier is equally valid for both shrouded and freestanding blades. For variable-speed, mechanical drive turbines, the fundamental blade frequency should be designed to be above the 4 × RS exciting force. For generator drive turbines (constant speed), the blade fundamental frequency is generally tuned between the harmonics of running speed if the blade frequency is below the fourth harmonic of running speed. The calculated magnification factor is reduced by a factor of 2 for tuned blades. When blade frequency tuning is not possible or the vibratory stresses are large, a lacing wire is used to help reduce the vibratory stresses. The calculated magnification factor can be reduced by a factor of 4 for blade design with a lacing wire.

8.6.2 Nozzle Passing Frequency (NPF) Excitation

Nozzle passing frequency excitation results from steam flow wakes that occur downstream of the nozzle vane trailing edges. The turbine blade receives a slight reduction in steam force as it passes through each of the nozzle vane trailing edge wakes. The frequency of nozzle passing excitation is calculated from

$$\text{NPF} = M(\text{RPM}/60) \tag{8.10}$$

where NPF = nozzle passing frequency
M = number of nozzles in 360° arc
RPM = turbine speed, rpm

The other harmonic of nozzle passing frequency excitation is equal to $n \times M(\text{RPM}/60)$, where n is the number of the harmonic of excitation. The severity of the nozzle passing frequency force depends upon the thickness of the nozzle vane trailing edge, the axial distance between the nozzle vanes and the rotating blades, and the velocity of the steam. Values of the resonant response magnification factor μ for true resonance conditions due to NPF and various packeted bladed disk natural frequencies are listed in Table 8.1 and also shown in Fig. 8.4. It shows the relative magnitude of the magnifier due to various combinations of modes and excitation. The magnifiers have been normalized with the magnifier due to 1 × NPF and the first tangential mode. The magnitude of magnifier due to 1 × NPF and the first tangential mode may vary between 20 and 30.

Condition	Magnifier
1 × NPF + 1st tang.	100.00
1 × NPF + 1st axial	33.33
1 × NPF + 1st axial rock	32.75
1 × NPF + 1st axial U	19.61
1 × NPF + 2d tang.	25.49
2 × NPF + 1st tang.	27.45
2 × NPF + 1st axial	11.76
2 × NPF + 1st axial rock	11.76
2 × NPF + 1st axial U	7.06
2 × NPF + 2d tang.	8.82

TABLE 8.1 Normalized Magnifiers for Resonance

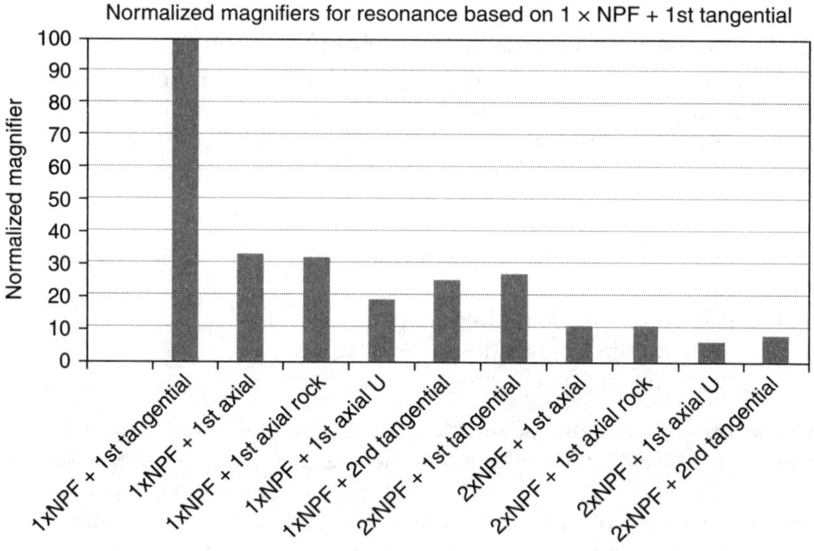

FIGURE 8.4 Magnifier vs. type of resonance.

If the SAFE diagram shows a coincidence of frequencies, but the mode shape does not match the shape of the force, the value of the magnifier should be reduced to one-third of the value shown in the chart. The reason for this is that a packeted bladed disk is a mistuned system, and the mode shapes are not of a pure sinusoidal form. Generally, these modes contain many different harmonics; therefore, some response is expected to occur away from the major harmonic component as in the forced vibration case.

For jobs such as failure analysis or new blade designs, a more advanced NPF resonant response analysis should be employed. A nozzle passing frequency-forcing function based on the exact steam conditions and the actual nozzle vane and rotating blade construction of the stage being analyzed should be available. A forced response calculation can then be done on the finite element model of the bladed disk using the results of the modal solution and assuming a value for system damping. This type of analysis is not done on a regular basis because experience shows that the above methods and guidelines provide good design. In the event that it becomes necessary to look into the design more closely, the FEA methods should be employed.

8.6.3 Partial Admission Excitation

Partial admission excitation results when the stationary nozzles do not cover the 360° arc of admission. Partial admission stages normally occur in the first stage of a turbine or next to an extraction opening in a turbine. Occasionally, a stage or two following one of these control stages may still have partial admission (less than 360°) nozzles. The turbine blades in a partial admission stage experience nonsteady forces as they go around from an unloaded region of arc to a fully loaded region and back to an unloaded region in each revolution. It is possible to have more than one group of nozzles in partial admission stages separated by an inactive arc. However, this feature exacerbates the unsteady blade loading in each revolution. A Fourier decomposition of the steam force loading per revolution is done to determine the harmonics of the running speed of concern. The frequency of concern of partial admission loading is generally the low harmonics of running speed. Based on the SAFE diagram plots and past operating experiences of short blades in many partial admission stages, they might be resonant at a rate greater than 15 times the running speed. Alternatively, a tall blade in partial admission stage could have a natural frequency low enough to be in a possible resonance with the low harmonics of the running speed.

The simplified equation explained earlier can be used to calculate the resonant response magnification factor for partial admission stages and is given below. The resonant response factor is a function of speed, number of separate groups of partial admission nozzles, and fundamental frequency of vibration.

$$\mu = \beta(\text{RPM})m/\omega_n \tag{8.11}$$

where μ = resonant response magnification factor
β = a constant
RPM = speed of turbine
m = number of separate groups of partial admission nozzles
ω_n = fundamental frequency of the blade, Hz

The magnification factor μ is directly dependent on the fundamental blade frequency ω_n, and the frequency will depend on the blade design. For a short first-stage blade, the fundamental blade frequency can range from 5000 to 10,000 Hz, which results in a magnification factor of about 5 or less depending upon the speed. The partial admission stage with a taller blade can have a fundamental blade frequency of 1000 to 2000 Hz, and the magnification factor can be as large as 20. These observations explain that partial admission stages having short blades can only experience forced response excitation. Alternatively, partial admission stages having taller blades can have a resonant response excitation.

It is complicated to look closely at the behavior of blading under a partial arc admission loading. For any natural mode of vibration as a single spring-mass system, the response can be looked at as below.

A transient component of excitation occurs when the blade enters and leaves the arc of loading. This analysis is generally performed in the time domain, and it becomes very complicated. Using the spring-mass concept and by following a method similar to that discussed by Kroon (1940), a magnification factor can be estimated, and the resonant response due to the various harmonics of force at many frequencies can be estimated by using a Fourier decomposition of the periodic force. A complete system model can be used, and a method similar to that described below can be employed to estimate resonant response.

The forcing function due to the partial arc admission is decomposed using a Fourier analysis. The frequency domain response analysis is performed on a bladed disk system for many harmonics used in the Fourier analysis of forcing function. The frequency domain response is then converted to a time domain solution. The response in the time domain is combined to find the nodal location with the highest response, and the stress analysis is performed at the right "time instant." Pigott (1980) described a more sophisticated method by utilizing generalized mass, generalized stiffness, and a generalized force concept for a cantilever packet of blades. Kroon's method, though simple, provides an insight to the response of the blade for a partial arc admission. Kroon's method and another simplified method utilizing the spring-mass-damper system are discussed in section to follow.

8.7 Factor of Safety Calculation

The traditional mechanical design evaluation method for a turbine blade is the factor of safety calculation. The factor of safety can be calculated by many methods, e.g., the Soderberg diagram, the Goodman diagram, or similar diagram. The failure line (when the factor of safety equals 1.00) on a Goodman diagram is defined by plotting a straight line connecting the fatigue strength with the ultimate strength of the blade material. The vibratory stress is plotted on the vertical axis and the mean stress on the horizontal axis. The equation used to calculate the factor of safety using the Goodman diagram is

$$1.0/\mathrm{FS} = \sigma_{\mathrm{vib}}/\sigma_{\mathrm{fs}} + \sigma_{\mathrm{mean}}/\sigma_{\mathrm{ult}} \tag{8.12}$$

where FS = factor of safety of blade
σ_{mean} = combined steady stresses in blade
σ_{vib} = vibratory stress in blade
σ_{ult} = ultimate strength of blade material
σ_{fs} = adjusted fatigue strength of blade material

A typical acceptable minimum value for the factor of safety can be 1.50 for turbine blades. If a change in design is required, the design change will be dictated largely by whether the steady stress or vibratory stress component is contributing most to the low factor of safety. A change in the natural frequency of the blade or the exciting frequency is often possible as a means to avoid a high response resonance resulting in high vibratory stresses. It is not possible to avoid all possible resonance in variable-speed turbines. In such cases the largest responses are avoided, and the blade is made strong enough to operate at the remaining resonances within the operating speed range. Another method of reducing resonant blade vibratory stresses is by increasing the blade chord width, which will lower the steam bending stress in the blade, thereby lowering the vibratory stresses in the blades. Another way to lower the vibratory stress level is by increasing the damping in the structure, which is sometimes achieved by utilizing lacing wires in the design.

When the steady stress is the cause for a low factor of safety, one method to improve the factor of safety is to increase the taper in the airfoil to reduce the centrifugal stress. This can also be achieved by reducing the blade height or pitch diameter. Another way to reduce steady stresses is by employing a lower-density or higher-strength blade material. Titanium alloy has been used successfully in exhaust end blades since it is both less dense and higher strength than the standard 403 stainless steel (SS) blade material.

The fatigue strength of a material depends on many factors, and fatigue properties are obtained through a test in a controlled laboratory environment on polished specimens. Therefore, these values for a real application should be adjusted to reflect the influence of those factors. The adjusted fatigue strength of a material is normally calculated from

$$\sigma_{\mathrm{fs}} = k_1 \times k_2 \times k_3 \times \cdots (\sigma'_{\mathrm{fs}}) \tag{8.13}$$

where σ_{fs} = adjusted material fatigue strength
$k_1, k_2, \cdots$ = correction factors
σ'_{fs} = fatigue strength (normally 50,000 psi for 403 SS at 10^6 cycles)

The typical correction factors used to calculate the blade material fatigue strength are

k_1 = surface finish factor (0.75 for machined surfaces)

k_2 = size factor (0.85 for large blades and 0.95 for small blades)

k_3 = reliability factor (0.816 for 99 percent reliability)

k_4 = temperature factor (includes this factor by using material creep-rupture properties above 550°F)

k_5 = notch sensitivity, which accounts for reduction in fatigue strength due to notches in material

For a typical root K_t ~8.0 (from Peterson "T" heads), making K_f ~5.9 and $k_5 = 1/K_f = 0.17$. For a blade K_t ~1.5 with K_f ~1.415 and $k_5 = 0.707$.

Using all the above correction factors would reduce the airfoil fatigue strength for a large blade made of SS 403 to σ_{fs} = (0.75) (0.85) (0.816) (0.707) (50,000) = 18,389 psi.

The typical adjusted values can be used as follows. However, these values must be adjusted based on a given construction and the manufacturers' experience.

Airfoil

$k_1 = 0.84$ surface finish

$k_2 = 0.90$ size factor

$k_3 = 1.00$ reliability factor (factor of safety will cover reliability)

$k_4 = 1.00$ temperature factor (included by reducing material strength)

$k_5 = 0.707$ notch factor

$K_f = 1.415$

Therefore,

$$k_{total} = k_1 \times k_2 \times k_3 \times k_4 \times k_5 = 0.5345$$
$$\sigma_{fs} = k_{total}\,\sigma'_{fs} = 0.5345\ (50{,}000) = 26{,}725 \text{ psi}$$

Root

$k_1 = 0.84$ surface finish

$k_2 = 0.90$ size factor

$k_3 = 1.00$ reliability factor (factor of safety will cover reliability)

$k_4 = 1.00$ temperature factor (included by reducing material strength)

$k_5 = 0.333$ notch factor

$K_f = 3.00$

Therefore,

$$k_{\text{total}} = k_1 \times k_2 \times k_3 \times k_4 \times k_5 = 0.2517$$

$$\sigma_{\text{fs}} = k_{\text{total}}\,\sigma'_{\text{fs}} = 0.2517(50{,}000) = 12{,}587 \text{ psi}$$

If a reliability study is performed on a row of blades in poor condition (pitted, dinged, etc.), an additional fatigue strength reduction factor k_6 should be used depending on the severity of the condition of the blade.

Steam Quality

Another correction factor to be considered that affects the fatigue strength of blade materials is the steam environment. The typical fatigue strength reduction values used for the steam environment are

$$k_7 = \text{superheated steam} = 0.85$$

$$k_8 = \text{transition steam (0 to 6 percent wet)} = 0.42$$

$$k_9 = \text{wet steam (more than 6 percent wet)} = 0.66$$

$$k_{10} = \text{corrosive steam} = 0.30$$

If one of the steam environment factors is combined with the previous fatigue strength reduction factors, the material fatigue strength can be reduced to a very low level.

A combined correction factor can be estimated for airfoil and root. These factors are listed in Table 8.2 and are plotted in Fig. 8.5.

	Correction Factor	
Steam	**Airfoil**	**Root**
Superheated	0.444	0.214
Transition	0.225	0.106
Wet	0.352	0.167
Corrosive	0.160	0.075

TABLE 8.2 Correction Factors for Different Steam Conditions

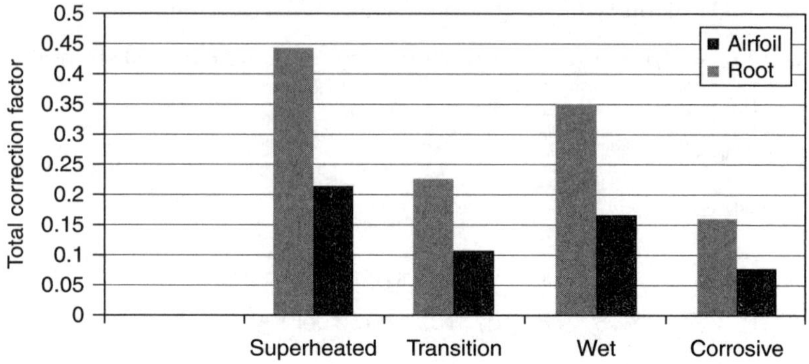

FIGURE 8.5 Total correction factor for airfoil and root section due to steam conditions.

8.8 Influence of Tolerance Stack Up in Root-Disk Attachment

The peak stress in both the root attachment and the disk slot is influenced by the stacking of machining tolerances between the root attachment and the disk slot. The assumption is made during the design phase that each hook shares the load equally for multihook (lands, lobes) root construction. It is prudent to assume, therefore, that this is not the case due to machining tolerances. It is more likely that either upper hook will be in contact or the lower ones. Therefore, there will be a gap between the blade root and the disk's land. Location of the gap, whether on the upper hook or the lower hook, depends on the manufacturing tolerances at the time of assembly. As a turbine starts, existing gap at assembly may close at a speed below the operating speed or it may not at all. This situation may pose a problem if the resulting stresses are larger than estimated based on contacts at each hook, which becomes evident even for a very small gap. This point can be clarified if the assumption is made that the closure of the gap will result from pure radial extension of the root section neglecting any bending displacement of the loaded hook. Assume that there is a chance of a 0.002-in gap at either of the hooks of the root attachment shown in Fig. 8.6 and the distance between the lower and upper hooks is 0.50 in. A strain equal to 0.004 in/in will result in order to close this starting gap. If the modulus of elasticity E has a value of 29.0×10^6 psi, then magnitude of stress equals 116,000 psi. To close the gap an average stress equal to 116,000 psi is required. Near the fillet, area stress will be further magnified due to the stress concentration. If there is sufficient centrifugal load to impose this magnitude of average stress, the gap will close before the operating speed is reached. The other hook starts to pick the load after closure of the gap, and the stress in the previous hook starts to decrease.

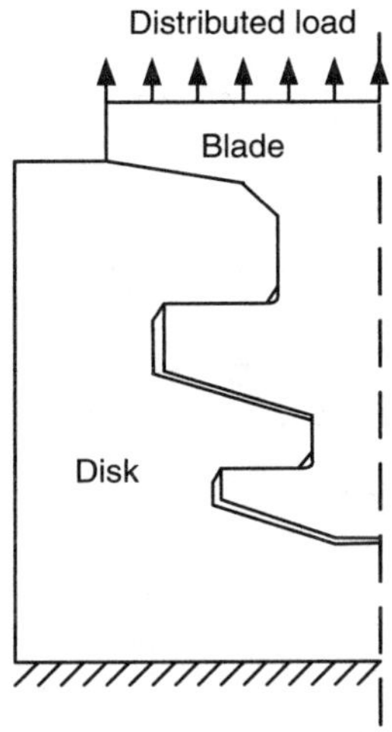

FIGURE 8.6 Blade load, root, and disk.

Lambert (1970) discussed the influence of tolerances on the load distribution. Singh (1998) examined the case of tolerance stacked up due to manufacturing tolerances on a steam turbine root disk assembly. The proposed method has the main ingredients for analyzing such a structure.

A possible concept to assess the influence is described below without specific numbers. The values of items such as stresses will vary on the design, displacement, and stress estimating methods, e.g., the finite element method (FEM). It is recommended to use modern tools such as FEM to analyze the influence of tolerances.

Figure 8.6 shows a root/disk attachment for a two-hook design commonly used in steam turbines. Three different conditions will be discussed that may exist depending on the manufacturing tolerances.

Condition 1: Perfect Contact on Both Hooks, No Gap

At the design stage it is convenient to assume perfect contact and the sharing of the load between both hooks to be equal. Figure 8.7 depicts the load sharing vs. the turbine speed, and it can be seen that as speed increases, the load at each hook moves toward 50 percent, but it does not become equal although the loads move closer together.

There will be variation in the magnitude of stress as speed is increased from rest to the operating speed. Figure 8.8 displays the possible variations at each hook location. Magnitudes of the maximum stress are different due to the stress concentrations, and it is possible that maximum stress at the lower hook may be lower than at the upper hook even though the average section stress at the lower

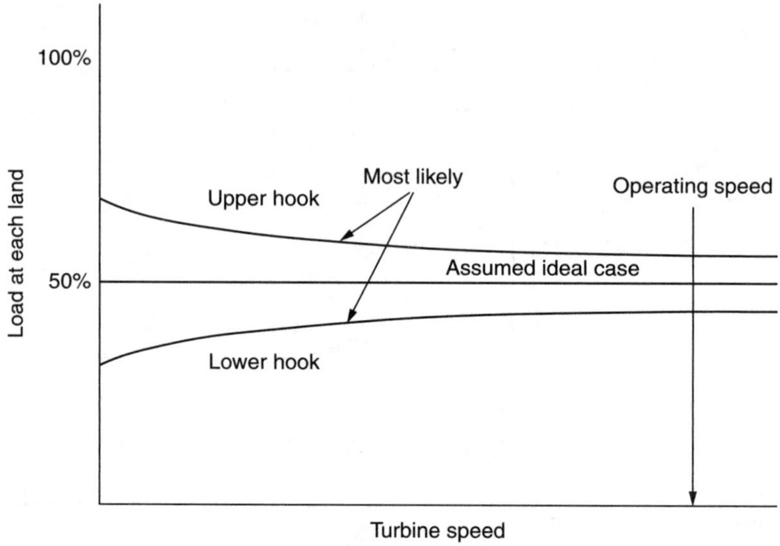

FIGURE 8.7 Distribution of load with turbine speed.

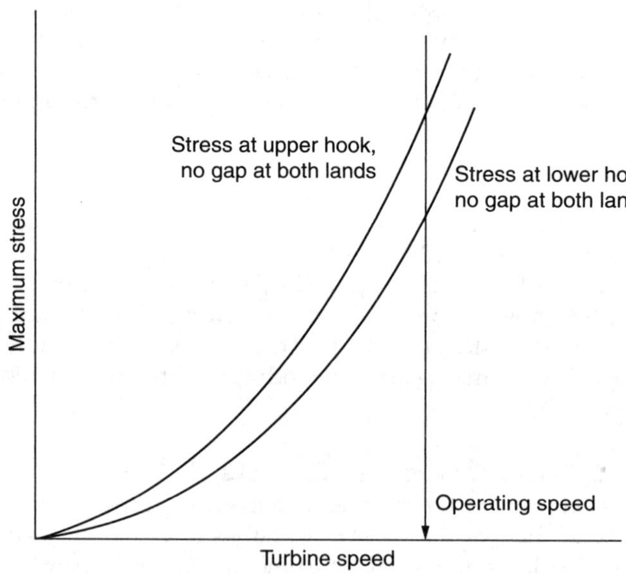

FIGURE 8.8 Variation in the magnitude of stress vs. speed.

hook may be larger than that at the upper one. The reason is that the cross-sectional area is less at the lower hook. It is possible that the stress concentration is less than at the upper hook location. The values of these stresses are the function of load distribution between the upper and lower locks.

Condition 2: Contact at Upper Hook with Gap at Lower Hook

As the turbine is started under this condition, 100 percent of the load is going to be carried by the upper hook. Assume that at a certain speed N_1 the lower hook gap closes and the lower hook starts sharing the load. A possible scenario is displayed in Fig. 8.9.

This analysis may be carried out with a different amount of the gap to determine the speed at which the lower hook may close and start sharing the load. Figure 8.10 is the depiction of such a result.

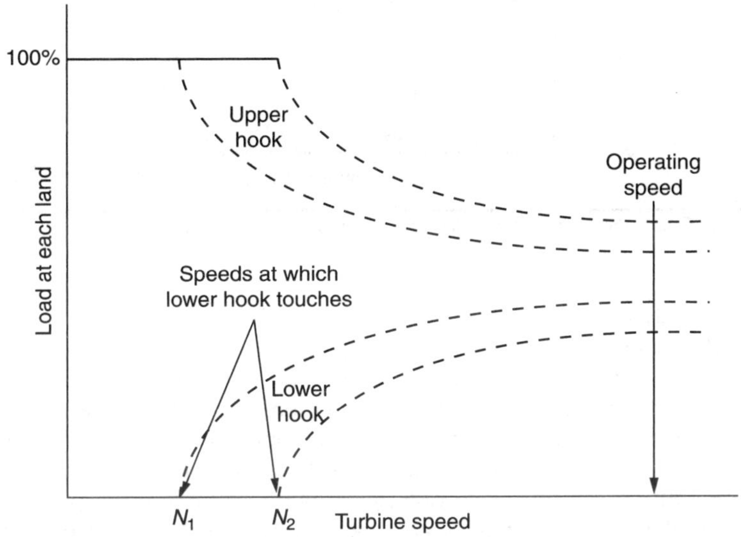

Figure 8.9 No gap at upper hook and gap at lower hook.

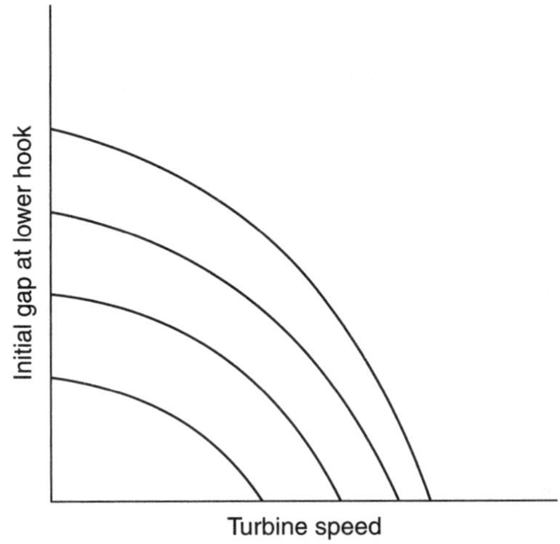

Figure 8.10 Speed at which lower hook closes.

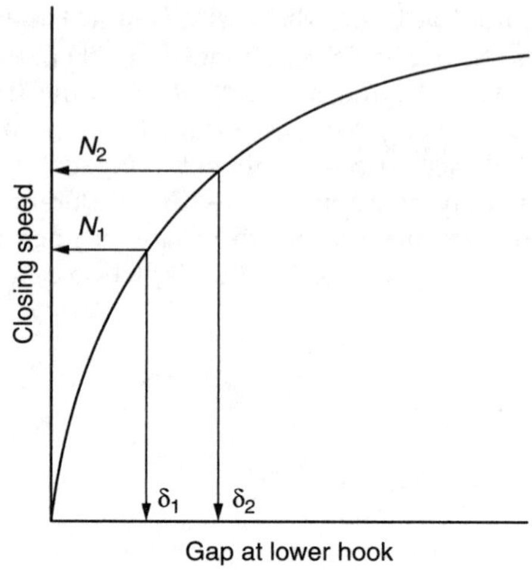

Figure 8.11 Speed at which lower hook closes.

Another plot to summarize the results shown in Fig. 8.10 may be displayed as in Fig. 8.11.

The results for maximum stress may be displayed as in Fig. 8.12.

Results of Fig. 8.12 show that with large gap at the lower hook it may not close before the operating speed and total load is carried by the upper hook that may result in high stress at the upper hook. However, if the gap at the lower hook closes before the operating speed, the load will be shared by the lower hook after the speed at which it closes.

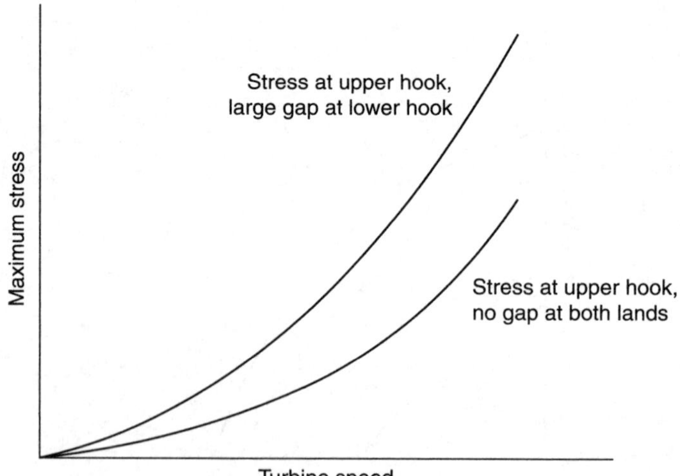

Figure 8.12 Maximum stress vs. turbine speed.

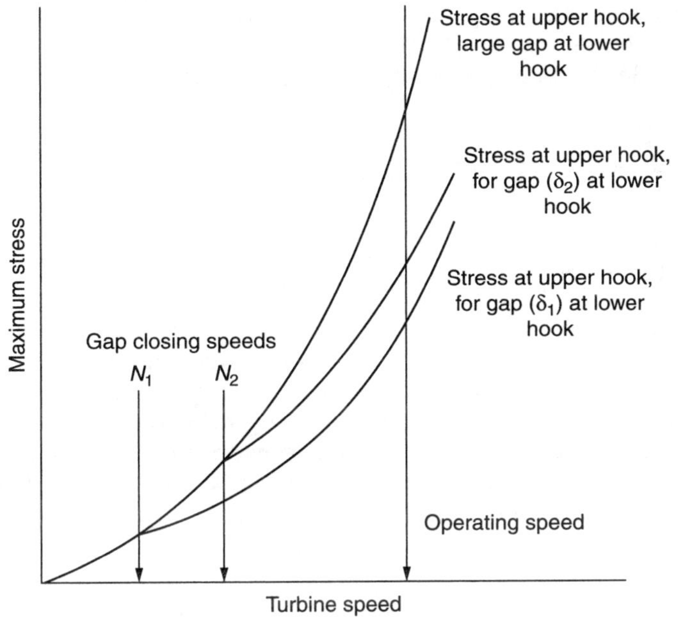

FIGURE 8.13 Reduction of stress at upper hook when lower hook closes.

It reduces the stress at the upper hook and there will be increase of stress in the lower hook. This scenario is depicted in Fig. 8.13.

Condition 3: Contact at Lower Hook, but a Gap at Upper Hook

If the starting gap exists at the upper lock, then a similar analysis may be performed as shown in condition 2. The behavior will be similar, but the magnitude of stress and speed to close a starting gap will be different.

8.9 Summary of Design Criteria

There are industry standards that manufacturers must comply with. One such standard is put forth by the American Petroleum Institute (API 612, 5th edition, 2003 or latest). This document states that the manufacturer shall verify by using the Campbell diagram or equivalent for each blade row to avoid following excitations after frequencies are corrected for speed and temperature.

$$15 \times \mathrm{RS}$$

$$1 \times \mathrm{NPF},\ 2 \times \mathrm{NPF}$$

Some of, but not all, the other excitations to be considered are the following: the first and second harmonics blade passing frequency stationary vanes frequency upstream and downstream of the blade row, steam passage splitters, irregularities due to vane pitch at horizontal split, casing openings (extraction or exhaust), partial arc, internal struts, meshing frequencies of gear mesh, etc.

Modes to be considered include the following:

In-phase tangential
Out-of-phase tangential
Axial
Torsional
Other response modes

Similar guidelines, rules, and criteria are followed by manufacturers. Excitations will depend on the design of the stage. The acceptable amount of the difference between excitation frequency and natural frequency is influenced by accuracy of the prediction, successful past experience, and by the features of blade construction. Thus, it is difficult to prescribe a numerical value to the difference between frequencies, but the above questions should be answered by the manufacturer and supported by analysis, test data, and past experience.

8.10 Checklist for Auditing a Blade Design

Some of, but not all, the questions should be satisfactorily answered for final decision as listed below. Depending on needs, questions may be added, or some may not be relevant for the project at hand.

Questions Pertaining to Performance of Turbine

What are the sizes of the inlet and exhaust openings?
What are the calculated inlet and exhaust velocities?
What are the implications if they are high?
Do they conform to relevant specifications?
Is there room for variation in steam condition, future up rate, and off-design condition?
How do they compare to similar steam turbines?

Is this a single-valve or multivalve turbine?
Does this conform to relevant specifications?
Is this appropriate to the service intended for this machine?

What type and/or manufacturer is the T&T (trip and throttle) and governor valves?
Does this conform to relevant specifications?
What are the fully open loss characteristics of these valves?
Are these loss characteristics substantiated by test data?

Is the bladed flow path characterized as "impulse," "reaction," or a combination of both?

Is the expected efficiency of this turbine better than that of the competitors, or better than earlier machines offered by this vendor?

Is new or different design or construction used for the flow path components?

Does the design appear to employ clean transitions, or "guided flow," between adjacent blade rows?

Do the airfoils in each blade row meet common expectations of angles and shapes?

Do any of the blade rows use special radial stacking, such as "leaning," "bowing," or "compound lean"?

Are special end wall shapes used, other than cylinders or cones?

Is the purpose of any unusual shapes explained?

Are claimed performance improvements substantiated by tests or credible analysis?

Is improved shaft or flow path sealing used to improve efficiency?

What is the principle behind the improved seal performance?

Does the seal design have proven value and reliability in this service?

Is moisture removal necessary to achieve efficiency and reliability expectations?

Is the scheme for trapping moisture outside the flow path rational?

Is trapped moisture expelled from the turbine case or passed from stage to stage?

Questions Pertaining to Structural Reliability of Turbine

Experience

Validated Blade

What is the experience with this blade?

How many times has it been used?

Is this a scaled derivative of a previously used blade?

What kind of service—constant load, variable speed, or cyclic—is the blade capable of?

What is the maximum temperature of its use?

What is the maximum speed at which it has been used?

What was the range of height of its use?

Were there any reported failures?

What was the reason for the failure?

Unvalidated Blade

Describe the reasons for selecting this blade.

Compare the prior successful experiences to this design (tip speed, hp/blade, maximum centrifugal stress, maximum bending stress, mean stress, alternating stress, and others as appropriate).

Acceptance Based on Stress

What is the maximum mean stress?

At what condition (maximum power, maximum continuous speed, trip speed)?

What is the maximum alternating stress?

At what condition (maximum power, maximum continuous speed, trip speed)?

How were the stresses calculated (FEA method, based on experience, etc.)?

How is reliability estimated (for fatigue damage, creep damage)?

Compare calculated results to similar successful stages in service.

Acceptance Based on Vibration Consideration

What was the analytical method used to estimate natural frequency?

Was the frequency analysis performed on a single blade?

Was the frequency analysis performed on a packet of blades?

Was the frequency analysis performed on a bladed disk?

What is the expected accuracy of the analytical results?

Are there any test data to correlate with analytical results?

What kind of test was used (bench test, running strain gage test, etc.)?

Was the influence of manufacturing tolerance considered?

What are the considered excitations?

Describe the method to ensure resonance-free operation.

If resonance is unavoidable, how is the response estimated?

Is it a partial admission stage?

If yes, how is the reliability estimated?

Describe the acceptance criteria.

Describe how the effects of erosion and corrosion have been taken into account.

How is the reliability of "locking blade" and "locking pin(s)" estimated?

Manufacturing

Method of manufacture
Drawn
Machined from bar stock
Cast
Forged
Envelope forge

Surface condition
Surface finish
Surface treatment if any and why (shot peening, hipping, etc.)
Erosion protection
Stellite
Flame hardening

Geometric tolerance on airfoil, length, and taper and twist
Root attachment

Tenon
Method of manufacturing
Method of pinning
Shape (round, kidney, sectional)
Shroud
Method of manufacturing
Method of putting hole
Assembly of shroud

Assembly
Blades on disk (caulking, shims, etc.)
Shroud on blades (control of peening operation)
Static weight balance (short- and moderate-length blades)
Moment balance (long or heavy blades)
Interim rotor balance
Blade row closure (pinning or other method)

Materials of Construction

List materials.
What is the specification of the material?
What is the experience of the material in a similar application?
How are the properties being determined?
Will the material certifications be supplied?

Quality

Characteristics to be controlled
- Material certifications
- Root dimensional tolerance
- Blade profile tolerance
- Surface finish
- Surface modification (e.g., shot peening, erosion shields, other)
- Tenon peening qualification (assembled stage)
- Trailing-edge area (assembled stage)
- Root to disk groove or firtree contact check
- Others identified as appropriate to specific blade and contract

Controlling document or standard for each controlled characteristic
- Acceptance criteria
- Methods
- Industry standards (API, ASTM, ANSI, etc.)
- Proprietary standards (manufacturer, A&E, end user)

Documentation requirements
- Certificate of compliance, MTRs, PQRs, process qualifications, calibrations, etc.
- Identify for each test or inspection

Witness requirements
- Identify affected test or inspection
- Responsible party
- Notification requirements
- Where (supplier, subvendor, independent test lab, etc.)

Review and approval requirements
- Document review—for approval and for review
- When: at time of inspection or at contract close

CHAPTER 9

Life Assessment Aspects for Blade

9.1 Assessment of Useful Life of Blade in Presence of High Cycle Fatigue

The concept, design, and manufacturing of rotating and reciprocating machinery were very much evident during the industrial revolution of the 1800s. The experience of the industry was that material was failing under repeated loading. This posed a real problem because design that was deemed safe failed under repeated loading. The surprise failure of ductile material seemed to be a brittle failure. Detailed subsequent investigations identified the concept of fatigue failures, where the failure is due to a combination of steady load and repeated load cycles.

Failure due to fatigue is the major reason for the vast majority of component failures in ground, air, and sea vehicles. Most turbine blade failures are the result of the application of a cyclic load. Thus, considerable attention in design is given to deal with and avoid/alleviate fatigue problems. This section describes some of the key aspects that need to be considered in blade design.

The fatigue data of the material are plotted on a diagram called the *S-N* curve. Out of this evolved the concept of the factor of safety based on many different diagrams, e.g., a Goodman diagram. Designers use these methods to estimate the reliability of the blades by establishing an acceptable factor of safety under combined loading of steady stress and alternating stress. A factor of safety above a certain predetermined number provides an acceptable design. However, this method does not yield a quantitative estimate of safe life, i.e., how many cycles are required for fatigue damage.

This section examines and takes advantage of the available theories of fatigue failure to estimate the number of cycles to failure during high cycle fatigue. Many equations and figures have been discussed in earlier sections. However, to make the current discussion complete, they will be included here.

The following terminology will be used in this section:

σ_a = alternating stress

σ_m = mean stress

σ_e = fatigue strength

σ_{ult} = ultimate strength

$\Delta\varepsilon$ = strain range

$\Delta\sigma$ = stress range

σ_f' = fatigue strength coefficient

b = fatigue strength exponent

d = fatigue ductility exponent

$A = \sigma_f'/E$ for zero mean stress

C = fatigue ductility coefficient

E = modulus of elasticity

FS = factor of safety based on stress

N_f = cycles to failure

The response of a material under a cyclic load where the amplitude of the stress is above the elastic limit or yield strength of the material is discussed. In the beginning when load is increased during a cycle, the material follows the monotonic stress-strain curve. After the load has exceeded the yield strength of the material, it is plastically deformed. After unloading, the material does not return to its original position, and there is a permanent deformation. The test data show a hysteresis loop when a material is subjected to repeated cycle loads. The area of the loop is proportional to the work that has not been recovered. The area under the hysteresis loop represents the energy per unit volume that has been stored in the material. Figure 9.1 shows such a hysteresis loop, and the plot of strain amplitude vs. cycles to failure is shown in Fig. 9.2, which is similar to the *S-N* curve for the material.

The total strain range consists of the elastic strain range (recovered after unloading) and an inelastic strain, or plastic range (not recovered after unloading), as shown in Fig. 9.1.

$$\text{Total strain} = \text{elastic strain} + \text{plastic strain}$$

As discussed above, the test data of many steels when plotted on a log-log scale will fall on a straight line for both strains (elastic and inelastic) vs. cycles to failure. It is equivalent to a power relationship when analyzed on a linear scale. Equation (9.1) is such a relationship,

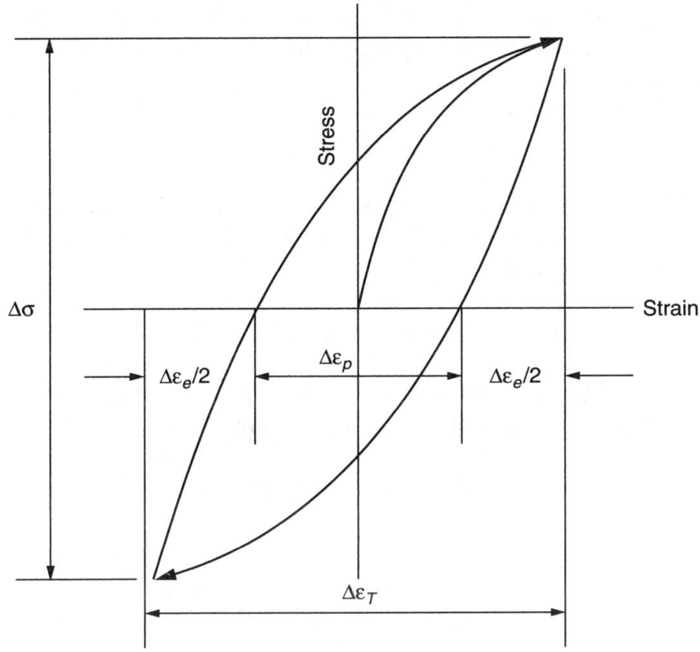

FIGURE 9.1 Hysteresis loop during cyclic loading.

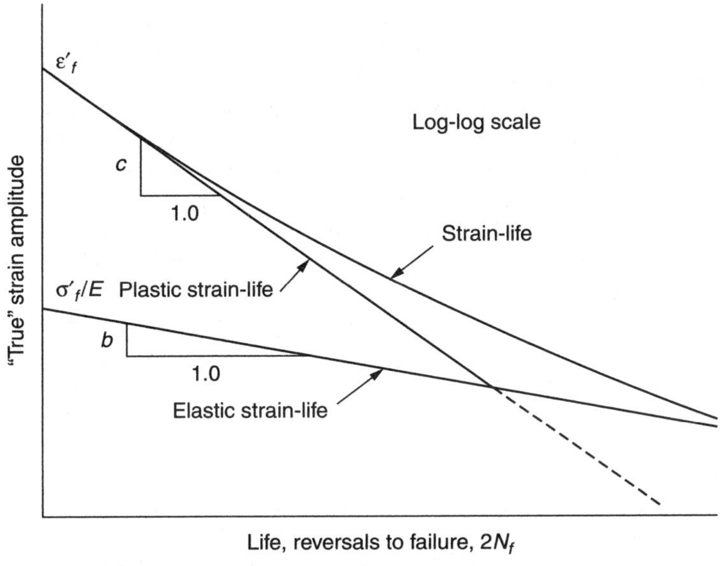

FIGURE 9.2 Strain vs. cyclic curve.

known as the Manson-Coffin relationship when the mean stress is zero. This relationship is expressed in total strain range or strain amplitude.

$$\Delta\varepsilon_{\mathrm{Total}}/2.0 = AN_f^b + CN_f^d \tag{9.1}$$

After reviewing and analyzing the test data of many materials, Morrow (in Graham et al., 1968) modified Eq. (9.1) to include the effect of mean stress as follows:

$$\Delta\varepsilon_{\mathrm{Total}}/2.0 = [(\sigma_f' - \sigma_m)/E]\, N_f^b + CN_f^d \tag{9.2}$$

In the case of high cycle fatigue (HCF), the inelastic strain can be considered to be negligible, i.e., plastic strain = 0.0.

$$\Delta\varepsilon_{\mathrm{Total}}/2.0 = [(\sigma_f' - \sigma_m)/E]\, N_f^b \tag{9.3}$$

This will be used as the basic relationship between the strain and the cyclic life when the inelastic strain is negligible. Thus, this relationship can be considered to represent the case for high cycle fatigue.

For a blade reliability evaluation, designers construct a Goodman diagram (or any other diagram) and estimate a factor of safety for the design. The need is to estimate safe design life of the blade in the event of high cycle fatigue. A process that may be used to estimate HCF life from test data is detailed in Figs. 9.3 to 9.6. The graphical process to create a Goodman diagram from test data starts from Fig. 9.3. This figure shows a sample under the action of an alternating load, and it includes the resulting alternating stress with zero mean stress. Figure 9.4 is a typical *S-N* diagram that is obtained by repeating the test on many samples with different magnitudes of imposed stress. It is the plot of applied stress on the vertical axis and the number of cycles to failure on the horizontal axis. Multiple tests are generally required to run at each stress level to estimate any statistical variability that may exist in the property. An average curve is plotted, shown in Fig. 9.4, and it is now straightforward to draw a Goodman diagram.

Figure 9.5 is the plot of the alternating stress on the vertical axis corresponding to any desired life. This point is connected by straight line to the ultimate stress on the horizontal axis. This line represents a life that is expected, if applied alternating stress is equal to the point on the vertical axis for zero mean stress. A point is located on the diagram based on the estimated alternating and mean stress of a design. If this point lies on the line, then the life of the design is given by the life representing that line. If the point falls below this line, then life should be greater. Each radial line represents a life corresponding to the fatigue strength on the vertical axis.

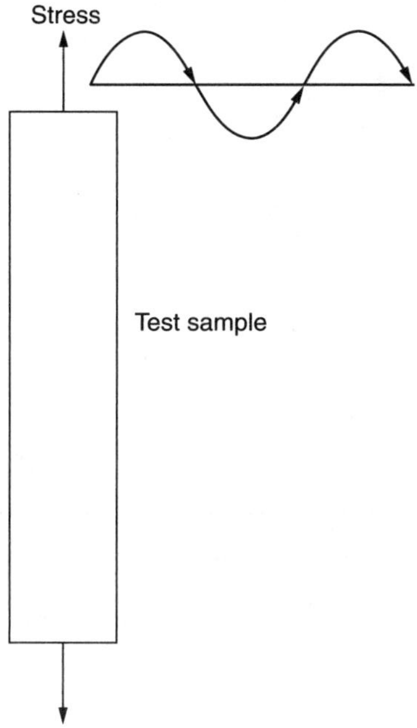

FIGURE 9.3 Test sample for fatigue test.

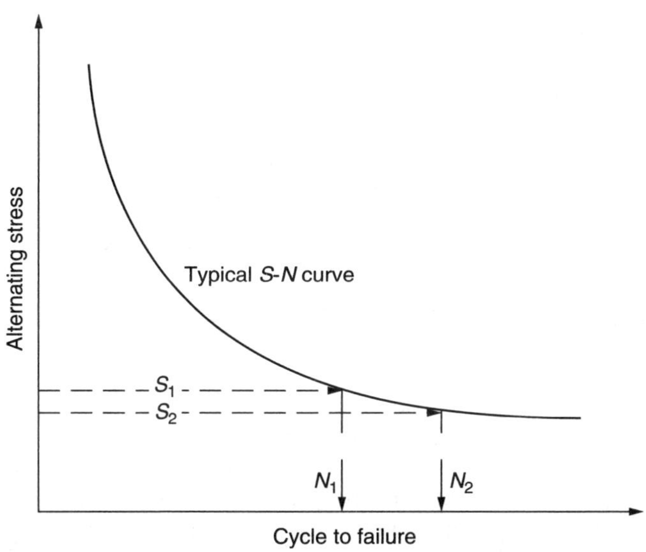

FIGURE 9.4 Typical *S-N* curve.

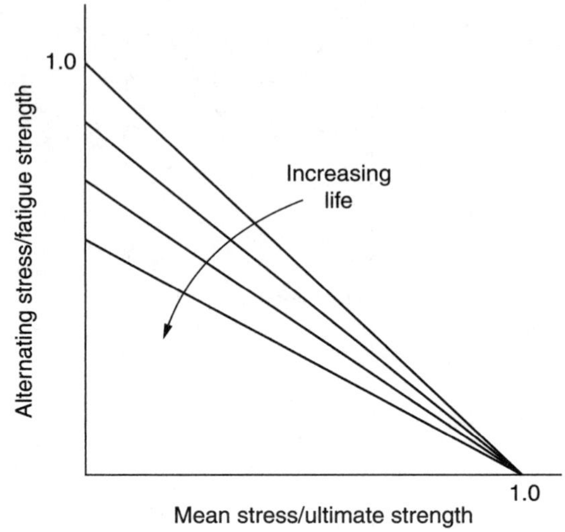

FIGURE 9.5 Goodman diagram with constant life lines.

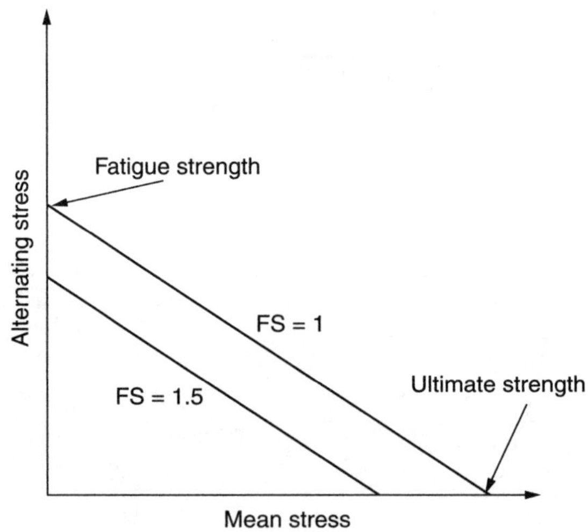

FIGURE 9.6 Goodman diagram with factor of safety line.

Traditionally, the use of a Goodman diagram is as shown in Fig. 9.6. A radial line corresponding to the fatigue strength represents a factor of safety equal to unity. A point is located on this diagram corresponding to alternating stress and mean stress estimated for the design in question. From this design point, a parallel line to the original radial line is drawn to estimate a factor of safety by using Eq. (9.4).

9.2 Factor of Safety Concept for High Cycle Fatigue

A factor of safety (FS), based on stress or a margin is estimated by

$$\sigma_a/\sigma_e + \sigma_m/\sigma_{\text{ult}} = 1/\text{FS} \tag{9.4}$$

The plot of σ_a vs. σ_m [Eq. (9.4)] for a given material's properties is a straight line, as shown in Fig. 9.6 where the top line is the plot when FS equals 1.00 and the bottom line corresponds to a factor of safety of 1.50. Any combination of alternating and mean stress that falls on this line will have a factor of safety equal to 1.50. Figure 9.5 represents a life concept, and Fig. 9.6 represents the margin in the design when the permitted stresses are limited to a lower magnitude.

The estimate of blade life due to high cycle fatigue can be made in conjunction with the Goodman concept. This method provides an estimate of life by taking advantage of stresses used in constructing a Goodman diagram (Singh, 2001).

9.3 Life Estimation

It is appropriate to use Morrow's equation which includes the effect of mean stress on cycles to failure, but neglecting the inelastic stress reduces Morrow's equation to the form of Eq. (9.3).

$$\Delta\varepsilon_{\text{Total}}/2.0 = [(\sigma_f' - \sigma_m)/E]\, N_f^b$$

$\Delta\varepsilon_{\text{Total}} = (2\sigma_a)/E$ where σ_a is alternating stress and one-half of stress range

$$\sigma_a = (\sigma_f' - \sigma_m)\, N_f^b \tag{9.5}$$

When mean stress is zero, i.e., $\sigma_m = 0$, then $\sigma_a = \sigma_e$. Substituting this into Eq. (9.5) yields

$$N_f^b = \sigma_e/\sigma_f' \tag{9.6}$$

Substituting Eq. (9.6) into Eq. (9.5) and rearranging yield

$$\sigma_a/\sigma_e + \sigma_m/\sigma_f' = 1 \tag{9.7}$$

For a constant life, the relationship between σ_a and σ_m is a straight line, as shown by Eq. (9.7). The straight line is defined by the *x* intercept of σ_f' and the *y* intercept of σ_e. This equation has the same form as the Goodman equation, which contains the quantity σ_{ult} in place of σ_f'. The Goodman line of Eq. (9.8) and the straight line of Eq. (9.7) are shown in Fig. 9.7. Morrow's equation in essence estimates a longer life than that given by the Goodman radial line and σ_f' (point *A*) as

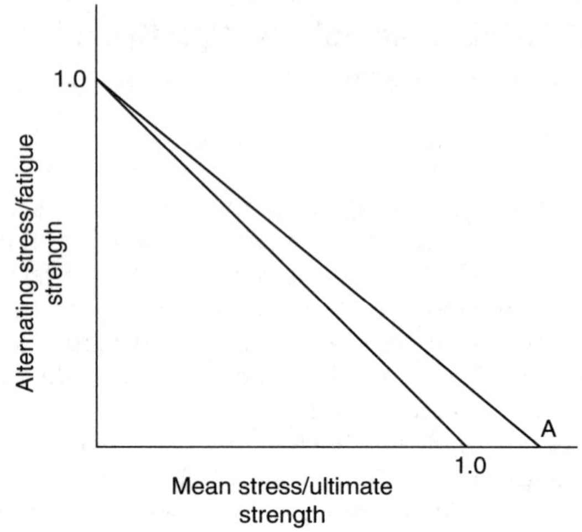

FIGURE 9.7 Comparison of Goodman and Morrow equations.

the failure point compared to σ_{ult} that is used in the Goodman diagram.

The stress at final failure σ_f' can be estimated from the cyclic stress-strain curve (as discussed earlier). The Goodman equation in Eq. (9.4) for FS = 1 is

$$\sigma_a/\sigma_e + \sigma_m/\sigma_{\text{ult}} = 1 \tag{9.8}$$

To be consistent with the Goodman equation and to insert more conservatism in the estimation of life, σ_f' is substituted with σ_{ult}. By combining the Goodman equation in Eq. (9.4) and the life equation in Eq. (9.5), an expression for the cycles to failure N_f for a specified FS is obtained. It can be expressed in terms of either mean stress or alternating stress for a given factor of safety. These expressions are given as Eqs. (9.9) and (9.10).

The expression for N_f when FS and the mean stress are known is

$$N_f = \{(1/\text{FS} - \sigma_m/\sigma_{\text{ult}})[\sigma_e/(\sigma_{\text{ult}} - \sigma_m)]\}^{1/b} \tag{9.9}$$

The expression for N_f when FS and the alternating stress are known is

$$N_f = \{[\sigma_{\text{ult}} - \sigma_{\text{ult}}(1/\text{FS} - \sigma_a/\sigma_e)]/\sigma_a\}^{-1/b} \tag{9.10}$$

The implication of Eqs. (9.9) and (9.10) regarding the life of blade design can be observed in Fig. 9.8. This shows a superimposition of Figs. 9.5 and 9.6. The radial lines *AF, BF, CF, DF,* and *EF* represent constant life in increasing order. Line *AG* is a Goodman-based factor of safety as defined by Eq. (9.4). The intersecting points *A, H, I, J,* and

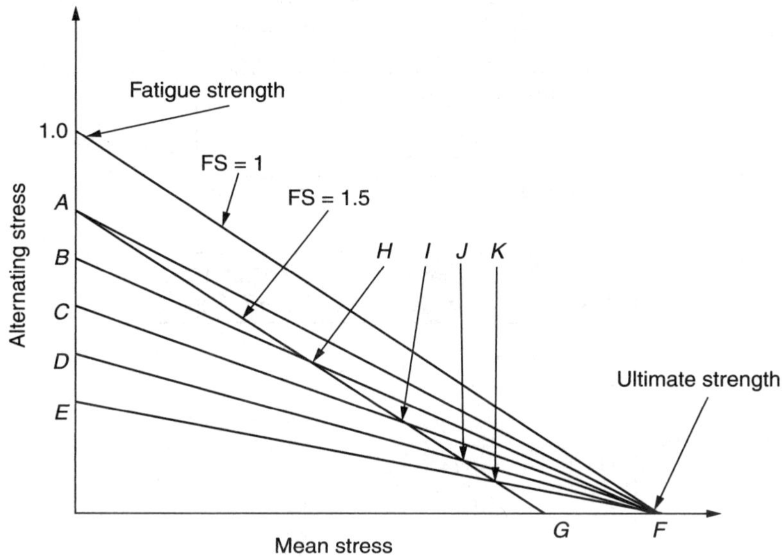

FIGURE 9.8 Superimposed Goodman lines.

K, although they represent the same factor of safety, lie on lines denoting different lives. The implication of this observation is that designs having the same factor of safety as defined by Eq. (9.4) will yield different lives. Life depends on the magnitude of the mean and alternating stresses.

The importance of the above observations from Fig. 9.8 is examined by an example, which will demonstrate the following points:

1. It is important to estimate the life of the design together with the factor of safety.
2. If the stresses that have been used to estimate the factor of safety are known, then the life estimation can be made by using either Eq. (9.9) or (9.10).
3. It might be possible for a design with a lower factor of safety to yield higher life than one that has a higher factor of safety.
4. Many times, difficulties arise in ranking designs for technical superiority, and the techniques shown so far may be used to gain insight into the reliability of various alternative designs.

Eight different designs of similar components and manufactured with identical materials were examined. Damage in these designs is considered to be due to high cycle fatigue. Table 9.1 lists the properties of the material used for this example. Table 9.2 lists the stresses used for this example. Table 9.3 lists the eight designs with their associated factor of safety estimated by using Eq. (9.4).

	Magnitude, ksi
Ultimate strength	125
Fatigue strength	78

TABLE 9.1 Properties of the Material

Design #	Steady Stress, ksi	Alternating Stress, ksi
A	30	20
B	35	18
C	32	17
D	31	18
E	50	12
F	60	10
G	40	16
H	65	8

TABLE 9.2 Steady and Alternating Stresses of Eight Designs

Design #	FS
A	2.014
B	1.958
C	2.110
D	2.089
E	1.806
F	1.644
G	1.904
H	1.606

TABLE 9.3 Eight Designs and Their Factor of Safety

The factor of safety is evaluated using the Goodman equation [Eq. (9.4)], and an examination of the data of Table 9.4 shows that design C ranks best while design H is rated the worst among the eight designs considered.

The important information to be derived from the data is the safe operating life of the component, and the factor of safety in itself does not provide this crucial information. Hence, the HCF life of each

Design #	FS	Ranking Based on FS
C	2.110	1
D	2.089	2
A	2.014	3
B	1.958	4
G	1.904	5
E	1.806	6
F	1.644	7
H	1.606	8

TABLE 9.4 Ranking of Designs Based on Factor of Safety

Design #	FS	Ranking Based on FS	Life Cycles	Ranking Based on Life
H	1.606	8	1.06E + 14	1
F	1.644	7	1.53E + 13	2
E	1.806	6	3.94E + 12	3
G	1.904	5	3.65E + 11	4
C	2.110	1	2.74E + 11	5
D	2.089	2	1.64E + 11	6
B	1.958	4	1.42E + 11	7
A	2.014	3	6.17E + 10	8

TABLE 9.5 Eight Designs and Ranking Based on FS and Estimated Life

design is estimated by using Eq. (9.9). Table 9.5 lists the estimated HCF life for each design and the ranking based on life. The surprising fact is that the worst design (H) based on the factor of safety becomes the best if ranking is done based on life. This is important information because the main concern must be safe operating life.

The mean and alternating stresses for each design are listed in Table 9.2. Through examination of this table it becomes clear that the major parameter impacting life is the alternating stress. The next deciding factor is the mean stress. Designs B and D have the same alternating stress, but design B has a larger mean stress, thus lower life.

The unexpected results shown in Table 9.5 indicate that a quick estimate of life may differentiate many designs. The assumption is that the variability in manufacturing, material properties, and stress is the same for each design. Alternatively, if the variations are different for each design, then a probability simulation study for life is recommended.

9.4 Process of Shot Peen and Laser Peen to Improve Fatigue Life

Surface treatment, e.g., shot peen and laser peen, induces compressive stress on the surface of a part that in turn increases the fatigue resistance capability of the part. For example, the root attachment and base of the airfoil have high stresses. These are the typical locations where fatigue cracks in turbine blades have been observed. These locations have both high steady stress and high alternating stress, and cracks initiate and propagate from these locations. Residual compressive stress is introduced by the process of shot peen and/or laser peen that reduces the applied stress, resulting in an increase of the fatigue life of a component. These methods of surface treatment have been widely used and have been successful in increasing the surface life of components.

John Almen is credited with discovering the shot peen method. The basic test strip to estimate the magnitude of compressive stress has been named after him and is called the *Almen strip*. Almen observed that when a metal sheet is blasted by shot, it tends to bend and to stretch the sheet, which essentially led to the standard of measuring intensity.

Shot peen is a process in which a compressive layer in the metal is introduced by plastically deforming the metal near the surface. It is achieved by impacting the metal surface by a round metallic glass or by ceramic particles called shots. The velocity of impact is such that it plastically deforms the metal at the site of impact. The plastic deformation at the surface introduces a compressive stress, but it also introduces a tensile stress in the substrate, which is due to the force balance required in the net section.

This is essentially a cold working process, and each shot strikes the metal as a hammer does, creating dimple like indentations on the surface (Fig. 9.9).

The creation of these indentations is the result of plastic deformation. The material in the substrate while trying to restore the original shape undergoes compression underneath the indentations or dimples (Fig. 9.10). Uniform compressive layer may result due to overlapping indentations.

Depending on the geometry of the real part, its material, shot material, shot intensity, and shot coverage of the surface, it has been observed that fatigue resistance of the material has increased up to 1000 percent in some cases.

9.5 Basic Explanation for Increase in Fatigue Life

The basic fatigue property of a material is depicted through a strain amplitude vs. life curve (similar to the *S-N* curve in Fig. 9.11). This shows the expected fatigue life of the part for any applied alternating load. The applied alternating stress is σ_1, and the corresponding fatigue

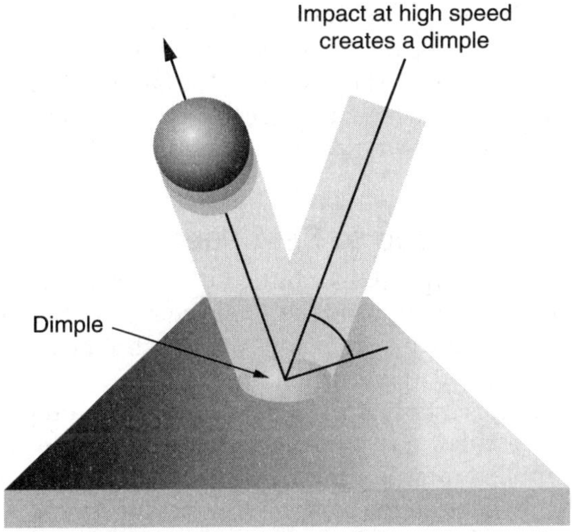

FIGURE 9.9 Plastic deformation at the site of impact. (*Courtesy of Metal Improvement Co.*)

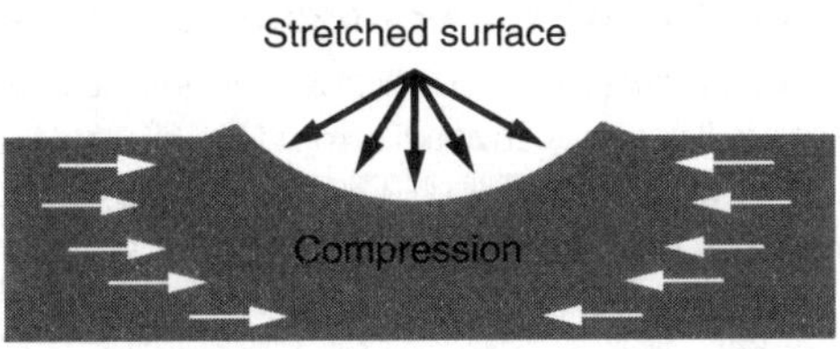

FIGURE 9.10 Residual compressive zone. (*Courtesy of Metal Improvement Co.*)

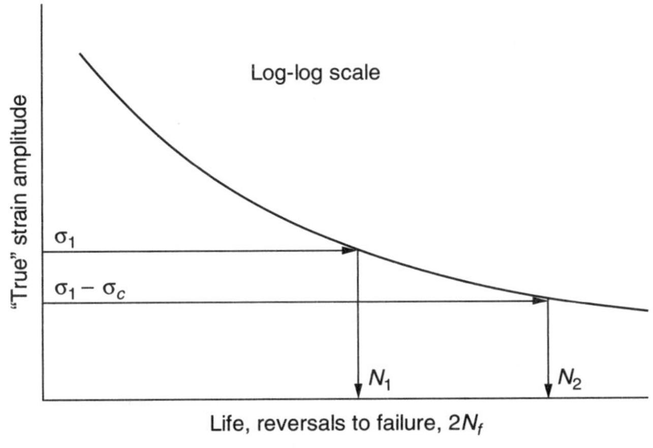

FIGURE 9.11 Strain amplitude vs. life curve.

life is denoted by N_1. To increase the fatigue life N_1 is to reduce the alternating stress, after shot peen compressive stress σ_c is introduced at the surface, resulting in a net $\sigma_2 = \sigma_1 - \sigma_c$. The fatigue life corresponding to σ_2 is denoted by N_2, which is larger than N_1. The improvement will be a function of factors of the process described earlier.

9.6 Residual Stress due to Shot Peen

Figure 9.12 shows a typical distribution of residual stress that results from shot peen. The magnitude at the surface and the depth of the compressive zone depend on factors such as shot medium, its intensity, size, and material of construction.

The tensile stress starting from the depth of the compressive layer results from the balance of force, which may be explained with the following simple example. The depth of the compressive layer is denoted by d. To balance the compressive residual stress, tensile stress is induced in the subsurface. The total compressive force is balanced by the resulting tensile force. Based on some shot peen data, it is assumed that the peak tensile stress diminishes to zero at the boundary.

9.6.1 Linear Approximation of Compressive Layer Profile

The magnitude of compressive stress at the surface is important in itself, and its profile becomes important in the estimation of the resulting stress distribution under load. The actual profile must be determined by a test. However, in most practical applications during design phase, the magnitude of the stress may not be available. In its

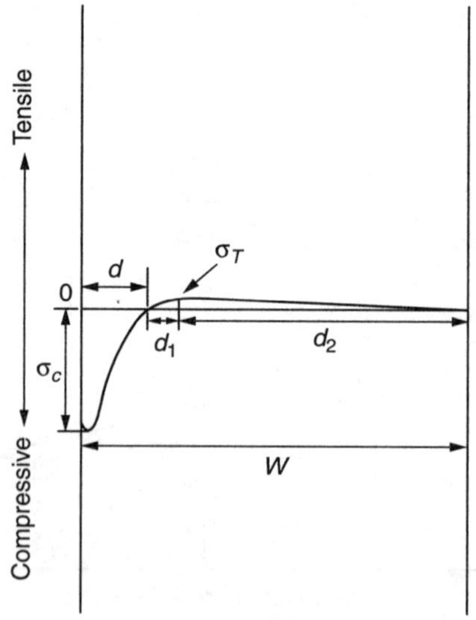

Figure 9.12 Typical residual stress profile after shot peen.

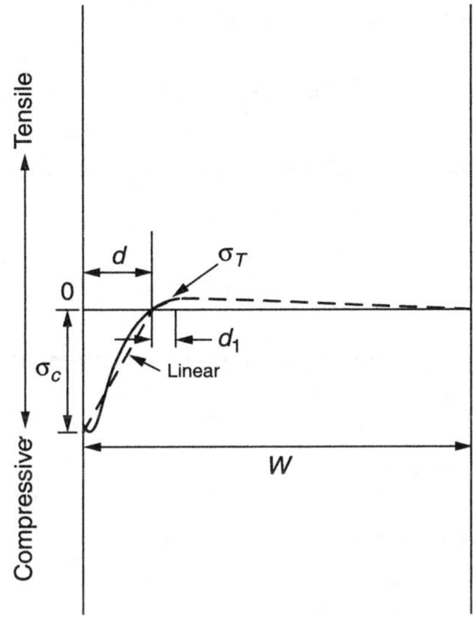

FIGURE 9.13 Linear approximation of compressive profile.

absence an approximation can be made. The simplest way is to assume that the profile must consist of two straight lines. The first line starts from the surface at the compressive stress, and it passes through the depth of the compression zone until it reaches a tensile stress to balance the forces created by the compression. The second line starts from this point, and it ends up at the boundary with zero stress. This approximation is depicted in Fig. 9.13.

The equations of these two straight lines are given below.

The equation for line 1 for the distance from the surface to $d + d_1$ is equal to

$$\sigma_1 = (\sigma_c/d)(x - d) \tag{9.11}$$

The equation for line 2 between $d + d_1$ and W is equal to

$$\sigma_2 = \sigma_T(W - x)/(W - D) \tag{9.12}$$

where σ_c = resulting peak compressive stress assumed at surface
σ_T = resulting peak tensile stress in body to balance compressive force
W = width of sample

Estimation of Location of Peak Tensile Stress

Compressive force = area under compressive layer multiplied by σ_c

$$\text{Compressive force} = (1/2)d\sigma_c \tag{9.13}$$

$$\text{Tensile force} = (1/2)(W - d)\sigma_T \tag{9.14}$$

These forces must be equal as stated below.

$$(1/2)(W - d)\sigma_T = (1/2)d\sigma_c \tag{9.15}$$

The ratio of the tensile peak stress to the peak compressive stress is given by

$$\sigma_T/\sigma_c = 1/(W/d - 1) \tag{9.16}$$

Figure 9.14 shows the plot of the percent tensile peak stress as a function of the width of the beam.

$$\text{Slope of compressive stress} = \sigma_c/d \tag{9.17}$$

$$\text{Slope of resulting tensile stress} = \sigma_T/d_1 \tag{9.18}$$

$$\sigma_T/d_1 = \sigma_c/d \tag{9.19}$$

$$d_1 = (\sigma_T/\sigma_c)d = d^2/(W - d) \tag{9.20}$$

Distance from surface D is

$$D = d_1 + d = (\sigma_T/\sigma_c)d + d = Wd/(W - d) \tag{9.21}$$

Figures 9.15 and 9.16 show D as a function of W for the compression depth, where $d = 0.005$ in and $d = 0.006$ in, respectively.

Equations (9.16) and (9.21) provide the estimation of the magnitude of the peak stress and its location from the surface for the estimated depth of the compressive layer.

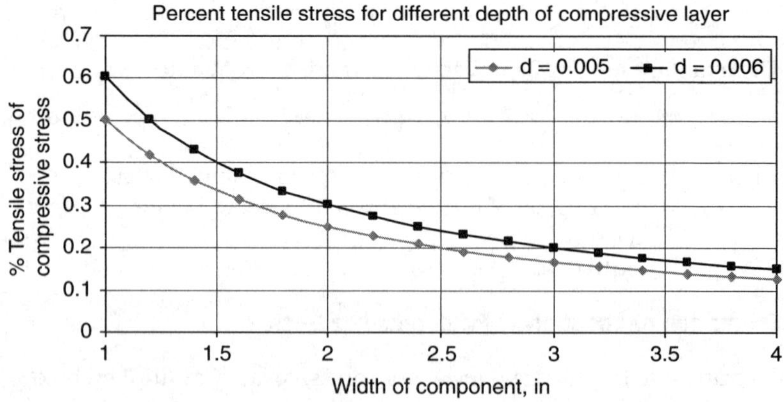

FIGURE 9.14 Tensile stress as percent of compressive stress due to shot peen. (*Courtesy of Safe Technical Solutions, Inc., Bethlehem, PA.*)

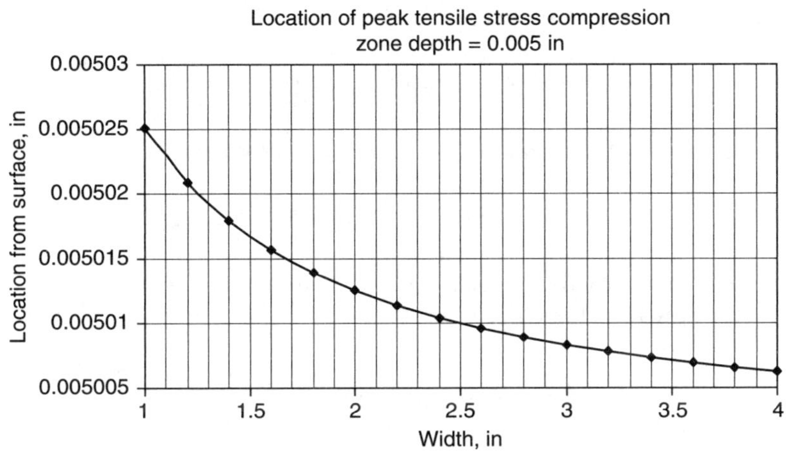

Figure 9.15 Location *D* of peak tensile stress for depth = 0.005 in. (*Courtesy of Safe Technical Solutions, Inc., Bethlehem, PA.*)

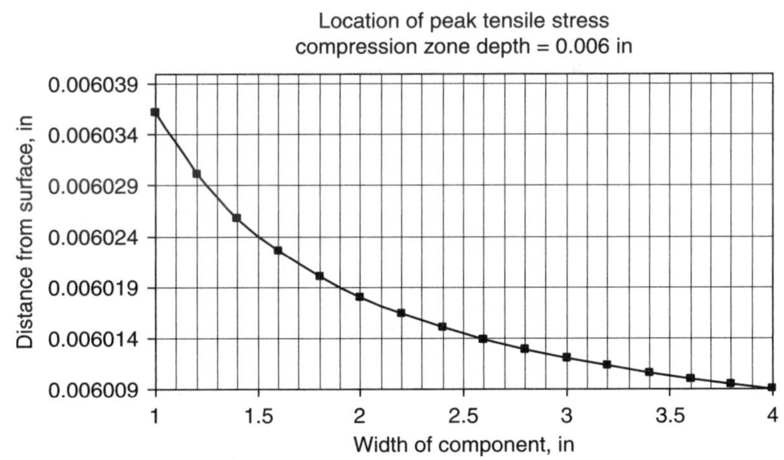

Figure 9.16 Location *D* of peak tensile stress for depth = 0.006 in. (*Courtesy of Safe Technical Solutions, Inc., Bethlehem, PA.*)

9.6.2 Exponential Approximation of Compressive Layer Profile

A better approximate equation may be an exponential equation from the surface to the resulting peak tensile stress point. From this tensile point to the other boundary of the beam, the same straight line is used as described for the linear case. Note that the magnitude of the tensile stress for this case might be different from that of the linear case.

The equation of first curve for $0 < x < (d + d_1)$ is

$$\sigma_1 = \sigma_c(e^{-d} - e^{-x})/(e^{-d} - 1) \tag{9.22}$$

The equation of the straight line for $(d + d_1) < x < W$ is

$$\sigma_2 = \sigma_T(W - x)/(W - D) \tag{9.23}$$

Now d_1 and σ_T can be estimated by performing a force balance.
Compressive force F_1 for $0 < x < d$ is

$$F_1 = \int_0^d \sigma_1 dx = \int_0^d [\sigma_c(e^{-d} - e^{-x})/(e^{-d} - 1)]\,dx$$

$$= \sigma_c[(1 + d)e^{-d} - 1]/(e^{-d} - 1) \tag{9.24}$$

The tensile force F_2 between $d < x < D$ is

$$F_2 = \int_d^D \sigma_1 dx = \sigma_c[e^{-D} + (D - d - 1)e^{-d}]/(e^{-d} - 1) \tag{9.25}$$

The tensile force F_3 between $D < x < W$ is

$$F_3 = \int_D^W \sigma_2 dx = \sigma_T(W - D)/2 \tag{9.26}$$

Then the compressive force must be equal to the tensile force.

$$F_1 = F_2 + F_3 \tag{9.27}$$

Assume $D \approx d$.

$$\sigma_T/\sigma_c = 2[(1 + d)e^{-d} - 1]/[(e^{-d} - 1)(W - d)] \tag{9.28}$$

Hence,

$$\sigma_T = 2\sigma_c[(1 + d)e^{-d} - 1]/[(e^{-d} - 1)(W - d)] \tag{9.29}$$

When $x = D$, the tensile stress σ_T can be estimated from Eq. (9.22) as follows:

$$\sigma_T = \sigma_c(e^{-d} - e^{-D})/(e^{-d} - 1) \tag{9.30}$$

Equate Eqs. (9.29) and (9.30) for $x = D$:

$$e^{-D} = e^{-d} - 2[(1 + d)e^{-d} - 1]/(W - d)$$

$$D = -\mathrm{Ln}\{e^{-d} - 2[(1 + d)e^{-d} - 1]/(W - d)\} \tag{9.31}$$

$$d_1 = D - d \tag{9.32}$$

Figures 9.17 and 9.18 shows D as a function of W for the compression depth, where $d = 0.005$ in and $d = 0.006$ in, respectively.

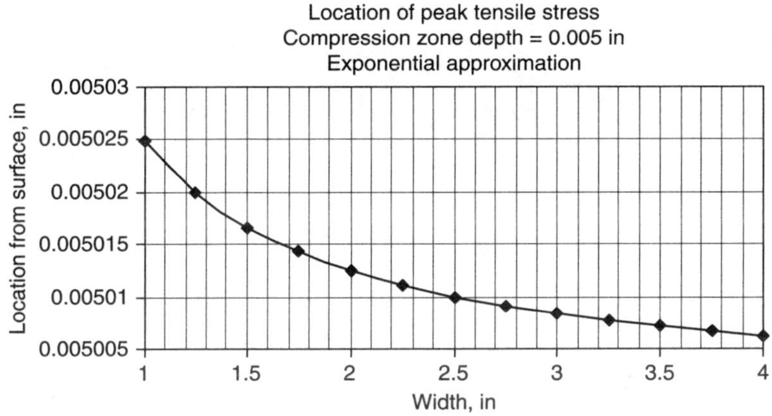

FIGURE 9.17 Location D of peak tensile stress for depth = 0.005 in. (*Courtesy of Safe Technical Solutions, Inc., Bethlehem, PA.*)

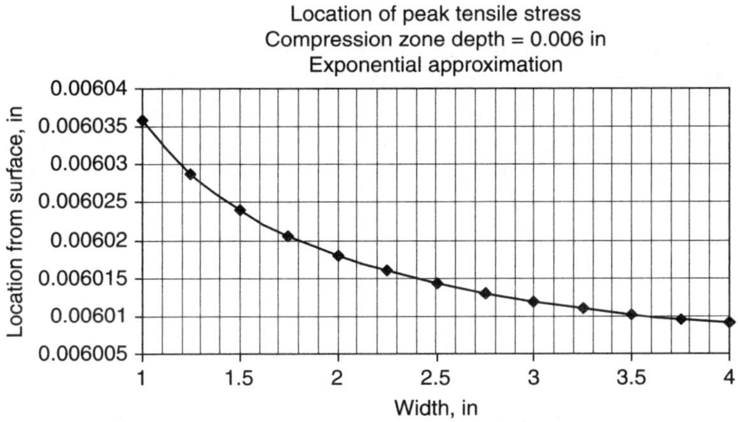

FIGURE 9.18 Location D of peak tensile stress for depth = 0.006 in. (*Courtesy of Safe Technical Solutions, Inc., Bethlehem, PA.*)

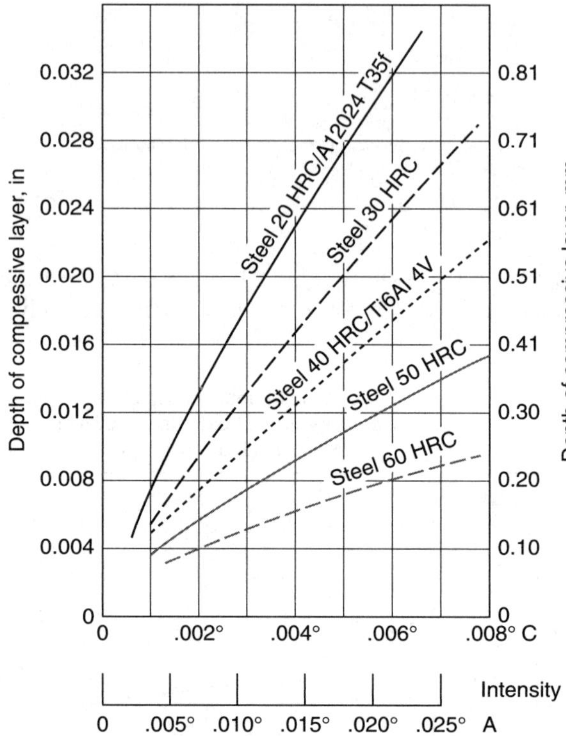

Figure 9.19 Depth of compression vs. Almen arc height. (*Courtesy of Metal Improvement Co.*)

The test data as shown in Fig. 9.19 can be used to estimate the depth of the compressive layer for various steels having different hardness.

The depth also depends on the peen parameters and the hardness of the material. It has been observed that the magnitude of compressive stress is directly related to the material's tensile strength. The higher amount of compressive stress may be achieved in the material with a higher tensile strength. This seems to be due to a rigid crystal structure of such materials. These rigid crystal lattices may withstand larger strains, thereby resulting in higher compressive stress.

9.7 Combination of Applied and Residual Compressive Stress

For the case of a reverse bending of a beam of zero mean stress, see Fig. 9.20.

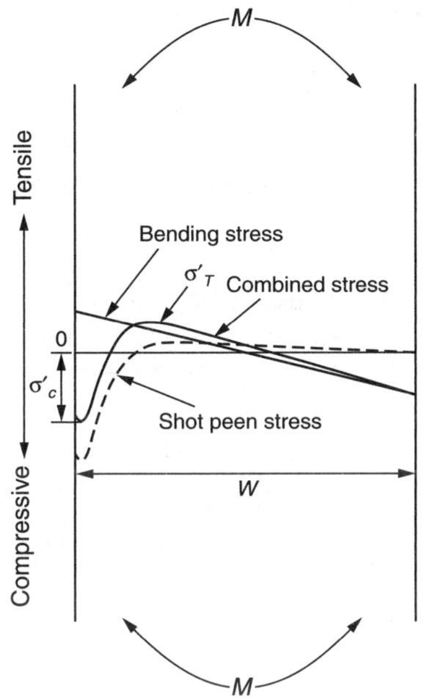

FIGURE 9.20 Combined stress after loading for reverse bending of a beam.

9.7.1 Linear Approximation of Combined Stress

When a load is applied to the shot peen component, the residual stress due to the shot peen modifies the stress due to the applied load. It is essential to estimate the stress under the combined effect of the shot peen stress and the stress generated by the imposed load. The equations for a residual stress profile were discussed earlier. To demonstrate the process, consider the case of pure bending (reversed bending) of a beam as in Fig. 9.20.

Let us assume that the reversed bending moment M generates a tensile bending stress σ_B at the surface. The stress at the opposite surface will be a compressive stress equal to σ_B.

$$\sigma_3 = \sigma_B(1 - 2x/W) \tag{9.33}$$

The resultant stress for $0 < x < D$ equals

$$\sigma_{R_1} = \sigma_3 + \sigma_1 \tag{9.34}$$

$$\sigma_{R_1} = \sigma_B(1 - 2x/W) + (\sigma_c/d)(x - d) \tag{9.35}$$

The resultant stress for $D < x < W$ equals

$$\sigma_{R_2} = \sigma_3 + \sigma_2 \tag{9.36}$$

$$\sigma_{R_2} = \sigma_B(1 - 2x/W) + \sigma_T(W - x)/(W - D) \tag{9.37}$$

9.7.2 Exponential Approximation of Combined Stress

The expression for the resultant stress for the exponential case is developed using the method described for the linear case.

The resultant stress for $0 < x < D$ equals

$$\sigma_{R_1} = \sigma_3 + \sigma_1$$
$$\sigma_{R_1} = \sigma_B(1 - 2x/W) + \sigma_c(e^{-d} - e^{-x})/(e^{-d} - 1) \tag{9.38}$$

The resultant stress for $D < x < W$ equals:

$$\sigma_{R_2} = \sigma_3 + \sigma_2$$
$$\sigma_{R_2} = \sigma_B(1 - 2x/W) + \sigma_T(W - x)/(W - D) \tag{9.39}$$

The approximated shot peen stress, applied stress for the reversed bending, and the combined stress are plotted in Fig. 9.21 for both the linear and exponential methods.

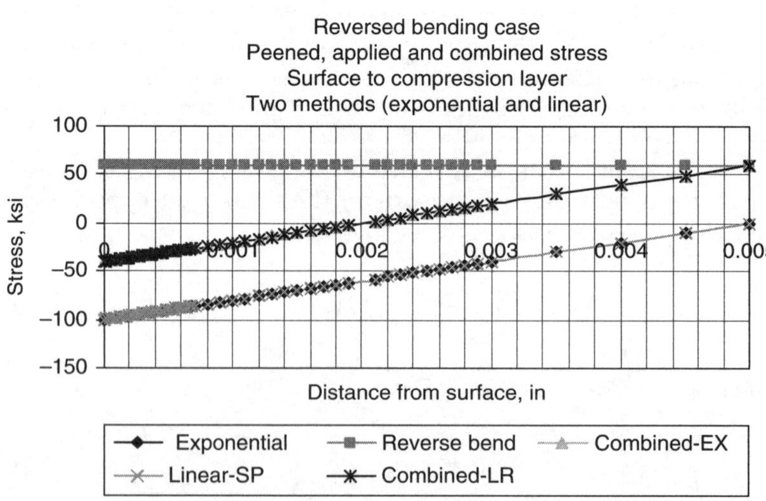

FIGURE 9.21 Shot peen, applied and combined stresses for two methods. (*Courtesy of Safe Technical Solutions, Inc., Bethlehem, PA.*)

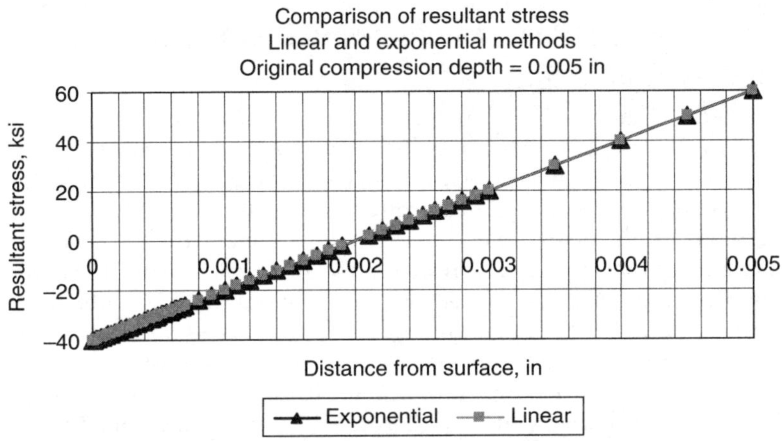

FIGURE 9.22 Combined stresses for two methods. (*Courtesy of Safe Technical Solutions, Inc., Bethlehem, PA.*)

Figure 9.22 shows the combined stress for the two approximations, and it appears that the difference between the two methods is very small. Note that the compression zone of the shot peen of 0.005 in reduces to 0.002 in after application of the load. This is an important aspect to be considered by the designers.

To further explore the difference of the combined stress, the percent difference between the results of the linear method and those of the exponential method from the surface to the compression layer has been plotted in Fig. 9.23.

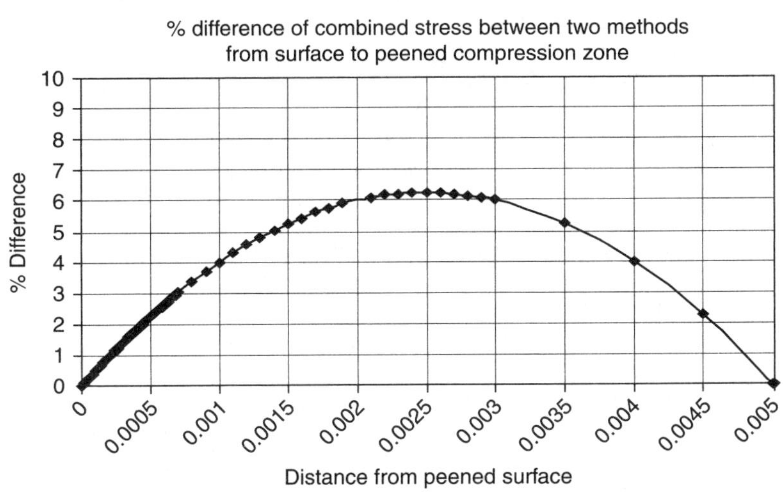

FIGURE 9.23 Percent difference of combined stress. (*Courtesy of Safe Technical Solutions, Inc., Bethlehem, PA.*)

9.8 Mechanistic View of Improvement in Fatigue Life due to Shot Peen

The fatigue damage in the metal may be due to low cycle fatigue and high cycle fatigue. The damage due to LCF happens when alternating stress is high enough to cause the material to yield. In this situation the effect of mean stress on the life is minimal. The strain amplitude (strain range) controls the expected life. However, damage due to HCF also depends on the imposed mean stress. The life estimation due to fatigue was discussed in detail earlier. The example of the reversed bending can be LCF or HCF. Depending on the applied stress and the magnitude of the compressive stress due to the shot peen situation, the damage mechanism can change from LCF to HCF. This will provide a large improvement in life. If the situation is HCF, there is likelihood that a compressive zone due to residual stress will yield greater benefit.

It is clear that the improvement in fatigue life is achieved by lowering the magnitude of the alternating stress when the mean stress is zero. However, an understanding of the situation is not clear, and it can be said that this poses complications in decision making for designers. It is known that the fatigue property of a material depends on the magnitude of the mean stress. The basic fatigue property of a material is obtained from a test for zero mean stress, and it is displayed on the *S-N* curve. The fatigue life of materials directly depends on the amount of residual plastic strains as determined by a hysteresis loop for a given type of cycle. This topic was covered in detail earlier. It is also known that the effect of mean stress in the HCF regime is large while in the LCF regime it is not that pronounced, and most of the time for design purposes it is assumed to be negligible.

The effect of the mean stress on the fatigue property is captured in the following equation, proposed by Morrow, which was discussed earlier.

$$\Delta\varepsilon/2 = [(\sigma_f' - \sigma_m)/E](2N_f)^b + \varepsilon_f'(2N_f)^c \tag{9.40}$$

9.8.1 Improvement during High Cycle Fatigue

Tensile Mean Stress

The Goodman diagram captures the effect of mean stress on fatigue life. Figure 9.24 shows a situation for constant alternating stress in the HCF regime. Three situations have shown the effect of shot peen. The resulting final points for zero mean stress, for applied mean stress, and combined with shot peen stress have been plotted. Radial lines passing through these points from the ultimate strength points estimate the life. Corresponding lives are denoted by N_0, N_1, and N_2 for the mean stress σ_m of 0, $\sigma_m = \sigma_1$, and $\sigma_m = \sigma_1 - \sigma_c$ respectively. Note that $N_0 > N_2 > N_1$.

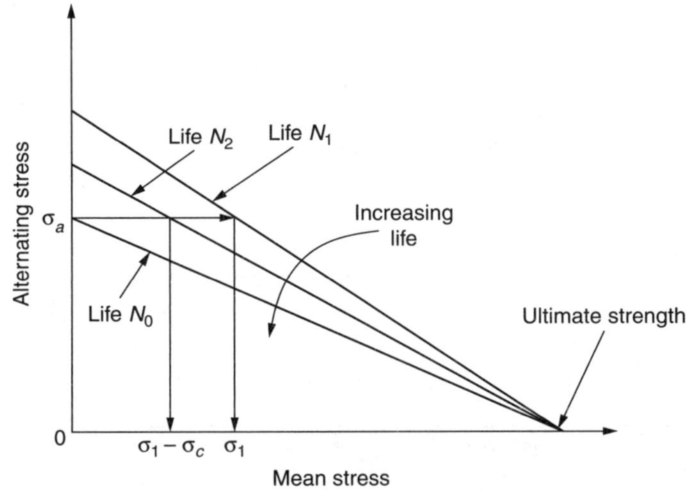

FIGURE 9.24 Goodman diagram showing improvement in life for reduced mean stress.

Another way to look at this situation is by using Morrow's equation. Figure 9.25 shows the same effect of the modification of mean stress on life. It directly shows the effect of material property on life, and the Goodman diagram has been derived from these properties.

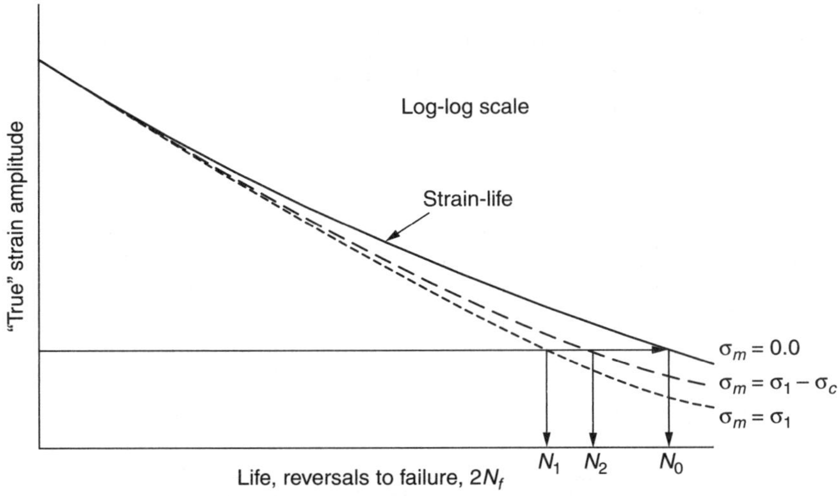

FIGURE 9.25 Improvement in life for reduced mean tensile stress.

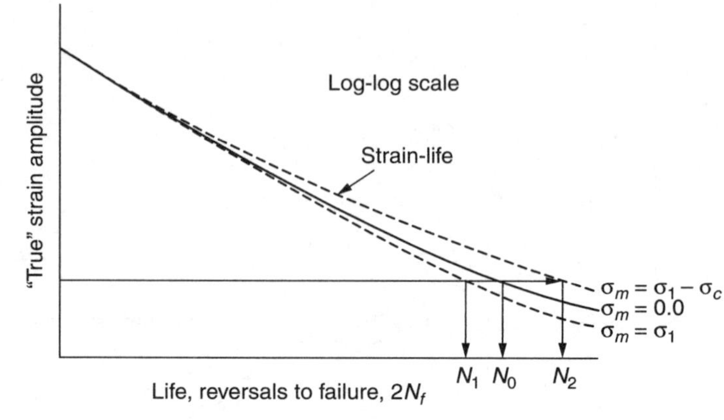

FIGURE 9.26 Improvement in life for reduced mean tensile stress to compressive stress.

Compressive Mean Stress

Figure 9.26 shows the effect of compressive mean stress and considers that the combined stress of applied mean stress and shot peen stress will result in a compressive stress. An assumption is made as before that the alternating stress is in the HCF regime and modification will occur in the mean stress first. The corresponding lives are denoted by N_0, N_1, and N_2 for the mean stress σ_m of 0, $\sigma_m = \sigma_1$, and $\sigma_m = \sigma_1 - \sigma_c$ respectively. It can now be noted that $N_2 > N_0 > N_1$. A large improvement in life is achieved.

9.8.2 Improvement during Low Cycle Fatigue—Zero Mean Stress

Figure 9.11 is reproduced as Fig. 9.27 with the magnitude of applied stress to be in the LCF regime (i.e., the material may yield due to applied load). This shows similar improvement in life due to shot peen.

9.9 Process of Laser Peen

The laser peen process was first developed at Battelle Laboratory, and the first application was for a fan blade of a military aircraft engine by GE Aviation. The commercialization of the process has been developed by Metal Improvement Company and Livermore National laboratory. Laser peen is capable of providing higher compressive stress and a deeper compressive layer. Even though it is more involved than the shot peen method, laser peen has been applied successfully in many practical applications.

The laser peen process uses a laser beam that is pulsed upon the metal surface, which produces a planar shock wave that travels through the metal. It in turn plastically deforms a surface layer of the metal.

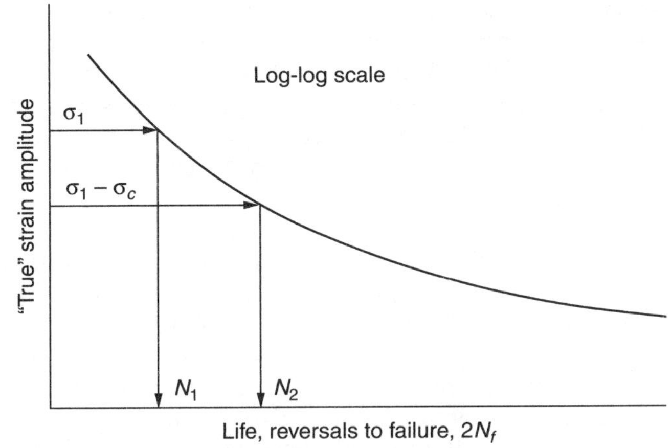

FIGURE 9.27 Improvement in LCF life.

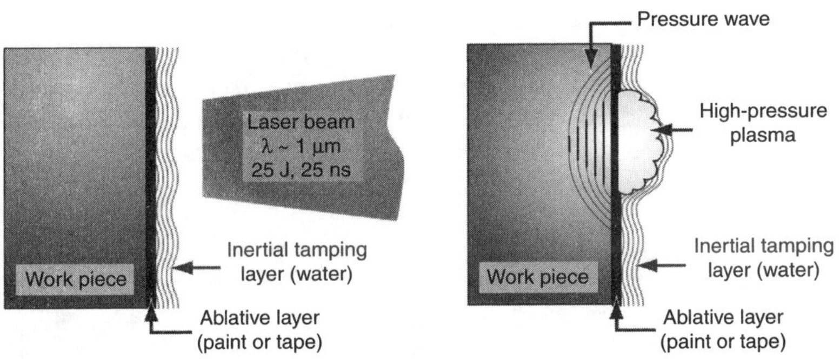

FIGURE 9.28 Laser peening concept in which ablation from a sacrificial surface creates a high-pressure plasma and consequent shock wave that results in a deep compressive stress in the workpiece.

The basics of laser peening are shown in Fig. 9.28. A roughly 25-J at 25-ns output beam from a Nd:glass laser is propagated onto the workpiece in which it is desired to induce residual compressive stress. The area to be peened has been covered with material to act simultaneously as an ablative layer and a thermal insulating layer. A thin stream of water is made to flow over the ablative layer. The laser light transparently passes through the water, and the leading temporal edge of the laser pulse is absorbed on the ablative layer. This absorption rapidly ionizes and vaporizes more of the ablative material to rapidly form a plasma that is highly absorbing for the rest of the laser pulse. The plasma pressure rapidly builds to approximately 100 kbar (1 million psi) with the water serving to inertially confine the pressure. This rapid rise in pressure effectively creates a shock wave that penetrates into the metal plastically straining the near surface layer. The plastic strain results in residual compressive stress that penetrates to a depth of 1 to 8 mm depending on the material and the processing conditions. This deep level of compressive stress creates a barrier to crack initiation and to crack growth and consequently enhances the fatigue lifetime and/or resistance to stress corrosion cracking of the material.

Figure 9.29 shows an example of the effectiveness of laser peening to enhance the fatigue strength or fatigue lifetime under cyclic loading. The data of a four-point loading fatigue tests on smooth samples, $Kt = 1$, for BSTOA titanium 6/4 are shown in Fig. 9.29. The deep levels of compressive stress retard crack initiation and thus increase the fatigue lifetime at a given stress loading or enable application of a greater stress loading at a given desired fatigue lifetime. The fatigue benefits of laser peening can be even greater when the geometric influence of real components is considered. As the local stress, Kt value associated with the geometric features increases and the benefit of deep residual stress becomes even greater as resistance to crack growth becomes a dominating factor.

Figure 9.30 shows fatigue results for a hole and keyway slot cut into an aluminum test sample. In this case, the laser peening benefits of very deep residual stress result in a 10-fold increase in fatigue lifetime over shot peening and a 20 times increase over machined parts.

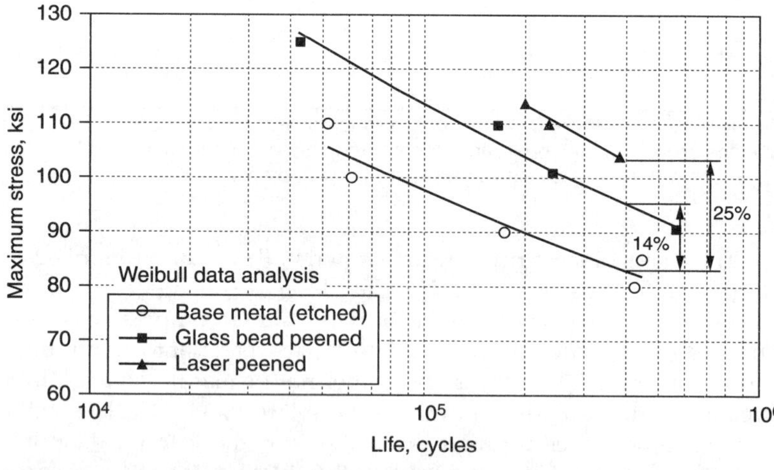

FIGURE 9.29 Stress-lifetime curve in BSTOA titanium 6/4 showing the increased fatigue strength enabled by shot and laser peening over base metal samples. Shot peening adds an excellent 14 percent increase to the fatigue strength, and the deeper compressive stress induced by laser peening provides a 25 percent increase in fatigue strength over the unpeened samples.

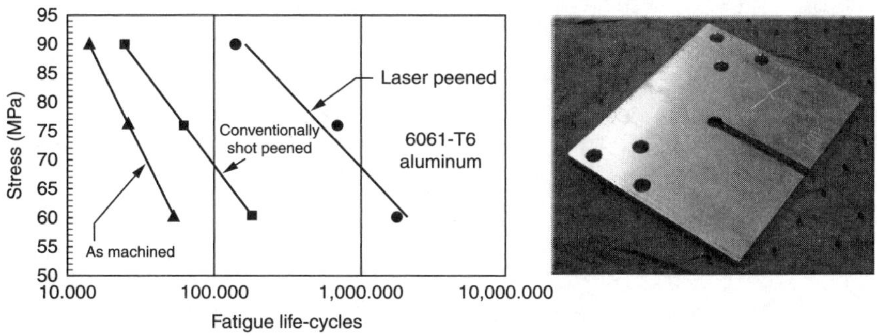

FIGURE 9.30 Benefits of the deep residual stress induced by laser peening are even more dramatic in components where geometric features create an increase in the stress intensity factor. A notched specimen of Al 6061 as expected shows a significant increase in fatigue life when shot peened, but an even greater 20 times lifetime improvement when laser peened. The deeper level of compressive stress is especially effective when the *Kt* factor is greater than 1 as generated by the central hole.

Production laser peening of steam turbine blades for plant deployment began in 2007. The requirement for improved efficiency in power generation is pushing the size of blades and hence the stress loadings. Laser peening blades of steel and titanium for the major gas and steam turbine manufacturers retard failure due to fatigue, fretting fatigue, and failure due to water droplet erosion.

CHAPTER 10

Estimation of Risk

10.1 Probabilistic Concept to Quantify Risk of a Proposed Design

It is a fact that design decisions are often made with insufficient information, and as a consequence, this method of making decision has risk associated with it. The claim can be made that sufficient margin in the design is made to ensure safe operation; however, the risk is not known.

The importance of this section is underscored by Siddall (1983) and by Singh and Drosjack (2008). Siddall stated that "Probabilistic design is the codification of risk judgment; and *is not* the statistical analysis of the design problem." Singh and Drosjack in relation to the design process stated that "there is a need for combining physics-based theory of material behavior with probabilistic methods that account for uncertainty (variability) in data in order (a) to achieve a robust design, and (b) to estimate the risk in accepting the resulting design."

Singh (2001) considered the design process in light of probabilistic analysis and risk estimation. The fact is that no design has a 100 percent guarantee, i.e., absolutely no potential for failure. There is always a finite risk, however small it may be.

Drosjack et al. (2005) described the need to move from today's concept of deterministic design to consider risk as a component in each design for tomorrow's process and pointed out the need to quantify risk. Total risk is the product of the likelihood of an event to occur by the consequence of the occurrence of the event. The likelihood of the occurrence of a design quantity or the probability of stress to be larger than a predetermined magnitude is engineering risk. In the deterministic approach to design, various parameters are treated as single-valued, i.e., as having zero uncertainty in the value. To accommodate this false assumption, an intended design margin is assigned during analysis, to cover the influence of this uncertainty, called the *factor of safety*. The basic fact is that if one has complete knowledge of the events and behavior of the component, there is no need for this margin. It is thus appropriate to term it a "factor of ignorance," where ignorance reflects the lack of knowledge of the uncertainty incorporated in the analysis. Larger safety factors reflect greater uncertainty on the part of the designer. By this act, engineers try to guarantee a safe design. If the margin is set to a high number, it might force an overdesign. The danger is that this approach is not always able to accommodate the lack of knowledge of the uncertainty, and it might in some cases result in underdesign. Failures verify this fear where the factor of safety didn't actually cover the variations in critical parameters.

It can be stated that the only true certainty in life is that there is uncertainty in life. That is the case in most engineering analyses whether quantum physics, allowable loading for turbine blades and material properties, or estimation of stresses. Even many times equations used to estimate engineering quantities have unreliability associated with them. The laws of physics may not be precise either. Thus, no matter what engineering design decisions are made, there is a risk associated with using the calculated results.

The theories of statistics and probability have evolved and grown, driven by needs and technical developments and accelerated by the fantastic advances in computers and computational capabilities. As a result, there are much more comprehensive analysis capabilities available today than in past years.

The basic appropriate theories were discussed earlier on the basis of deterministic physics where the basic equations are developed and explained. In this section a method is discussed in which parameters and properties are not single-valued. Rather, all real values have an inherent variability or uncertainty. As a result, solutions are not single-valued—rather they are best represented as probabilistic. That is, answers are described as a range with the likelihood for each answer. Each value can then be described as having a technical risk associated with the likelihood of the answer occurring. The designer can select an answer based on the acceptable level of risk. This section provides sufficient detail for designers to consider how a "true" analysis must incorporate the uncertainty inherent in any calculation. It is possible to quantify the risk associated with a design. Thus, design decisions do not need to incorporate factors of safety (factors of ignorance) to accommodate the uncertainty. Rather the risk may be quantified and design decisions made on the acceptable level of risk.

The risk associated with a design can be managed when risk is estimated. Enhancements in analytical tools and capabilities make that determination a reality. The needs are real, but require greater application to grow and flourish. This method has the promise to push designs into new frontiers from the basis of size, loads, speeds, pressures, materials, costs, and delivery times.

The premise of classical analysis is that each variable, e.g., force, dimension, and material properties contributing to the response of a mechanical system, is single-valued. This is almost always far from the truth. A very good treatment of uncertainty is given in Siddall (1983), Melchers (1987), Singh (1985), Singh and Ewins (1988), Singh (1992), Singh (2001), and Singh and Drosjack (2008). Some pertinent information is taken from these references, and the following discussion is influenced by the work cited in the references mentioned above.

For example, every dimension on a drawing has a tolerance band as does every manufactured component. Material properties are based on tests of representative samples, and published values follow statistical protocols. Tensile strength is a value with a mean and standard deviation associated with it, developed by the statistical analysis of the data.

It is imperative to incorporate the effect of uncertainty into the design process in a rational manner that is supported by mathematics, physics, and general good sense. To do that, the uncertainty must be measured or calculable. The measure of the degree of uncertainty of the likelihood of an event occurring is probability. For many years, it was not possible to perform this analysis, but today the rapid growth of the availability and capability of computer tools makes this analysis very practical.

A dominant characteristic of engineering design is uncertainty—uncertainty about, among other things, operational loads, manufacturing dimensions, material properties, and measured data. The estimation or prediction of the behavior and/or response of a structure based on measured data will also exhibit uncertainty (randomness). There is an assumption of uncertainty in all these situations. There might also be uncertainty about the validity of a hypothesis or theory that is to be used to predict the performance of the design. The concern is whether a model that was valid in certain circumstances is applicable to a new situation in which the circumstances are uncertain. Most modeling protocols involve linearization of nonlinear phenomena, e.g., bearing stiffness characteristics used in rotor dynamics.

It is a reality in the process design that says, "the only place the equipment is not expected to be operated is at the design point!"

As engineering knowledge and tools become more sophisticated, the factor of ignorance approach will become unsatisfactory. Uncertainty should be applied in the design analysis, and decisions should be made based on a calculated response that incorporates the uncertainty in the result.

The process discussed in this section, although aimed at blade design, is general in nature; i.e., it may be applicable in the design processes of other mechanical components.

Certain benefits of probabilistic analysis can be stated as follows:

- It considers uncertainties with appropriate physics-based law(s).
- It provides an "estimate of risk."
- It shows that finite life and an acceptable level of risk may be economic realities.

The reasons for adopting a probabilistic approach during the design phase can be summarized by three items (Siddall, 1983):

1. Theoretical validity
2. Reliability
3. Economic considerations

Theoretical validity deals with many phenomena encountered in design with multivalued quantities. The correspondence between predicted and actual performance requires the consideration of variability. Furthermore, even using statistical strength and stress models having admissible ranges of values from 0 to infinity assumes abnormally large stresses and abnormally low strengths that rarely occur. There are functions whose admissible ranges are finite, e.g., among others, the diameter of a shaft, thickness of a shell, temperature variation, and pressure variation.

An achievable reliability goal should be decided before designing, or it should be incorporated into the design methodology with defined levels of performance and of safety.

Economic benefits will be realized when improved design is expected to minimize costly testing programs, reduce total time from design to test, optimize design based on a reliability target, and meet other requirements. It will have appropriate weight, size, volume, etc. A rational approach to design should focus on total life cost rather than be based on a policy of replacements and warranties. It lends itself to overdesign and unachievable manufacturing tolerances without the estimation of reliability and a goal of reliability in design. It directly relates to higher cost without any idea of risk.

Figure 10.1 shows a blade from which a test sample is machined. Many samples are tested to estimate variation in material properties.

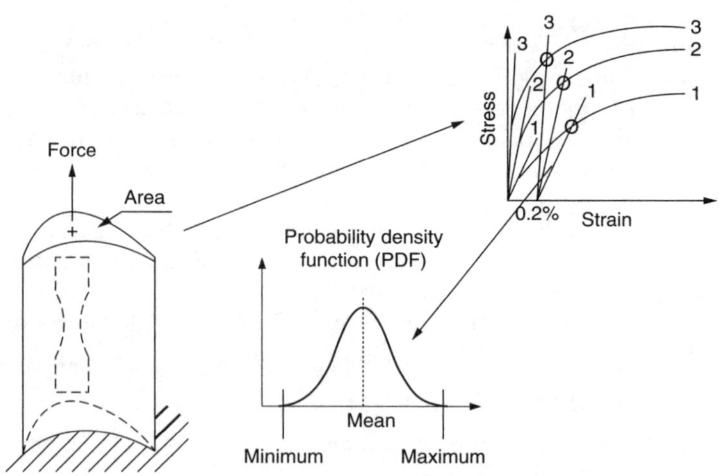

FIGURE 10.1 Test samples of parts showing variability in data.

Variation in the test data will depend on the number of samples tested. Some stress-strain curves are shown to depict the possible differences. When the data, e.g., among others, yield strength and ultimate strength of the material, are plotted as shown, clearly there is a statistical distribution for any property, e.g., yield strength and ultimate strength among others.

In the deterministic design evaluation of a mechanical component, parameters such as loads, material properties, and geometric dimensions are assumed to be known quantities, i.e., single-valued. The assumption that there is no uncertainty about them is not realistic in any practical application.

Due to uncertainty in the variables, the response (displacement, stress) is also expected to be uncertain, multivalued. The estimation of the occurrence of a response above the "limit" is a measure of the probability of the risk. Alternatively, the number of times a response meets the criteria is a measure of the probability of success, i.e., a measure of the technical reliability of the structure.

Many probabilistic methods can be employed to estimate reliability, e.g., among others, FORM (First Order Reliability Method) and SORM (Second Order Reliability Method). In this particular case, the Monte Carlo simulation method has been used to obtain the solution.

Mathematically, the response of a structure depends on the interaction between the imposed stress S and the component's resistance R. The deterministic method defines the margin by the ratio R/S. There are uncertainties in the values of both S and R, and these are represented in statistical terms by a probability density function (PDF). The probability of success is expressed as

$$P_s = P(R \geq S) \tag{10.1}$$

where P_s is the probability of success.

In general terms, Eq. (10.1) takes the following form:

$$P_s = P[H(R, S) \geq 0] \tag{10.2}$$

where $H(R, S)$ is a limit function and defines the relationship between R and S. For example, if R is the yield strength of the material and S is the imposed stress, H might be expressed as

$$H(R, S) = 0.67R - S \tag{10.3}$$

The above relationship states that the function H is equal to 67 percent of the yield strength less the imposed stress. It has been stated that for reliable operation H must be positive, but this is a deterministic statement.

The probabilistic aspect of the problem can be stated as follows:

$$P_s = P[(0.67R - S) \geq 0] \tag{10.4}$$

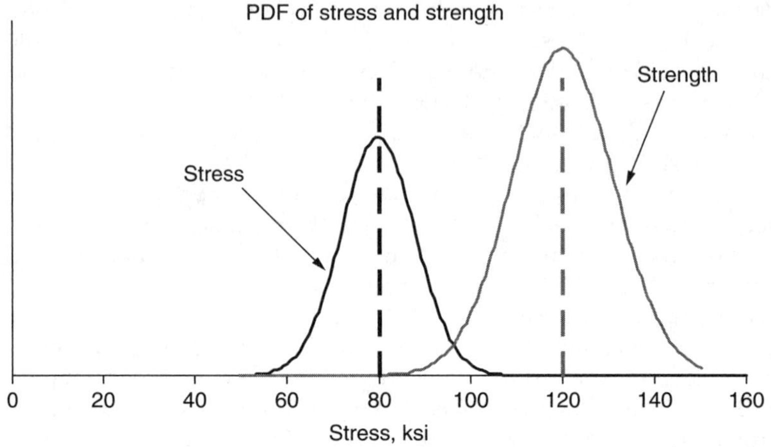

FIGURE 10.2 Description of probability of failure when 67 percent of the yield strength is less than or equal to the applied stress.

A typical representation of the probability density functions for R and S is shown in Fig. 10.2. It can be seen that the average magnitude of S (applied stress) is less than the average strength (say, 67 percent of the mean value of the yield strength of the material). It satisfies the condition for safe design from a deterministic point of view with the estimated factor of safety equal to 1.5. However, when the possible uncertainties in R and S are considered, it becomes clear that there is still a possibility of S being larger than R in the area in the middle of the graph where the stress and strength curves cross each other. Hence, there exists a chance of failure due to the statistical nature of R and S, but the deterministic approach indicates success.

The following example of a beam under a tensile load demonstrates the importance of the need for probabilistic analysis. A deterministic limit analysis is performed, and an assumption is made for the minimum and maximum values of the loads. Also the minimum and maximum possible cross-sectional areas of the beam are considered due to the tolerance on the diameter of the beam. The yield strength will have minimum and maximum values.

$$\text{Factor of safety} = \text{yield strength}/S$$

where

$S = P/A$
P = applied load
A = cross-sectional area
Min FS = minimum value of yield strength/maximum value of S
Max FS = maximum value of yield strength/minimum value of S
Mean FS = average value of yield strength/average value of S

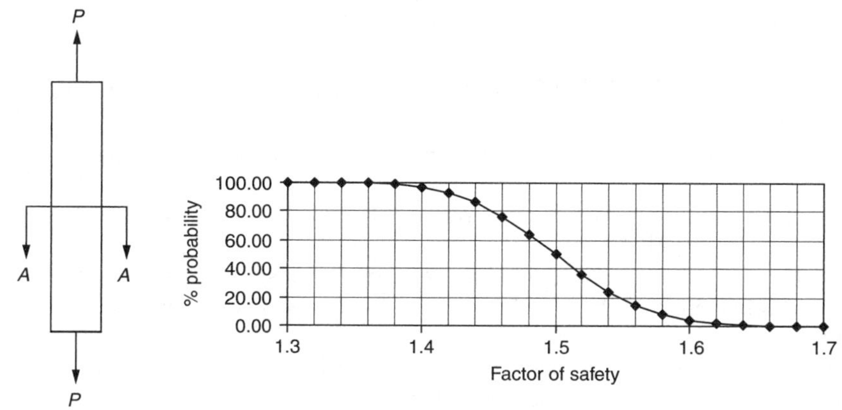

Figure 10.3 Factor of safety vs. % probability for a beam in tension.

The result of the probabilistic analysis (a brief description of the Monte Carlo simulation is provided in the Appendix), which provides the probability of a certain magnitude of FS considering the statistical variation of the parameters, is shown in Fig. 10.3. The analysis is based on the assumption that material properties and applied load are not single-valued, but these have statistical variations as described in Fig. 10.1. Also, the dimension of the cross section of the beam has some tolerances. During manufacturing the only guarantee is that the specified dimension stays within the tolerance band. When many beams are manufactured in a similar way, the final dimension will have a statistical dimension, even though these are expected to be within the tolerance band and the FS will range from 1.3 to 1.7.

Limit type of deterministic analysis for a beam where FS = strength/stress *below* is given below.

Min	FS	1.03
Max	FS	2.18
Mean	FS	1.50

The factor of safety equation given above has been used to conduct a Monte Carlo type of analysis, and the equation may be rearranged as follows:

$$\mathrm{FS} = (\sigma_a/\sigma_e + \sigma_m/\sigma_{\mathrm{ult}})^{-1}$$

The limit function H can be written as

$$H(\mathrm{FS}_T, \sigma_a, \sigma_e, \sigma_m, \sigma_{\mathrm{ult}}) = (\sigma_a/\sigma_e + \sigma_m/\sigma_{\mathrm{ult}})^{-1} - \mathrm{FS}_T \qquad (10.5)$$

where FS_T is a desired value for FS.

The probabilistic mathematical statement is

$$P_s = P[H(\mathrm{FS}_T, \sigma_a, \sigma_e, \sigma_m, \sigma_{\mathrm{ult}}) \geq 0] \tag{10.6}$$

where P_s represents the probability of occurrence for a desired value of FS_T.

The magnitude of P_s for each value of FS_T has been estimated and is shown in Fig. 10.3.

This simple example demonstrates that there is approximately a 100 percent probability of achieving a factor of safety equal to 1.34 while the probability of achieving a factor of safety of 2.18 (maximum by limit analysis) is approximately zero. The probability of achieving the average value of 1.5 is 50 percent, as expected.

The probabilistic calculation becomes more complicated, particularly when the response is a function of many variables. To mitigate this complexity, techniques have been developed that perform many calculations (often in the thousands) rapidly. One such method is the Monte Carlo method. In a Monte Carlo simulation, e.g., as the analysis proceeds forward, a picture of the probability of the response being larger than the limit value emerges. The results of Monte Carlo simulations can prove that the level of risk associated with the response is larger than the limit value. Based on this information, better decisions can be made regarding the risks of using a component or operating a component with a suspected defect as well as the risks associated with using the component up to a particular number of load cycles.

10.2 Probabilistic Treatment of Factor of Safety Based on Goodman Equation

Singh (2001) has demonstrated the probabilistic reliability assessment of mechanical components. Figure 10.4 helps explain the reasoning behind this evaluation and shows an *S-N* diagram of a material with a band of curves. The band represents the scatter of the test data. The estimated stress applied to a component is also shown to have variation in the *y* axis. This variation results from the influence of variations in operating loads, and geometric tolerances, and includes the inaccuracy in the estimation procedure.

The resulting assessment of the operating life is shown on the *x* axis. It is clear that due to variations in material property and stress, the operational cycle is not single-valued but varies within a range. The band of cycles represents the potential life of the blade. The implication of this can be seen in the probabilistic Goodman diagram shown in Fig. 10.5. The factor of safety is not single-valued but has a range of values; therefore, the operating life will also have a range.

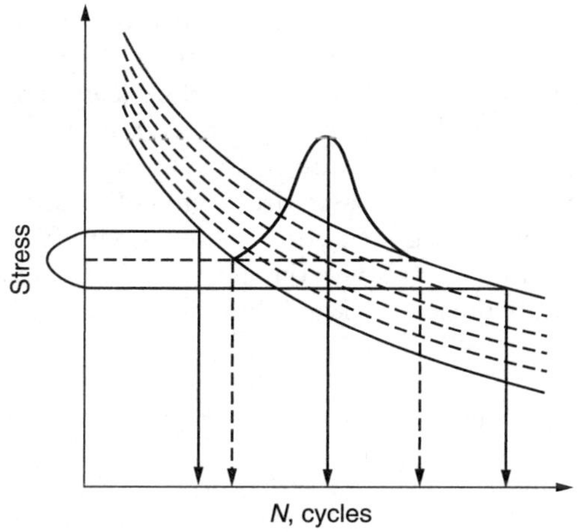

Figure 10.4 Depicting scatter of data in S-N curve.

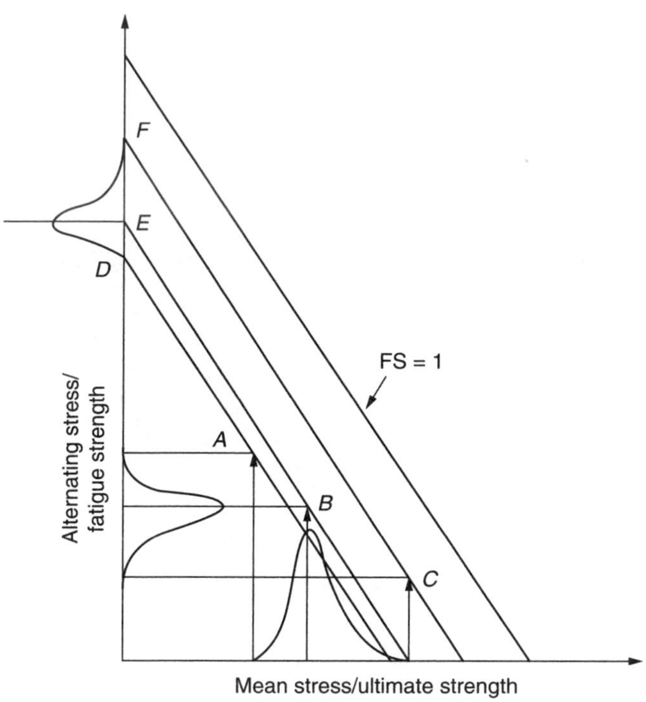

Figure 10.5 A probabilistic aspect of the Goodman diagram.

The factor of safety based on life (FN) was discussed in the previous section and is restated below in Eq. (10.7). Both factors of safety, FS and FN, depend on variables, such as the mean stress, alternating stress, ultimate strength, fatigue strength, and fatigue strength exponent of the material. Uncertainty in the stress due to variations in loading and the observed scatter in the material properties can be described by statistical distributions. After quantifying this uncertainty or scatter in the variables, one needs to estimate the chance of the factor of safety (based on stress) having a value larger than a prescribed limit.

$$\mathrm{FN} = [(1/\mathrm{FS} - \sigma_m/\sigma_{\mathrm{ult}})/(1 - \sigma_m/\sigma_{\mathrm{ult}})]^{1/b} \tag{10.7}$$

The limit function H can be written as

$$H(\mathrm{FS}, \sigma_m, \sigma_{\mathrm{ult}}, b) = [(1/\mathrm{FS} - \sigma_m/\sigma_{\mathrm{ult}})/(1 - \sigma_m/\sigma_{\mathrm{ult}})]^{1/b} - \mathrm{FN} \tag{10.8}$$

The probabilistic mathematical statement can be stated as

$$P_s = P[H(\mathrm{FS}, \sigma_m, \sigma_{\mathrm{ult}}, b) \geq 0] \tag{10.9}$$

The above equation has been used to estimate the probability of HCF life for a steam turbine blade (Fig. 10.6) described in the

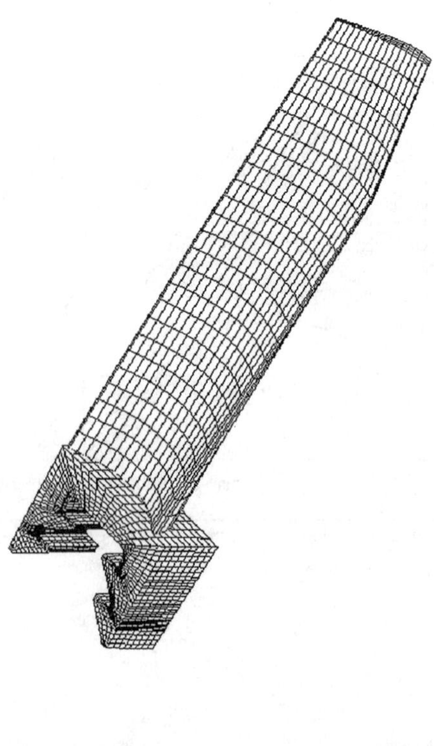

FIGURE 10.6 A typical blade.

Mean Stress, psi	Alternating Stress, psi	FS	Cycles
60,000	10,000	1.644	1.530E+13

TABLE 10.1 Goodman Factor of Safety and Estimated Life

earlier section. Table 10.1 lists stresses, the calculated factor of safety, FS, based on the Goodman equation, and the estimated life.

The life estimate was made using the same stress magnitude, and the estimated lives are listed in the last column of Table 10.1.

The results presented were obtained using a traditional deterministic type of analysis. Since the stress magnitude and material properties are not single-valued, a probabilistic analysis was performed to account for the variability. Table 10.2 lists the assumed variations of these parameters to be used in the probabilistic analysis.

The blade design was reanalyzed using a probabilistic analysis that incorporated the variations in properties identified in Table 10.2. The plot in Fig. 10.7 shows the probability vs. life cycle for that design

	Total Variations, % of Mean
Ultimate strength	20
Fatigue strength	16
Mean stress	20
Alternating stress	16
σ_p	20
b	20

TABLE 10.2 Assumed Variation in Properties (Typical of Turbine Blade Material)

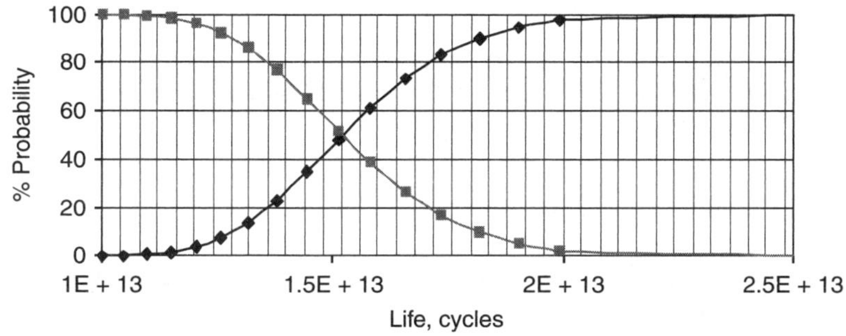

FIGURE 10.7 Probabilistic life assessment for blade.

evaluation, which can be compared with the deterministic result in Table 10.1. The probabilistic analysis gives a range of life from 1E+13 to 2.5E+13. Identification of the acceptable level of risk can greatly improve the accuracy of the design life and also provide one with more design options. The curve with decreasing probability and increasing life depicts reliability, and the other curve shows unreliability.

Singh (1985) demonstrated the implications of geometric variations on the randomness of blades' natural frequency. The large available computing capabilities of computers help designers to apply advanced statistical and probabilistic techniques. It is instructive to use the transformation of random variables to demonstrate the process and the many nuances in applying probabilistic techniques. For simplicity, assume that the geometric variation may be reflected in the cross-sectional property of the blade, e.g., the moment of inertia of the cross section. The basic equation of vibration of a cantilever beam can be used to demonstrate the method. Any one mode of vibration may be assumed as a single-DOF spring-mass-damper system.

10.3 Transformation of Random Variables

The functional relationship between the natural frequency and the moment of inertia of the cross section may be expressed by

$$\Omega = g(I) \tag{10.10}$$

The statistical expression for the change in Ω due to random variation in I is expressed as follows:

$$\omega_n = g(i) \qquad \omega_n = \text{natural frequency} \tag{10.11}$$

or

$$i = g^{-1}(\omega_n) \tag{10.12}$$

Following the rules of transformation of random variables, the probability density function (PDF) of the natural frequency may be written as follows:

$$f_\Omega(\omega_n) = f_I\,(I)\ |di/d\omega_n| \tag{10.13}$$

where $f_I\,(i)$ is the PDF of the moment of inertia and the second term is called the *Jacobian matrix.*

$$f_\Omega(\omega_n) = f_I\,[g^{-1}(\omega_n)]\ |d[g^{-1}(\omega_n)]/d\omega_n| \tag{10.14}$$

The probability of the natural frequency ω_n to have a value greater than or equal to a targeted frequency ω can be computed as follows:

$$P(\omega_n \geq \omega) = \int_{\omega}^{\infty} f_{\Omega}(u)\, du \qquad (10.15)$$

The above expressions have been developed for statistical variation in only one independent variable. Such expressions for multiple variables may also be developed, but these become complicated.

10.3.1 Single Cantilever Beam

The expression for the natural frequency of a cantilever beam is

$$\omega_n = C_n[(Ei)/(\rho l)]^{1/2} \qquad (10.16)$$

where E = modulus of elasticity
i = moment of inertia
ρ = mass
l = length of the beam
C_n = constant for a mode (3.52 for mode 1)

$$di/d\omega_n = (2\rho l\omega_n)/(C_n^2 E) \qquad (10.17)$$

$$f_{\Omega}(\omega_n) = f_I(I)\,|(2\rho l\omega_n)/(C_n^2 E)| \qquad (10.18)$$

For simplicity, the PDF for the moment of inertia is taken as the normal distribution, and the expression is given by

$$f_I(I) = \{1/[\sigma_I(2\pi)^{1/2}]\} \exp[-(i - i_0)^2/(2\sigma_I^2)] \qquad (10.19)$$

where i_0 = mean value of moment of inertia
σ_I = standard deviation of moment of inertia

$$f_{\Omega}(\omega_n) = \{1/[\sigma_I(2\pi)^{1/2}]\} \exp[-(i - i_0)^2/(2\sigma_I^2)]\ \ |(2\rho l\omega_n)/(C_n^2 E)| \qquad (10.20)$$

where

$$i = \omega_n^2 \rho l / C_n^2 E$$

$$i - i_0 = \omega_n^2 \rho l / C_n^2 E - \omega_{n0}^2 \rho l / C_n^2 E$$

$$(i - i_0)^2 = (\rho l / C_n^2 E)^2 (\omega_n^2 - \omega_{n0}^2)^2$$

Finally,

$$f_{\Omega}(\omega_n) = \{1/[\sigma_I(2\pi)^{1/2}]\} \exp[-(\rho l/C_n^2 E)^2(\omega_n^2 - \omega_{n0}^2)^2/(2\sigma_I^2)]\,|(2\rho l\omega_n)/(C_n^2 E)| \qquad (10.21)$$

To appreciate the implications of Eqs. (10.16) through (10.21), a single blade is chosen. It is assumed that there will be variations between blade to blade in a manufactured lot. This comes about due

	Total % Deviation	Mean	SD	Minimum	Maximum
Moment of inertia, in^4	12	0.0021	0.000021	0.001974	0.002226
Modulus, psi	15	3.00E+07	375,000	27,750,000	32,250,000
Density, lb/in	20	5.00E–02	0.0008333	0.045	0.055
Length, in	2	2	0.0033333	1.98	2.02
Damping ratio	15	0.003	0.0000375	0.002775	0.003225
Constant C_1		3.52			

TABLE 10.3 Variation in Geometric and Material Properties

to geometric tolerances. Material properties also vary among blades. Mean values and variation in properties for an assumed blade design are listed in Table 10.3. Minimum and maximum values are estimated based on plus/minus six sigma from mean values. Variation in frequency of the first mode due to variation in the moment of inertia of the cross section is estimated.

The probability density function (PDF) of frequency for the first mode is given by Eq. (10.20). A graphical representation is provided in Fig. (10.8).

Figure 10.9 is a Campbell diagram that shows a band of frequencies as it is clear from Fig. 10.8. The nominal frequency based on mean values is also shown. The implication is that the response due to excitation will also be registered by the blade not only at the operating

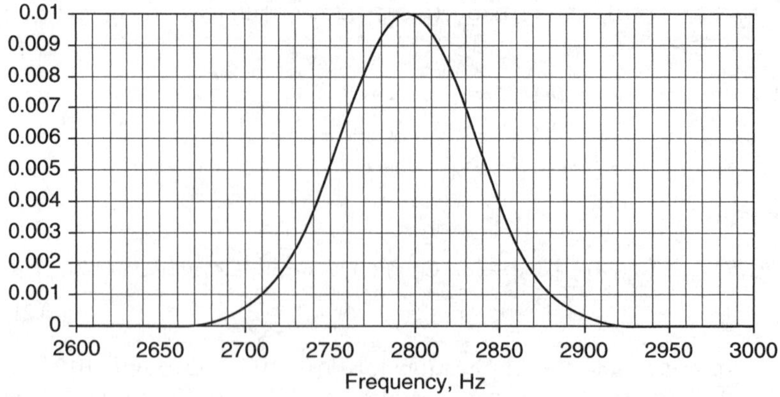

FIGURE 10.8 Probability density function.

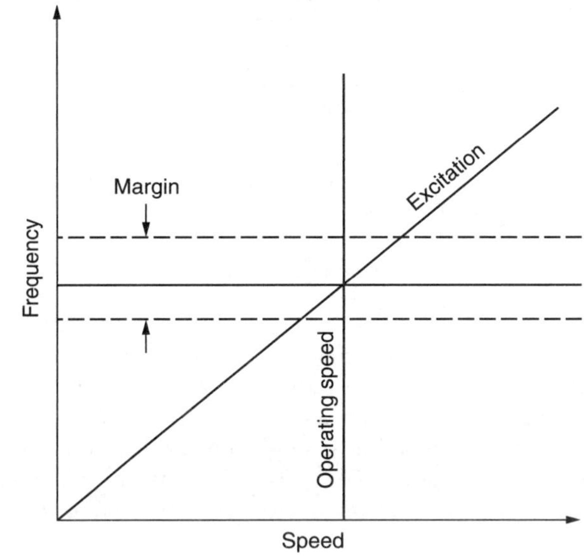

FIGURE 10.9 Variation in natural frequency.

speed but it will also occur at speeds less than and greater than the operating speed.

Probability of the occurrence at any value of frequency has been estimated through a Monte Carlo simulation using Eq. (10.15). This depicts two curves, one is for percent probability for the occurrence for less than a frequency while the other one is for greater than a frequency. It should be noted that they cross each other at probability equal to 50 percent because a normal distribution was assumed for the analysis. A graphical representation is provided in Fig. 10.10.

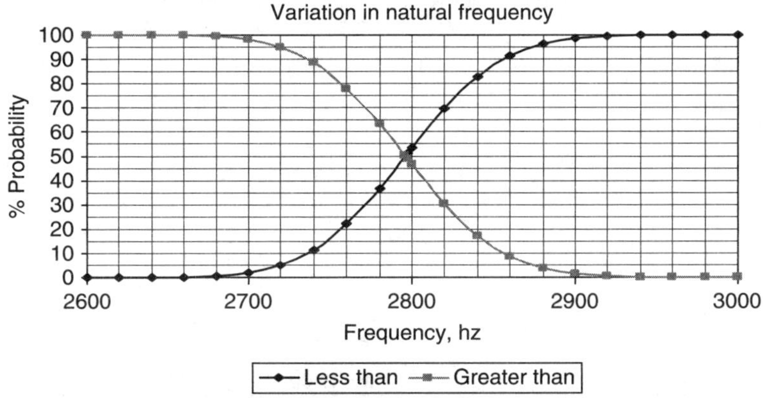

FIGURE 10.10 Probability of frequency occurrence.

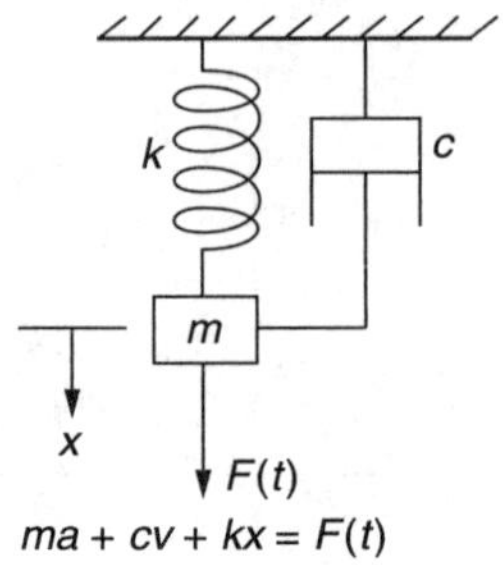

FIGURE 10.11 Spring-mass damper.

Due to variation in frequency, value of response will be different at each frequency within the band. A single spring-mass-damper (SMD) system is used to explain this feature. Any single mode of vibration may be estimated as a single spring-mass-damper system. Figure 10.11 shows a single SMD. The response of this system was developed in Chap. 5. A simplified form of equation for magnifier is given by Eq. (10.22).

$$\text{Abs}(\mu) = 1/\{[1 - (\omega/\omega_n)^2]^2 + [2\zeta(\omega/\omega_n)]^2\}^{1/2} \tag{10.22}$$

The response curve for each natural frequency shifts from that for the nominal natural frequency. Graphical plot of Eq. (10.22) for three natural frequencies is shown in Fig. 10.12.

Figure 10.13 is a plot of multiple response curves, each representing a natural frequency within the band as explained earlier. The probability of final response of the system at any exciting frequency

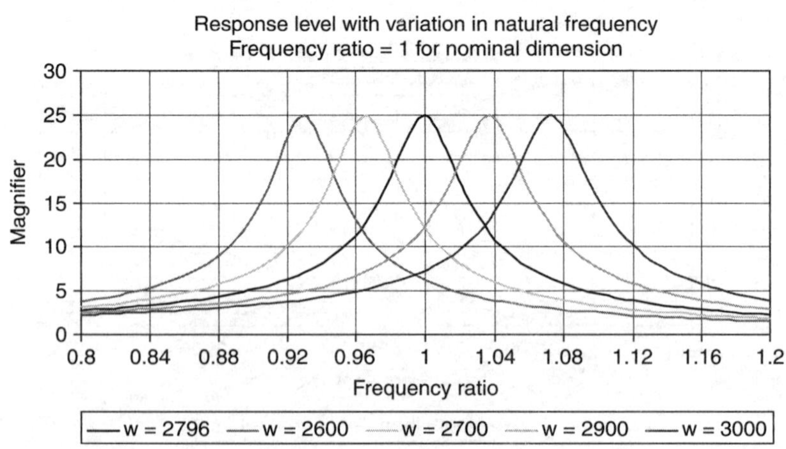

FIGURE 10.12 Response levels for variation in natural frequency.

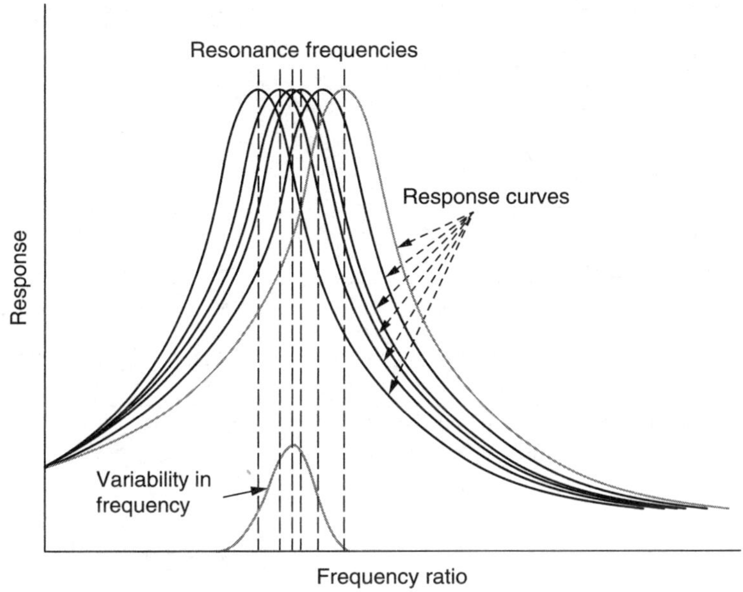

Figure 10.13 Response at different frequency.

depends on the probability of occurrence of the natural frequency. The curve at the bottom part depicts a possible scenario of the variation of response for exciting frequency equal to the natural frequency estimated with average properties of the blade.

A system will generally respond to both sides of a frequency ratio equal to 1, and the range will depend on damping present in the system. There is a variation in the magnitude of natural frequency due to uncertainty in the input variables, e.g., dimension and material properties.

Table 10.4 lists variation in natural frequency and it lists also assumed variation in the damping for further analysis.

Percent probability as a function of the magnifier at four different frequencies is given in Fig. 10.14.

Figure 10.15 shows the percent probability as a function of magnifier at the nominal natural frequency.

	Total % Dev.	Mean	SD	Minimum	Maximum
Natural freq. Hz	20	2794.00	46.57	2514.60	3073.40
Damping ratio	15	3.00E–03	3.75E-05	2.78E–03	3.23E–03

Table 10.4 Variation in Frequency and Damping

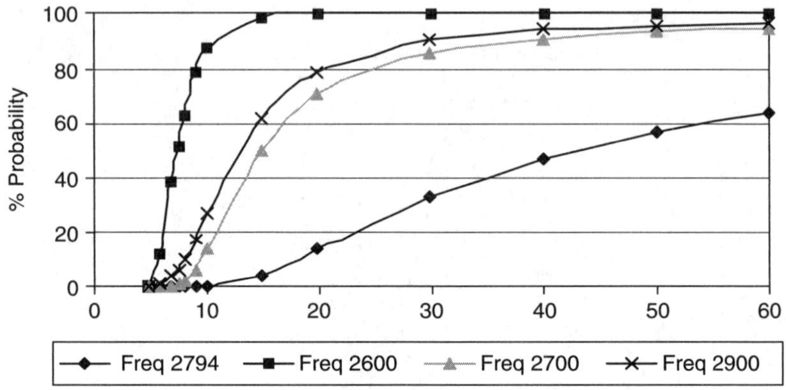

FIGURE 10.14 Percent probability vs. magnifier at various frequencies.

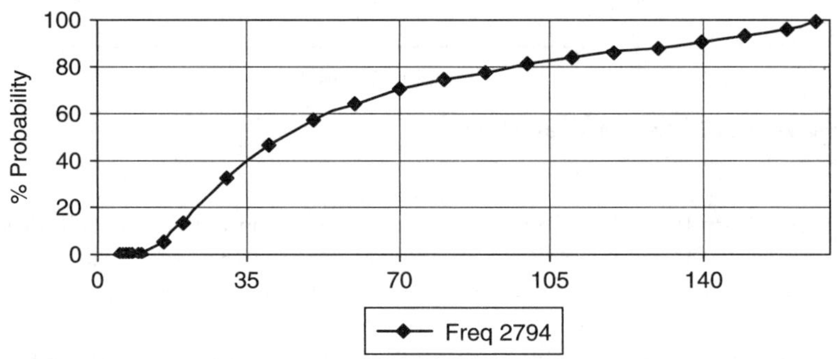

FIGURE 10.15 Percent probability vs. magnifier at 2794 Hz.

10.4 Probabilistic Low Cycle Fatigue Concept

To account for the variability in material properties, it is customary to test multiple samples. Figure 10.16 shows the results depicting variations in *S-N* data for the material. It shows that different samples might fail at different numbers of applied cycles even when the applied stress is identical.

The strain vs. life relationship was discussed earlier.

$$\Delta\varepsilon_{\text{Total}}/2.0 \qquad \text{or} \qquad \text{strain amplitude } \varepsilon_{\text{Total}} = AN_f^b + CN_f^d$$

The second term is controlled by plastic strains, and this strain contributes mainly to LCF-type damage. Even though the total strain should be considered for the estimate, for demonstration purposes

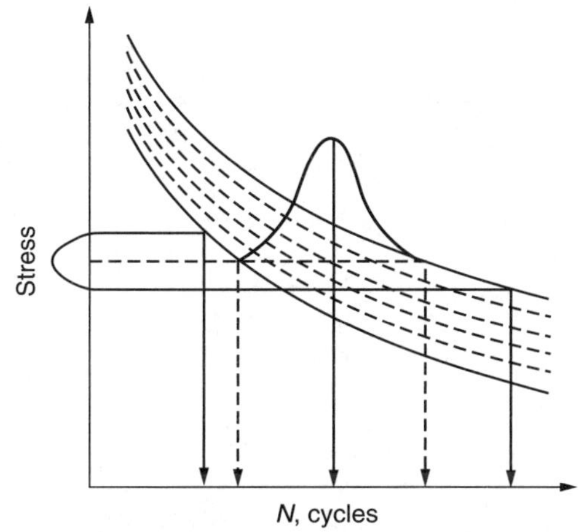

FIGURE 10.16 Depicting scatter of data in S-N curve.

the second term will be used to estimate the probabilistic LCF cycles.

$$\varepsilon_{\text{plastic}} = CN_f^d \tag{10.23}$$

$$N_f = (\varepsilon_{\text{plastic}}/C)^{1/d} \tag{10.24}$$

The limit function is given by

$$H(\varepsilon_{\text{plastic}}, C, d) = (\varepsilon_{\text{plastic}}/C)^{1/d} \qquad \text{cycles} \tag{10.25}$$

The equation to estimate the probability of certain life is

$$P_s = P[H(\varepsilon_{\text{plastic}}, C, d) \geq 0] \tag{10.26}$$

CHAPTER 11
Summary

In several sections, the path followed in the design of turbine blades was discussed. This technology has evolved and grown, driven by needs and technical developments and accelerated by the fantastic advances in computers and computational capabilities. The estimation of static stress has advanced from basic force by area analysis or bending moment by moment of inertia of a beam analysis to more sophisticated finite element analysis. The localized stresses can now be estimated with greater accuracy. Alternatively, vibratory stress is estimated more accurately by using advanced methods aided by a better prediction of forces. The analysis for the vibration of a steam turbine blade system has gradually advanced from the concept of spring-mass-damper model to beam and plate analysis and from a cantilever beam to packets of beams. Finally, the use of packeted bladed disk analysis to portray a more accurate estimation can now be made, and it is being done. The development of new materials and advancements in manufacturing capabilities has increased the reliability of the product, but it has also helped in the design for higher temperature and forces. As a result of the work of many researchers and designers of related disciplines, there are much more comprehensive analysis capabilities available today than in past years.

11.1 Deterministic Reliability Estimation

The theory is first explained on the basis of deterministic physics where the basic equations are developed and explained. Examples of typical results are shown in a stepwise fashion that would be followed in a comprehensive analysis of real turbine blades. However, throughout this analysis, it is assumed that parameters are deterministic, single-valued, and constant.

11.1.1 Stress and Fatigue Analysis

Given steady stress (centrifugal + bending) and alternating stress (equation of stimulus) and after adjusting for fatigue strength, a

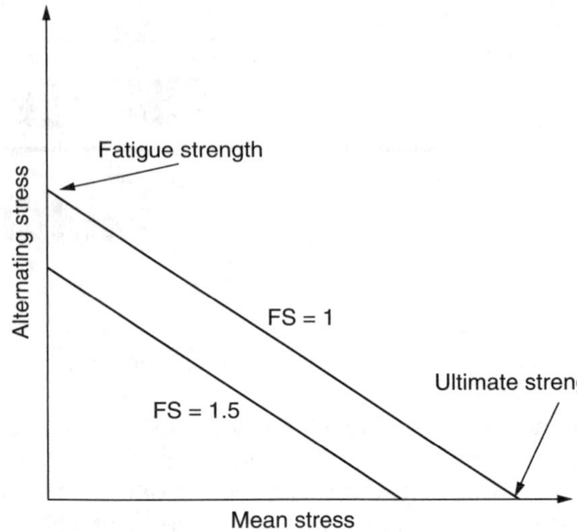

FIGURE 11.1 Basic design reliability estimation.

Goodman diagram (Fig. 11.1) is created. A predetermined factor of safety defines acceptance of the design.

$$\sigma_{\text{mean}} = \sigma_{\text{CF}} + \sigma_{\text{bending}}$$

$$\sigma_{\text{vib}} = S(\pi/\delta)(\text{RRF})\sigma_{\text{bending}}$$

$$1.0/\text{FS} = \sigma_{\text{vib}}/\sigma_{\text{fs}} + \sigma_{\text{mean}}/\sigma_{\text{ult}}$$

$$\sigma_{\text{fs}} = k_1, k_2, k_3, \text{etc. } (\sigma'_{\text{fs}})$$

11.1.2 Creep Analysis

$$\sigma_{\text{avg}} \rightarrow \text{LMP} \rightarrow (T + 273)(\log t_r + C)$$

If temperature T is known, then the rupture time t_r can be estimated (Fig. 11.2).

11.1.3 Modal Analysis

The descriptions of Campbell and SAFE diagrams have been given in earlier sections. The analysis based on a single packet of blades that are cantilevered at the base is given in the left of Fig. 11.3. It assumes that the disk has infinite stiffness. The right side of the figure shows a SAFE diagram for tangential vibration plotted using the results of a bladed disk analysis. Note that there is a small difference in frequencies between these two analyses. The disk seems to have nearly infinite stiffness in the tangential direction. Also, there is a direct correspondence of mode shapes to the type of analysis. However, during

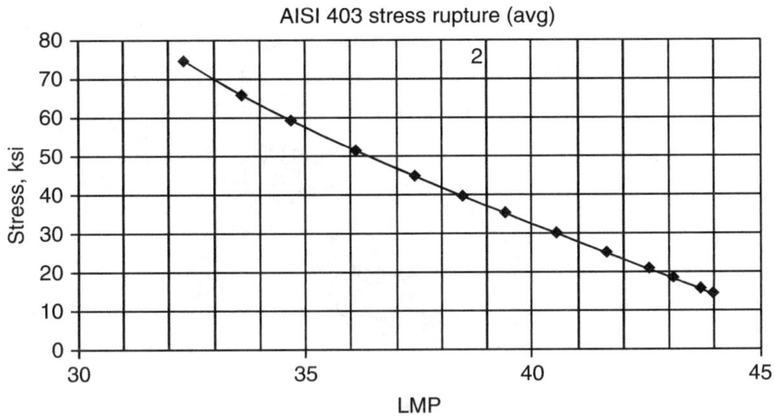

FIGURE 11.2 Basic design creep rupture estimation.

a vibration in the axial direction (out of plane of the disk) there is expected to be a difference in frequencies between the two types of analyses due to the disk's flexibility in the out-of-plane direction.

11.1.4 Response Analysis

The next step is to estimate the vibratory stress imposed upon the blade due to any interruption in the flow. A forced response analysis should be performed to estimate this stress (Fig. 11.4). Finite element codes or user's developed codes allow this determination.

11.1.5 Goodman Factor of Safety Based on Above Analysis

The Goodman diagram (Fig. 11.5) is plotted, and a factor of safety is estimated as discussed earlier.

$$\sigma_{\text{mean}} = \sigma_{\text{CF}} + \sigma_{\text{bending}}$$

Vibratory stress σ_{vib} is used as estimated above.

$$\sigma_{\text{fs}} = k_1, k_2, k_3, \text{etc. } (\sigma'_{\text{fs}})$$

$$1.0/\text{FS} = \sigma_{\text{vib}}/\sigma_{\text{fs}} + \sigma_{\text{mean}}/\sigma_{\text{ult}}$$

11.2 Deterministic Life Estimation

The factor of safety does not provide any indication of the life of the blades. The real question is how long the blade will operate safely, and it is important to have the answer to this question so that the real cost of the stage can be ascertained. This information will also help in the repair and replacement strategy.

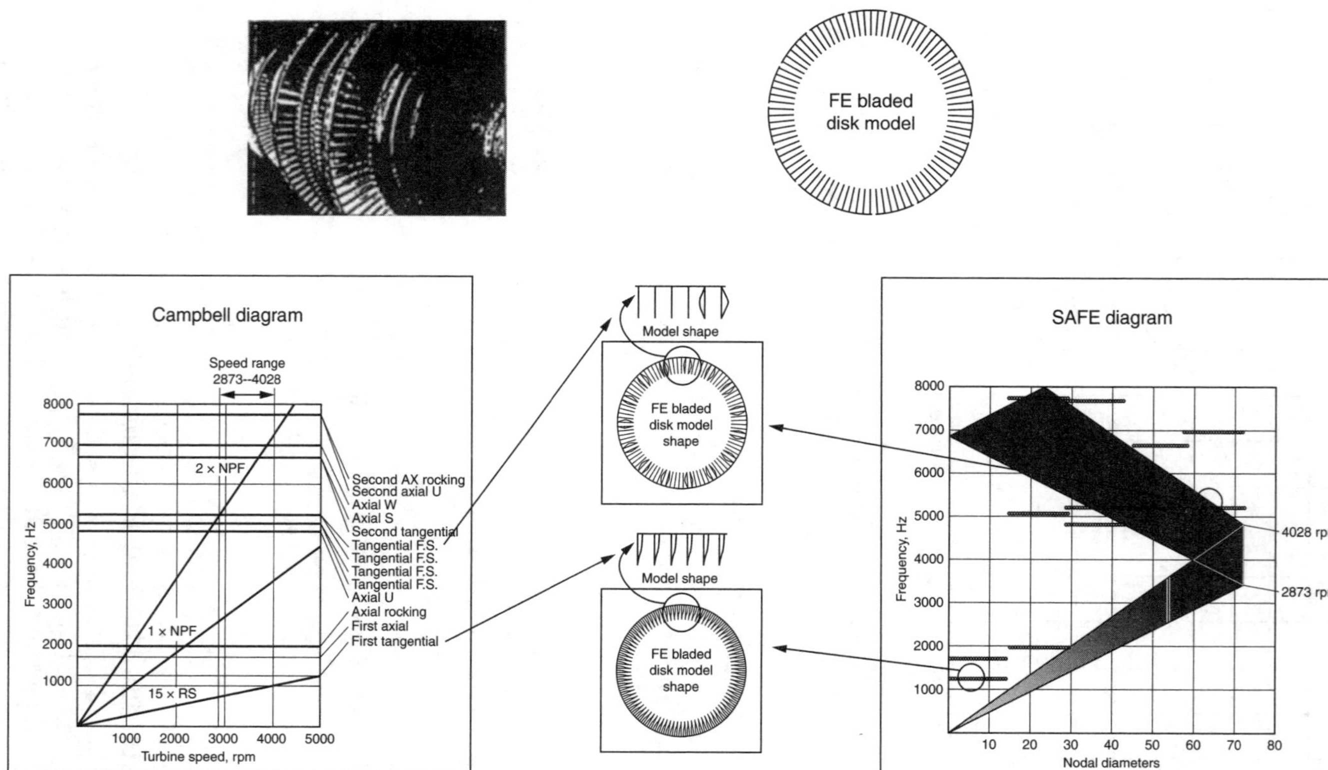

FIGURE 11.3 Modal analysis.

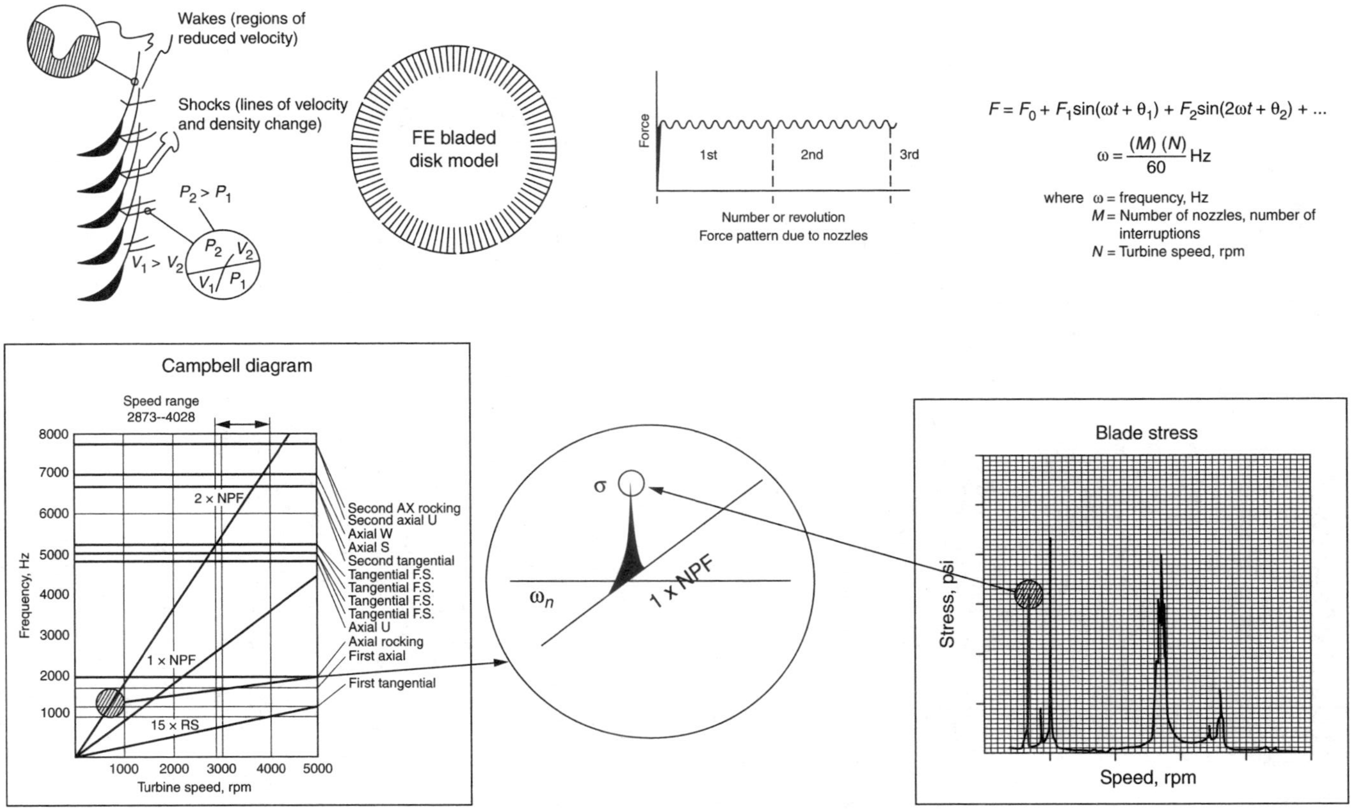

FIGURE 11.4 Estimation of response.

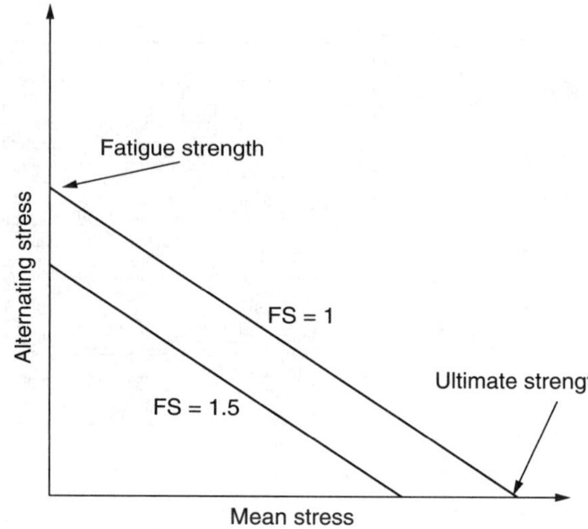

FIGURE 11.5 Goodman factor of safety.

Once the operating cycle is known, an analysis can be performed to estimate life, and equations are listed for both high cycle and low cycle fatigue (Fig. 11.6).

11.3 Probabilistic Reliability Analysis

It can be stated that the only true certainty in life is that there is uncertainty in life. That is the case in most engineering analyses whether they are based on quantum physics or turbine blade allowable loading. Thus, no matter what engineering design decisions are made, there is a risk associated with using the calculated results.

The theory is briefly explained and physics is discussed in which parameters and properties are not single-valued or simply deterministic. Rather, all real values have an inherent variability or uncertainty. As a result, solutions are not single-valued—rather they are best represented as probabilistic. That is, answers are described as a range with a given likelihood for each answer. Each value can then be described as having a risk associated with it—the likelihood of its occurring. The user can select an answer based on the acceptable level of risk.

11.3.1 Probabilistic Goodman Analysis

Normally, the Goodman line is drawn based on the average values of fatigue strength as well as ultimate strength. It has been pointed out that there is a statistical distribution of these properties. The average values signify a 50 percent chance of occurrence. Figure 11.7 schematically shows the possible variations. Once the stress is applied, then a variation in the factor of safety can be estimated.

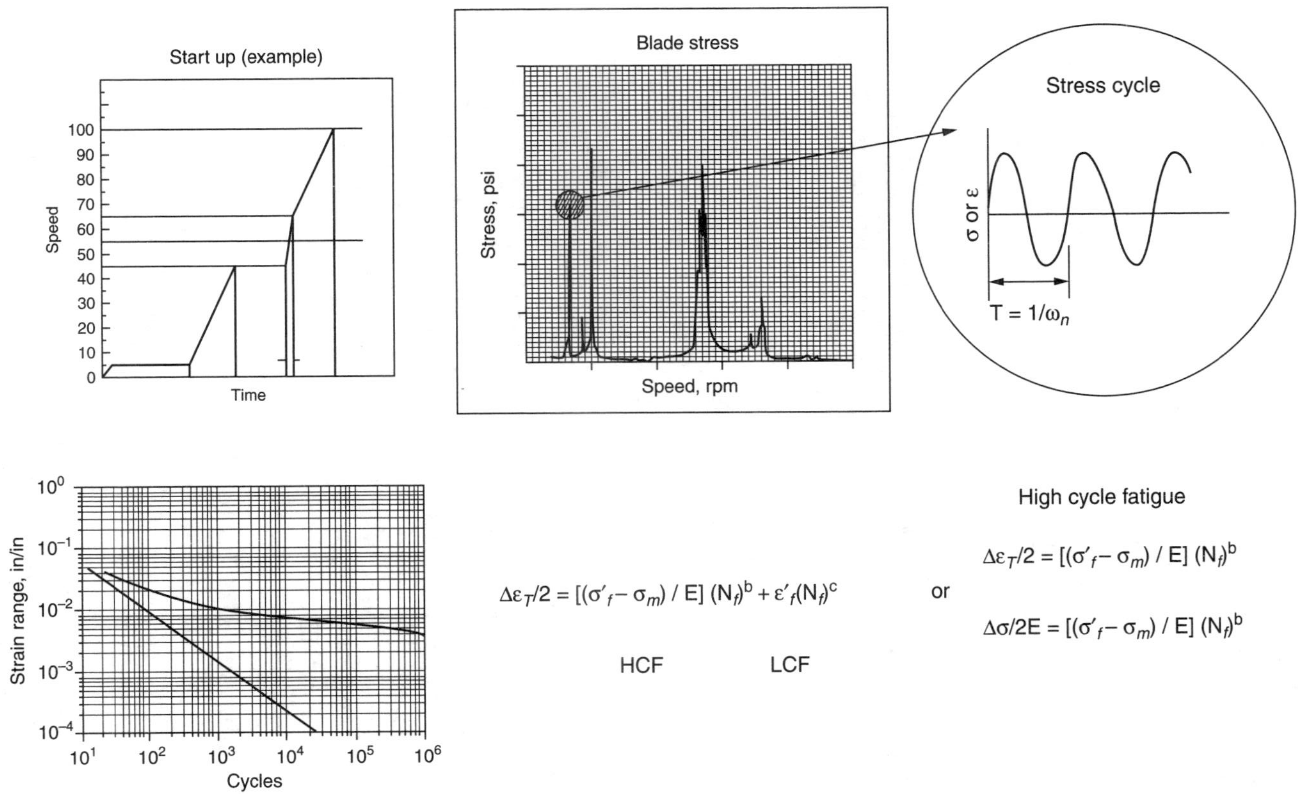

FIGURE 11.6 Deterministic life estimation.

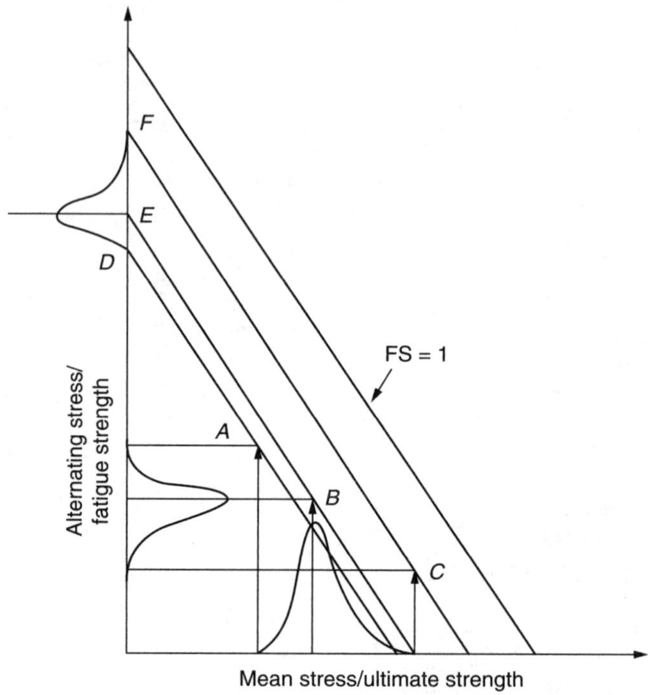

FIGURE 11.7 A probabilistic aspect of the Goodman diagram.

11.3.2 Probabilistic Frequency Analysis

There are variations in mechanical properties (E, ν, etc.) and the dimension (blade to blade) even within the tolerance band of blades. This will provide a statistical variation in the natural frequency. The Campbell diagram in Fig. 11.8 shows the implication of this aspect of variation. Figure 11.9 was created based on a Monte Carlo simulation to demonstrate the percentage likelihood of a frequency of a given magnitude. The average magnitude as used in design should have a 50 percent probability.

11.4 Probabilistic Life Estimation

The S-N curve or more accurately the $\Delta\varepsilon$ vs. N_f curve will have statistical variation. As explained in the earlier section, Monte Carlo simulation provides an estimate of the likelihood to achieve certain life (Fig. 11.10).

The basic theory has been explained to permit the reader to consider how a "true" analysis must incorporate the uncertainty inherent in any calculation. It is possible to quantify the risk associated with a design analysis. Thus, design decisions do not need to incorporate the factors of safety (factors of ignorance) to accommodate the

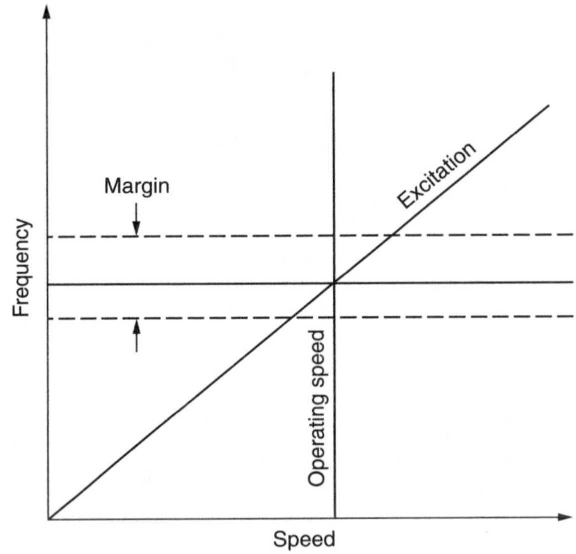

Figure 11.8 Variation in natural frequency.

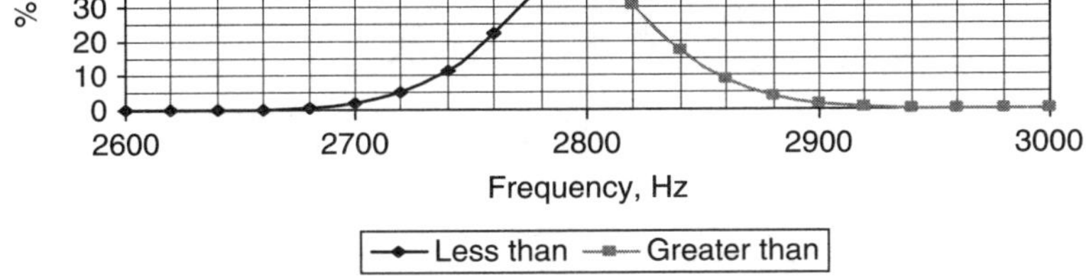

Figure 11.9 Probability of occurrence of frequency.

uncertainty. Rather, the risk may be quantified and design decisions made on the acceptable level of risk.

Risk can be managed when risk is quantified and enhancements in analytical tools and capabilities make that determination a reality. The probabilistic analysis tools are available and should be used. They are real, but require greater application to grow and flourish.

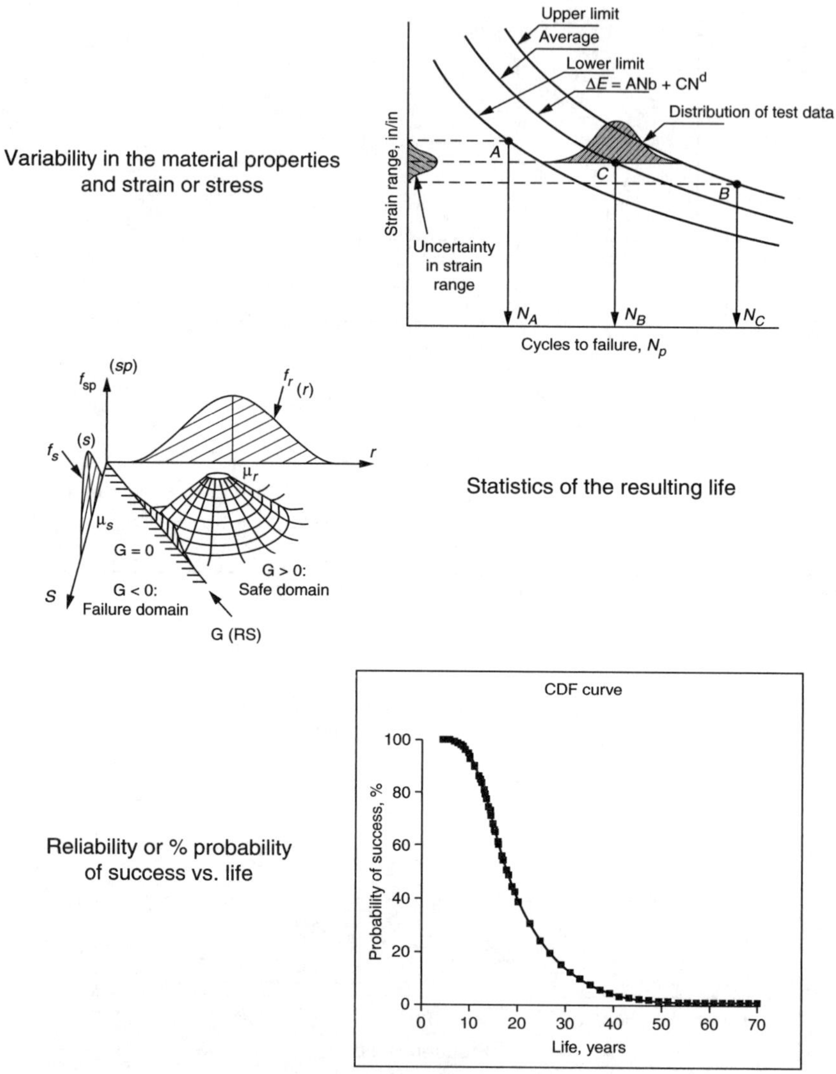

FIGURE 11.10 Probabilistic life estimation.

The time is right and the needs are there. New frontiers of design are being sought on the basis of size, loads, speeds, pressures, materials, costs, and delivery times.

It is believed that this represents a significant change in demands on our design approach. The time seems to have arrived to integrate the probabilistic analysis into our design tool kit.

APPENDIX

Fourier Series

It was discussed earlier that a mechanical system can be idealized by a collection of spring, mass, and dampers. During vibration, response contains a function that includes sin (ωt) and cos (ωt). These functions are continuous and can be evaluated for any value of ω. During each revolution of the turbine rotor, the turbine blades pass through steam as it exits the nozzle passage. Each blade experiences a pressure fluctuation as it passes from one nozzle opening to the next. This pattern of fluctuation in pressure is repeated for each complete revolution, and time of this pattern is a function of turbine speed. The pressure variations act on the blade as a variable periodic load. One period is one revolution; i.e., in terms of space it is 360° of rotation. It is evident, however, that the force, although periodic, is by no means as simple as a pure sine or cosine pattern. To aid in understanding the influence of this force on the blade vibration and to convert the shape of the force, a Fourier series analysis is conducted. A Fourier series analysis is capable of converting a periodic function to a summation of sine and cosine terms. This method facilitates the estimation of the response of the vibrating blades passing through pressure variations.

Without going into mathematical details, some basic equations of Fourier series are given below.

If a function $f(x)$ is bounded in a period $2L$ [i.e., $f(x + 2L) = f(x)$] and satisfies the following conditions

1. In any period $f(x)$ is continuous, except for a finite number of discontinuities.
2. In any period $f(x)$ has only a finite number of maxima and minima.

then $f(x)$ may be expressed by a series (Fourier series) as follows:

$$f(x) = a_0/2 + \sum_{n=1}^{\infty} a_n \cos(n\pi x/L) + b_n \sin(n\pi x/L) \qquad \text{(A.1)}$$

where a_n and b_n are constants, which depend on the function $f(x)$ to be evaluated. The series converges at $[f(x^+) + f(x^-)]/2$ at discontinuous points.

The following integral expressions help in the determination of constants a_n and b_n.

$$\int_{d}^{d+2L} \sin(n\pi x/L)\,dx = 0 \tag{A.2}$$

$$\int_{d}^{d+2L} \cos(n\pi x/L)\,dx = 0 \qquad n \neq 0 \tag{A.3}$$

$$\int_{d}^{d+2L} \sin(m\pi x/L)\,\sin(n\pi x/L)\,dx = \begin{cases} 0 & m \neq n \\ \pi & m = n \end{cases} \tag{A.4}$$

$$\int_{d}^{d+2L} \cos(m\pi x/L)\,\cos(n\pi x/L)\,dx = \begin{cases} 0 & m \neq n \\ \pi & m = n \end{cases} \tag{A.5}$$

$$\int_{d}^{d+2L} \sin(m\pi x/L)\,\cos(n\pi x/L)\,dx = 0 \tag{A.6}$$

The expressions for a_0, a_n, and b_n are given below:

$$a_0 = (1/L)\int_{d}^{d+2L} f(x)\,dx \tag{A.7}$$

$$a_n = (1/L)\int_{d}^{d+2L} f(x)\cos(n\pi x/L)\,dx \tag{A.8}$$

and

$$b_n = (1/L)\int_{d}^{d+2L} f(x)\,\sin(n\pi x/L)\,dx \tag{A.9}$$

Note that the series contains infinite terms, and the inclusion of more terms will make the approximation more accurate.

A.1 A Brief History and Description of the Monte Carlo Method

The method is named after a city in Monaco, Monte Carlo. Specifically, it was named after a game of roulette and a random number generator and dates back to the mid-1940s when a systematic development occurred.

In the development of the atomic bomb during World War II, the Monte Carlo method was used for the probabilistic simulation of the random diffusion of neutrons in fissile materials. A Polish-born mathematician named Stanislaw Ulam has been credited with the invention of the Monte Carlo method while he was working on the Manhattan project. According to the story, Stanislaw came up with the method while trying to estimate the probabilities of winning a card game of solitaire. "In 1946 during my illness I was playing solitaire. I spent lot of time to figure the chance of Canfield solitaire laid out with 52 cards," he is believed to have said.

Stanislaw recognized the potential to automate the statistical sampling procedure on the newly invented electronic computer. He worked with other scientists to develop the algorithms that were implemented on the computer. This showed the way to transform nonrandom problems into a random form and facilitated the solution they were looking for through statistical sampling. Finally, this made it possible to transform mathematical statistical sampling into a method that could be applied to a wide variety of problems.

The Monte Carlo method provides an approximate solution by conducting experiments on the computer and is a general method that can handle problems with and without probabilistic content. The error of approximation decreases as the number of samples is increased.

The following list comprises the foundation of Monte Carlo applications. A description of each item is beyond the scope of this book.

- Probability distribution function (PDF)—variables must be defined by a set of PDFs.
- Random number generator—an algorithm to generate uniformly distributed random numbers.
- Sampling rule—a method of sampling from specified PDFs.
- Tallying—results must be accumulated into overall scores for the quantities of interest.
- Error estimation—an estimate of error as a function of the number of trials.
- Variance reduction techniques—method that reduces computational time by reducing the error in the estimated solution.
- Algorithm—computer algorithm to be implemented on a computer.

Consider a deterministic model given in Fig. A.1. The X's are input parameters and the Y's are the responses of the system. For any set of values of X's, the Y's can be determined. No matter how many times the response is evaluated, it will have the same values for a given input.

Now consider the system of Fig. A.2. The input variables (X_1, X_2, and X_3) have uncertainty, and the response likewise is also random.

Typically, the analysis is conducted in following steps.

Step 1. Create a transfer function

$$Y = f(X_1, X_2, X_3, \ldots, X_n)$$

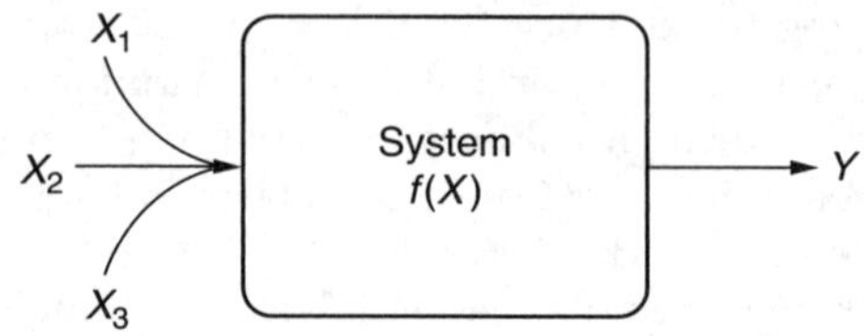

FIGURE A.1 Deterministic system model.

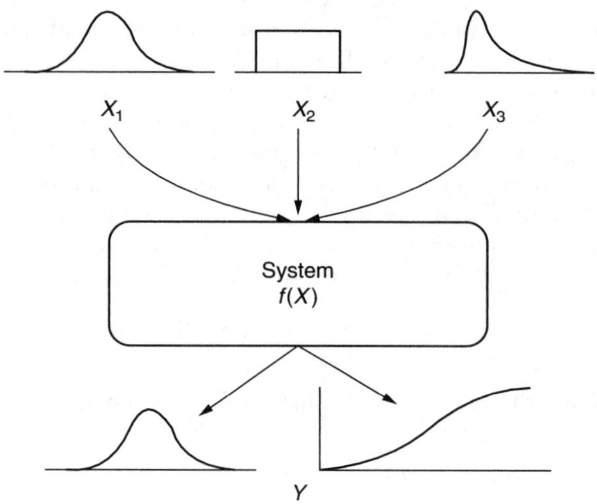

FIGURE A.2 Probabilistic system model.

Step 2. Generate a set of random inputs

$$X_{j1}, X_{j2}, X_{j3}, \ldots, X_{jn}$$

Step 3. Evaluate the response and store the results as Y_i.

Step 4. Repeat steps 2 and 3 for $j = 1, N$.

Step 5. Analyze the results using histograms, statistics, and confidence intervals.

Bibliography

Allen, R. C., "Steam-Turbine Blading," *Trans. ASME*, pp. 689–710, Nov. 1940.

API (American Petroleum Institute) Standard 612, 5th ed., API Publishing, Washington DC, paragraph 8.3, 2003.

Arkadyev, D. A., "Effect of Ties on the Vibrational Endurance of Turbine Blades," *Power Equipment Manufacture*, No. 3, pp. 6–9, 1968.

Bloch, H. P., and Singh, M. P., *Steam Turbines: Design, Applications, and Rerating*, 2d ed., McGraw-Hill, New York, 2009.

Cameron, T. M., Griffin, J. H., Kielb, R. E., and Hoosac, T. M., "An Integrated Approach for Friction Damper Design," *ASME Bound Volume DE-3*, pp. 205–212, 1987.

Campbell, W., "The Protection of Steam-Turbine Disk Wheels from Axial Vibration," *Trans. ASME*, vol. 46, pp. 31–160, 1925.

Chubb, S. B., "Evaluation of Wire Lacing for the Control of Gas Turbine Blade Vibration," ASME Paper 67-VIBR-47, 1967.

Coffin, L. F., "A Study of the Effects of Cyclic Thermal Stresses on a Ductile Metal," *Trans. ASME*, vol. 76, pp. 931–950, 1954.

Conway, J. B., Stentz, R. H., and Berling, J. T., "Fatigue, Tensile, and Relaxation Behavior of Stainless Steels," Technical Information Center, Office of Information Services, U.S. Atomic Energy Commission, 1975.

Crick, V., "Independent Self Adjusting Vibration Damper," Patent 3,723,023, Mar. 1973.

Davis, C. C., "Blade Vibration Damping Device," Patent 3,181,835, Mar. 1973.

Dello, J., "Frequency Evaluation of a Steam Turbine Bladed Disk," *Turbomachinery International*, Jan/Feb. 1987.

Den Hartog, J. P., "Forced Vibrations with Combined Coulomb and Viscous Friction," *Trans. ASME*, vol. 53, pp. 107–115, 1931.

Dowling, N. E., *Mechanical Behavior of Materials*, Prentice-Hall, Upper Saddle River, N.J., 1999.

Drosjack, M., Singh, M. P., and Rutan, C., "Determining Risk Associated Inspection and Overhaul of Turbomachinery: Yesterday, Today, and Tomorrow," NPRA, Reliability and Maintenance Conference and Exhibition, New Orleans, La., May 24–27, 2005.

Ewins, D. J., "The Effects of Detuning upon the Forced Vibrations of Bladed Disk," *Journal of Sound and Vibration*, vol. 9, no. 1, pp. 65–79, 1969.

Ewins, D. J., "A Study of Resonance Coincidence in Bladed Discs," *Journal of Mechanical Engineering Science*, vol. 12, no. 5, 1970.

Ewins, D. J., "Vibrational Characteristics of Bladed Disk Assemblies," *Journal of Mechanical Engineering Science*, vol. 15, no. 3, 1973.

Ewins, D. J., "Studies to Gain Insight into the Complexities of Blade Vibration Phenomena," Conference on Vibration in Rotating Machinery, ImechE, C184/76, pp. 165–172, 1976.

Graham, J. A., Marsh, K. J., and Appl, F. J. (eds.), *Fatigue Design Handbook*, SAE Pub. No. AE-4, Society of Automotive Engineers, Warrendale, Pa., 1968.
Griffin, J. H., "Friction Damping of Resonant Stresses in Gas Turbine Engine Airfoils," *ASME Journal of Engineering for Power*, vol. 107, pp. 329–333, 1980.
Griffin, J. H., and Menq, C. H., "Friction Damping of Circular Motion and Its Application to Vibration Control," *ASME Journal of Vibration and Acoustics*, vol. 113, pp. 225–229, 1991.
Hager, R. D., Lewis, G. W., and Wagner, J. M., "Use of Lacing Wire to Reduce Blade Vibration in an Axial Flow Compressor Rotor," NASA TM X-1603, 1965.
Halford, G., "The Strain Hardening Exponent—A New Interpretation and Definition," *Trans. Quarterly*, vol. 56, no. 3, 1963.
Johnson, D. C., and Bishop, R. E. D., "The Modes of Vibration of a Certain System Having a Number of Equal Frequencies," *Journal of Applied Mechanics, Trans. ASME*, pp. 379–384, Sep. 1956.
Jones, D. I. G., Nashif, A. D., and Stargardter, H., "Vibrating Beam Dampers for Reducing Vibrations in Gas Turbine Blades," ASME Paper 74-GT-95, Apr. 1974.
Kearton, W. J., *Steam Turbine, Theory and Practice*, First Published by Sir Isaac Pitman and Sons, London, 1922, Indian edition, 2004, 7th edition, CBS Publishers and Distributors, New Delhi, India.
Kielb, R., Griffin, J., and Menq, C., "Evaluation of a Turbine Blade Damper Using an Integral Approach," *Structures, Structural Dynamics and Materials Conference*, Williamsburg, Va., AIAA 88–2400, 1988.
Kroon, R. P., "Influence of Lacing and Centrifugal Force on Turbine Blade Stresses," *Journal of Applied Mechanics, Trans. ASME*, vol. 1, pp. 109–118, 1934.
Kroon, R. P., "Turbine-Blade Vibration due to Partial Admission," *Journal of Applied Mechanics*, pp. A-161, 1940.
Lambert, J. J., "The Influence of Assembly Tolerances on the Load Distribution of a Statistically Indeterminant Structure," M.S. thesis, Worcester Polytechnic Institute, Worcester, Mass., 1970.
Larson, F. R., and Miller, J., "A Time-Temperature Relationship for Rupture and Creep Stresses," *Trans. ASME*, vol. 74, pp. 765–775, 1952.
Lazan, B.J., *Damping of Materials and Members in Structural Mechanics*, Pergamon Press, New York, 1968.
Mallik, A. K., and Ghosh, A., "Fatigue Strength of Structural Members with High Damping Elastic Inserts," *Proceedings of First Design Engineering Conference*, Washington, pp 449–453, Oct. 5–9, 1974.
Manson, S. S., and Halford, G. R., "A Method for Estimating High-Temperature Low-Cycle Fatigue," *Proceedings of International Conference of Thermal and High-Strain Fatigue*, The Metals and Metallurgy Trust, London, pp. 154–170, 1967.
Manson, S. S. and Halford, G. R., *Fatigue and Durability of Structural Materials*, ASM International, Materials Park, Ohio, 2006.
Manson, S. S., "NASA TN2933," *National Aeronautics and Space Administartion*, 1953.
Manson, S. S., *Experimental Mechanics*, vol. 5, no. 7, pp. 193, 1968.
McKnight, R. L., Laflen, J. H., Halford, G. R., and Kaufman, A., "Turbine Blade Nonlinear Structural and Life Analysis," *Journal of Aircraft*, vol. 20, no. 5, pp. 475–480, May 1983.
Melcher, R. E., *Structural Reliability Analysis and Prediction*, Ellis Harwood Ltd., West Sussex, UK, 1987.
Menq, C. H., Chidamparam, P., and Griffin, J. H., "Friction Damping of Two-Dimensional Motion and Its Application in Vibration Control," *Journal of Sound and Vibration*, vol. 144, pp. 427–447, 1991.
Mitchell, M. R., "Fundamentals of Modern Fatigue Analysis for Design," in *Fatigue and Microstructure*, ASM International, Materials Park, Ohio, pp. 385–437, 1979.
Pigott, R., "Turbine Blade Vibration due to Partial Admission," *International Journal of Mechanical Sciences*, vol. 22, pp. 247–264, 1980.

Prohl, M. A., "A Method for Calculating Vibration Frequency and Stress of a Banded Group of Turbine Buckets," *Trans. ASME*, Paper 56-A-116, 1956.

Provanzale, G. E., and Skok, M. W., "A Cure for Steam Turbine Blade Failure," ASME Paper 73-PET-71, 1973.

Saint Venant, B. de, Historique Abrège in C. L. Navier, "Resume des lecons . . . . .," Carilian-Goeury, Paris, 1833.

Sanliturk, K. Y., and Ewins, D. J., "Modeling Two-Dimensional Friction Contact and Its Application Using Harmonic Balance Method," *Journal of Sound and Vibration*, vol. 193, no. 2, pp. 511–523, 1996.

Siddall, J. N., *Probabilistic Engineering Design*, Marcel Dekker, New York, 1983.

Singh, M. P., "Turbine Blade Dynamics: A Probabilistic Approach," *In the Vibrations of Blades and Bladed Disk Assemblies*, ASME book no. H000335, pp. 41–48, 1985.

Singh, M. P., "Predicting the Randomness of Forced Vibration Response of a Bladed Disc," ImechE Conference, Bath, United Kingdom, C432/128, 1992.

Singh, M. P., "Probabilistic Estimation of the Effect of Dimensional Tolerance for Turbine/Compressor Blade Attachment," *Recent Advances in Solids and Structures*, H. H., Chung, and Y. W., Kwon (eds.), ASME Book no. G01095, Pressure Vessel Publication, vol. 381, 1998.

Singh, M. P., "Probabilistic HCF Life Estimation of a Mechanical Component," *Proceedings of 2001 ASME International Mechanical Engineering Congress and Exposition*, New York, Nov. 11–16, 2001.

Singh, M. P., "Reliability Evaluation of Steam Turbine for Process Drive," A course sponsored by Texas A&M, 2005.

Singh, M. P., and Drosjack, M., "Emerging Advanced Technologies to Assess Reliability of Industrial Steam Turbine Blade Design," *Proceedings of the 37th Turbomachinery Symposium*, Turbomachinery Laboratory, Texas A&M University, College Station, Tex., pp. 169–188, 2008.

Singh, M. P., and Ewins, D. J., "A Probabilistic Analysis of a Mistuned Bladed Turbine Disc," ImechE Conference, Edinburgh, United Kingdom, C229/88, 1988.

Singh, M. P., Matthews, T., and Ramsey, C., "Fatigue Damage of Steam Turbine Blade Caused by Frequency Shift due to Solid Buildup—A Case Study," *Proceedings of the 23d Turbomachinery Symposium*, Turbomachinery Laboratory, Texas A&M University, College Station, Tex., pp. 107–114, 1994.

Singh, M. P., and Schiffer, D., "Vibrational Characteristic of Packeted Bladed Disc," Presented at *ASME Design Engineering Technical Conf.*, Washington, DC, 82-DET-137, Sep. 1982.

Singh, M. P., Thakur, B. K., and Sullivan, W. E., "Assessing Useful Life of Turbomachinery Components," *Proceedings of the 34th Turbomachinery Symposium*, Turbomachinery Laboratory, Texas A&M University, College Station, Tex., pp. 177–192, 2005.

Singh, M. P., and Vargo, J. J., "Reliability Evaluation of Shrouded Blading Using the SAFE Interference Diagram," *Journal of Engineering for Gas Turbine and Power*, vol. 111, pp. 601–609, 1989.

Singh, M. P., Vargo, J. J., Schiffer, D., and Dello, J., "SAFE Diagram: A Design and Reliability Tool for Turbine Blading," *17th Turbomachinery Symposium*, Dallas, Tex., Nov. 8–10, 1988.

Soo, S. L., *Thermodynamics of Engineering Science*, Prentice-Hall, Englewood Cliffs, N.J., 1958.

Srinivasan, A. V., "Flutter and Resonant Vibration Characteristics of Engine Blades," *Journal of Engineering for Gas Turbines and Power, Trans. ASME*, vol. 119, pp. 741–775, 1997.

Stodola, A., *Steam Turbine*, D. Van Nostrand Co., New York, 1905.

Thompson, W. T., *Vibration Theory and Applications*," Prentice-Hall, Englewood Cliffs, N.J., 1965.

Trumpler, W. E., and Owens, H. M., "Turbine Blade Vibration and Strength," *Trans. ASME*, vol. 77, pp. 337, 1955.

Vargo, J. J., "Blading," Publication ST-3, Turbodyne Corp., New York.

Viswanathan, R., *Damage Mechanisms and Life Assessment of High Temperature Components*, ASM International, Metals Park, Ohio, 1989.

Viswanathan, R., and Dooley, R.B., "Creep Life Assessment Techniques for Fossil Plant Pressure Parts," *Proceedings of International Conference on Creep*, JSME-IME-ASTM-ASME, Tokyo, pp. 349–359, Apr. 1986.

Walker, L. I., and Kushner, F., "Experimental Investigation of the Vibratory Characteristics of an Exhaust End Steam Turbine Rotating Blade and Wheel Assembly," ASME Paper 74-PET-38, 1974.

Weaver, F., and Prohl, M., "High Frequency Vibration of Steam Turbine Buckets," ASME Paper 56-A-119, 1956.

Williams, E. J., and Earles, S. W. E., "A Method of Damping Vibration of Gas Turbine Compressor Blading," *Vibration Problems in Industry, International Symposium, Session 2*, Paper 221, Apr. 10–12, 1973.

Williams, E. J., and Earles, S. W. E., "Optimization of Response of Frictionally Damped Beam Type Structures with Reference to Gas Turbine Compressor Blading," *ASME Journal of Engineering for Industry*, pp. 471–476, May 1974.

Wilson, D. G., *The Design of High-Efficiency Turbomachinery and Gas Turbines*, MIT Press, Cambridge, Mass., 1984.

Yampalskaya, R. G., and Arkadyev, D. D., "Determining the Vibration Characteristics of Blade with Damper Connections," *Power Equipment Manufacture*, no. 11, pp. 5–7, 1965.

Zemansky, M. W., and Van Ness, H. C., *Basic Engineering Thermodynamics*, McGraw-Hill, New York, 1966.

Index

Note: Page numbers followed by *f* denote figures; page numbers followed by *t* denote tables.

C